INTEGRATIVE ORGANISMAL BIOLOGY

INTEGRATIVE ORGANISMAL BIOLOGY

Edited by

Lynn B. Martin
Department of Integrative Biology
University of South Florida

Cameron K. Ghalambor
Department of Biology and Graduate Degree Program in Ecology
Colorado State University

H. Arthur Woods
Division of Biological Sciences
University of Montana

WILEY Blackwell

Published by John Wiley & Sons, Inc., Hoboken, New Jersey

Published simultaneously in Canada

Library of Congress Cataloging-in-Publication Data:

Martin, Lynn Bloxom, 1974–
Integrative organismal biology / Lynn B Martin, Cameron K Ghalambor, H. Arthur Woods.
pages cm
Includes bibliographical references and index.
ISBN 978-1-118-39878-4 (pbk.)
1. Variation (Biology) 2. Nature and nurture. I. Ghalambor, Cameron K. II. Woods, Harry Arthur. III. Title.
QH438.5.M37 2014
576.5′4 – dc23

2014033465

Cover image: Cover image provided courtesy Martyna Boruta

Woods – To Creagh, Nicholas, and Eliot, for their support, love, and boundless curiosity.

Ghalambor – To Dionna and Evie for all their love, support, and patience.

Martin – To Melissa, Arryson, Alex and AJ for their love, support, and sacrifices.

CONTENTS

ACKNOWLEDGMENTS

Woods – HAW thanks the National Science Foundation, the University of Montana for support, and the faculty and staff of l[1]Institut de Recherche sur la Biologie de l[1]Insecte, Université François Rabelais, Tours, France, for supporting a sabbatical stay, during which a significant portion of his work on this book was done.

Ghalambor – CKG thanks the National Science Foundation and Colorado State University whose generous support allowed for the writing and editing of the chapters making up this book. He would also like to thank the authors for their efforts in taking an integrative perspective on their research, and Wiley for providing the opportunity to publish this collection of chapters.

Martin – LBM thanks the National Science Foundation and the University of South Florida for support during the writing of chapters and editing of the book. He also thanks the authors of chapters for their efforts, as well as the Anna Ehler, Mindy Okura-Marszycki, Michelle Kwok, and especially Stephanie Dollan and Martin Noble at Wiley for their help with the book, and Krupa Muthu at Laserwords for help with proofing. Last, he thanks the Fall 2013 USF class in Physiological Ecology and the Martin lab of USF Integrative Biology for thoughtful comments on book contents.

NOTES ON CONTRIBUTORS

James S. Adelman, Department of Biological Sciences, Virginia Polytechnic Institute and State University, Blacksburg, VA, USA

Joshua Banta, Department of Biology, University of Texas, Tyler, Tyler, TX, USA

Jennifer A. Brisson, Department of Biology, University of Rochester, Rochester, NY, USA

Francisco Bozinovic, Departamento de Ecología and Center of Applied Ecology and Sustainability (CAPES), Facultad de Ciencias Biológicas, Universidad Católica de Chile, Santiago 6513677, Chile

Vincent Careau, Centre for Integrative Ecology, School of Life and Environmental Science, Deakin University, Waurn Ponds, Victoria, Australia

Alan A. Cohen, Groupe de Recherche PRIMUS, Department of Family Medicine, University of Sherbrooke, Sherbrooke, QC, Canada

Robert M. Cox, Department of Biology, University of Virginia, Charlottesville, VA, USA

Goggy Davidowitz, Department of Entomology, University of Arizona, Tucson, AZ, USA

Sarah Diamond, Department of Biology, Case Western Reserve University, Cleveland, OH, USA

Niels J. Dingemanse, Max Planck Institute for Ornithology, Seewiesen, Germany

Cameron K. Ghalambor, Department of Biology and Graduate Degree Program in Ecology, Colorado State University, Fort Collins, CO, USA

Richard Gomulkiewicz, School of Biological Sciences, Washington State University, Pullman, WA, USA

Thomas P. Hahn, Department of Neurobiology, Physiology and Behavior, University of California, Davis, Davis, CA, USA

Mark F. Haussman, Department of Biology, Bucknell University, Lewisburg, PA, USA

Bryan R. Helm, Department of Biological Sciences, North Dakota State University, Fargo, ND, USA

Shaun S. Killen, Institute of Biodiversity, Animal Health and Comparative Medicine, College of Veterinary and Life Sciences, University of Glasgow, Glasgow, Scotland, UK

Joel Kingsolver, Department of Biology, University of North Carolina, Chapel Hill, Chapel Hill, NC, USA

Scott A. MacDougall-Shackleton, Department of Physiology, Advanced Facility for Avian Research, University of Western Ontario, London, ON, Canada

Lynn B. Martin, Department of Integrative Biology, University of South Florida, Tampa, FL, USA

Paul R. Martin, Department of Biology, Queen's University, Kingston, ON, Canada

Kimberley J. Mathot, Max Planck Institute for Ornithology, Seewiesen, Germany

Daniel E. Naya, Departamento de Ecología y Evolución, Facultad de Ciencias, Universidad de la República, Montevideo, Uruguay

Daniel H. Nussey, Institute of Evolutionary Biology, University of Edinburgh, Edinburgh, Scotland, UK

Mihaela Pavličev, Cincinnati Children's Hospital, Cincinnati, OH, USA

Christina L. Richards, Department of Integrative Biology, University of South Florida, Tampa, FL, USA

Aaron W. Schrey, Department of Biology, Armstrong State University, Savannah, GA, USA

Lisa A. Treidel, Department of Biology, Bucknell University, Lewisburg, PA, USA

Haruka Wada, Department of Biological Sciences, Auburn University, Auburn, AL, USA

Günter P. Wagner, Department of Evolutionary Biology, Yale University, New Haven, CT, USA

Heather E. Watts, Department of Biology, Loyola Marymount University, Los Angeles, CA, USA

J. Keaton Wilson, Division of Biological Sciences, University of Montana, Missoula, Missoula, MT, USA

H. Arthur Woods, Division of Biological Sciences, University of Montana, Missoula, MT, USA

Anthony J. Zera, School of Biological Sciences, University of Nebraska, Lincoln, Lincoln, NE, USA

1

PLASTICITY, COMPLEXITY, AND THE INDIVIDUAL

Cameron K. Ghalambor[1], Lynn B. Martin[2], and H. Arthur Woods[3]

[1]*Department of Biology and Graduate Degree Program in Ecology, Colorado State University, Fort Collins, Colorado, USA*
[2]*Department of Integrative Biology, University of South Florida, Tampa FL, USA*
[3]*Division of Biological Sciences, University of Montana, Missoula, Montana 95812, USA*

INTRODUCTION

The first half of the 20th century was a pivotal time for biology. The different branches of evolutionary biology, from genetics to paleontology, operated independently, each with its own debates and guiding theories. A major shift started in the 1930s, when theoretical and empirical work in population genetics (1) reconciled Mendelian inheritance with natural selection, (2) demonstrated that microevolution and macroevolution were compatible, and (3) elevated Darwin's views of descent with modification and evolution by natural selection as unifying theories in biology (Mayr & Provine 1998). These achievements, now referred to as the *Modern Synthesis* (Mayr & Provine 1998), enabled researchers to delineate a set of principles that could explain variation in genetic and phenotypic patterns over space and time. For example, we can understand the maintenance and spread of selfish genetic elements within the genome (e.g., transposable elements) or selfish individuals in a population (e.g., cheaters) using the same conceptual framework and principles of evolution by natural selection. Nevertheless, the *Modern Synthesis*, in its original conception, was limited: its application and appeal were largely limited to branches of biology interested in population-level processes. While there were attempts to incorporate a role for individual development and physiology into this broader evolutionary synthesis by Goldschmidt (1940), Schmalhausen (1949), and Waddington (1942), such efforts remained

Integrative Organismal Biology, First Edition.
Edited by Lynn B. Martin, Cameron K. Ghalambor, and H. Arthur Woods.

outside the focus of mainstream evolutionary biology. As a result, the *Modern Synthesis* left out areas of biology concerned with the structure and function of individual organisms, such as physiology, development, functional morphology, neuroscience, and behavior (Gottlieb 2001). These organismal branches recognized and embraced evolutionary concepts such as homology in comparative studies (Hall 2012) or how "ontogeny recapitulated phylogeny" (Kalinka & Tomancak 2012), but they largely operated outside an evolutionary framework when studying how organisms work today. Subsequently, much of biology below the level of individuals came to favor reductionism, assuming that greater deconstruction would explain best how organisms, tissues, and cells work (Bartholomew 1986; Strange 2005).

Today, however, biology finds the barriers between fields breaking down; there is an increasing integration of evolutionary principles into organismal biology (Carroll 2005; Jablonka & Lamb 2005; Harrison *et al.* 2012; Nesse *et al.* 2010; Perlman 2013; Westneat & Fox 2012), a growing appreciation that organismal function might inform evolutionary theory (Schlichting & Pigliucci 1998; Bell 2009; Flatt & Heyland 2011), and a desire to understand how complex integrated systems, such as whole organisms, function and evolve (Wagner & Altenberg 1996; Stern 2010; Strange 2005; Martin *et al.* 2011; Noble 2013). The next few years may therefore become another pivotal time for biology, as technical and mathematical advances are diffusing shared approaches across biological levels of organization. Moreover, new, extended syntheses are emerging that are truly integrative in their views on interactions among environments, genomes, and phenotypes (e.g., Schlichting & Pigliucci 1998; West-Eberhard 2003; Jablonka & Lamb 2005; Davidson 2001; Pigliucci & Muller 2010). Whereas the *Modern Synthesis* simplified biological processes by developing a general theory grounded in mathematical models, these new approaches are playing out against the backdrop of rapid progress toward understanding the mechanisms underlying the complexity of life. In the process, we are faced with what at first glance appears as an overwhelming degree of information and contingency. Our goal here is to argue that as we begin to open the black box linking genotypes, phenotypes, and their natural environments, an evolutionary perspective of how whole organisms function is needed. In doing so, mechanistic studies of form and function will be infused and guided by evolutionary theory, evolutionary biology will incorporate an understanding of how underlying mechanisms constrain or facilitate certain ecological and evolutionary outcomes, and collectively biologists working across scales will be motivated by a shared perspective that promotes developing and testing general theory.

We feel that these goals toward an integrative organismal biology will be served best by elevating the roles of individuals in biology. Already, there is a strong tradition of research on the structure and function across levels of organization (see also Wake 2008; Satterlie *et al.* 2009; Mykles *et al.* 2010; Noble 2013). But only recently has this work started to become firmly grounded in evolutionary theory and in natural (as opposed to laboratory) contexts. In this chapter, we focus on two themes that reveal the importance of understanding individuals: (1) the concept of *phenotypic plasticity*, or the capacity for a given genotype to produce different phenotypes in response to different environments and (2) the concepts of *complexity and integration*, or how suites of interacting traits across levels of organization respond to, and evolve in response to, environmental variation. Below, we first provide a brief history of progress toward an integrative organismal biology. We then review why the concepts of plasticity, complexity, and integration have such unifying power. We end with a case study of how a focus on understanding mechanisms within individuals can influence biology across scales.

BRIDGING THE CONCEPTUAL DIVIDE

Conceptual unification occurs when disciplines come to share underlying theories and questions but use different approaches. The promise of unification is that it will reveal new insights not achievable otherwise. Conceptual unification in biology remains a work in progress, and a key barrier has been the division between wanting to understand *how* organisms work versus *why* they work one way as opposed to another (e.g., Mayr 1961; Orians 1962; Dobzhansky 1964). Mayr (1961) referred explicitly to how (functional biology) vs why (evolutionary biology) as complementary but not alternative explanations for pattern and process. In practice, asking *how* biological systems work lends to reductionism because we often assume (at least implicitly) that insight stems from decomposing wholes into parts. In contrast, evolutionary biology is based on a set of conditions; the presence of heritable variation, the historical and contemporary processes generating and acting upon that variation, and the outcomes and patterns of these processes. These conditions were first fully articulated by Darwin (1859) without any knowledge of DNA or a mechanistic understanding of how organisms were built, and they eventually became codified in the population genetic models developed during the *Modern Synthesis*. Yet, we now are rapidly moving into an era where the divide between functional and evolutionary biology is eroding.

Evolutionary theory provides context and explanations for patterns that functional biology cannot. For example, understanding the details of transcription and translation does not tell us why genes and proteins are conserved across related taxa, or why they predictably change in response to certain environmental pressures. Similarly, a functional biologist can study the biomechanics of how bill size influences the ability to crush a seed, but an evolutionary biologist is more likely to explain why populations of birds on different islands have larger or smaller bills. Yet, there is a desperate need for unification – because the power of functional or evolutionary biology to explain biological phenomenon is limited when they operate alone. Evolutionary theory's ability to predict how a given population will respond to natural selection is often limited, because genotypes and phenotypes are generally black boxes, as are the developmental and physiological pathways that translate genotypes into phenotypes. Such limitations are pervasive in optimality and game theory models, where the constraints on the response to selection are simply inferred from deviations from a predicted optimum (Parker & Maynard-Smith 1990). Indeed, many evolutionary biologists would argue that a mechanistic understanding is not needed to predict evolutionary outcomes because, over time, natural selection will find optimal solutions. By contrast, others argue we need mechanism if we are to understand the trade-offs that constrain responses to selection. These constraints may be genetic in the form of pleiotropic relationships, physiological or developmental in the form allocation trade-offs between traits competing for the same pool of resources, or functional when selection on one trait or function negatively impacts another. Mechanism therefore provides insight into why one optimal solution rises to the top, and is something that optimality approaches are unable to do. For example, in their review of the regulatory mechanisms underlying polyphenisms, Zera and Brisson (Chapter 5, this volume) demonstrate how the genetic and endocrine basis of different phenotypes act via developmental modules to constrain or bias the evolution of phenotypes. Similarly, an understanding of the physiological trade-offs associated with tolerance to different environmental conditions can predict where a species lives far more effectively than a strictly ecological approach (Chapter 18, this volume). Within this framework, unification acknowledges that there is a deep reciprocity: how can inform why, and why can inform how.

A skeptic may ask: why should functional biology be integrative? Can't it advance on its own, outside the umbrella of evolution? We feel that, ultimately, progress will be slower in nonevolutionary strands of functional biology. First, we would remind functional biologists of Dobzhansky's (1964, 1973) dictum that "nothing in Biology makes sense except in the light of evolution." Indeed, evolutionary theory is the only accepted unifying framework for organizing and understanding the hierarchy of life. Second, if we accept descent with modification, and all the genetic, developmental, and functional constraints that go with it, then we must recognize that even reductionist approaches to function are not divorced from the broader evolutionary context focusing on the unity of life (Hochachka & Somero 2002; Sharan *et al.* 2005; Carroll 2005; Gerhart & Kirschner 2007). For example, it is this evolutionary unity or conservation that serves as a key justification for using model organisms to infer general function. Indeed, most biomedical research uses model organisms as diverse as bacteria, yeast, plants, nematodes, flies, fish, and mice to better understand and improve human health precisely because all the taxa share many genes and proteins by virtue of their evolutionary history. However, despite such conservation, it is increasingly recognized that the same genes and proteins have been coopted over time to perform diverse functions. Thus, understanding the mechanism at the molecular and cellular level ultimately depends on how past selective pressures and phylogenetic history have shaped the organismal context and systems in which these processes are embedded.

A skeptic might also ask, conversely, why evolutionary biology should pay attention to mechanism, as prominent scientists have recently done (see Flatt & Heyland 2011: chapter 28 exchange with Stearns). We answer that organisms are the conduits through which genes and their products interact with each other and with their environments. Thus, knowledge of mechanism allows evolutionary theory to move beyond models of how selection acts on variation among "genotypes," to models of how selection acts on pathways, networks, and interactions (e.g., Proulx *et al.* 2005). Indeed, we now know that the genotype-phenotype map, and selection's ability to act on particular pathways, is significantly more complicated than either classic Mendelian genetics or the central dogma of molecular biology have led us to believe. In a real sense, organisms are emergent Gordian knots of complexity, all tied up by pleiotropic linkages and trade-offs that cascade in unexpected ways and at different levels of organization in response to their environments. Such integrative complexity requires thinking about evolutionary theory in a way that explicitly acknowledges the integrated character of organisms and that takes mechanism seriously. This rethinking is already underway, with progress in areas of evolutionary theory as diverse as stasis, evolvability, constraints, and the origins of complex traits (e.g., Moran 1994; Wagner & Altenberg 1996; Kirschner & Gerhart 1998; West-Eberhard 2003; Hansen & Houle 2004; Hansen 2006). We present detailed examples of these ideas below and other authors present similar arguments throughout the book (Chapters 2, 6, & 9, this volume).

INTEGRATIVE ORGANISMAL BIOLOGY: PROGRESS TO DATE

Despite the practical and philosophical barriers in place, the conceptual divide between functional and evolutionary biology has varied over time. During the 1950s and 1960s, the biological sciences started transitioning from the organism-based disciplines of zoology and botany to more specialized groupings of cellular and molecular biology, physiology, and ecology and evolution (Huey & Bennett 2008). Increasing knowledge of biological complexity also caused greater specialization within subdisciplines, as fields like animal physiology split into the ever more specialized topics such as endocrinology, neurobiology, and functional morphology (Huey & Bennett 2008). Although prominent physiologists, like George Bartholomew (1958, 1966), and evolutionary biologists, like Theodosius Dobzhansky (1964), argued against the gathering reductionism, these views found little traction. Serious erosion of the divide between functional and evolutionary biology started in the early 1980s when Arnold (1983) proposed the morphology-performance-fitness framework. This framework introduced physiologists and functional morphologists to quantitative genetics and provided the conceptual and statistical tools to study how selection acts upon variation in performance and fitness. Arnold's (1983) framework also catalyzed the development of evolutionary physiology (Garland & Carter 1994), and enabled physiologists and evolutionary biologists to interact productively (Kingsolver & Huey 2003). Building from the foundations laid out by Bartholomew (1964), Huey and Pianka (1981) and Huey and Bennett (1986) introduced approaches that explicitly dealt with the problem of proper phylogenetic inertia in comparative studies – approaches that are now standard in comparative physiology (Garland *et al.* 2005). Although the importance of these evolutionary ideas was openly debated among physiologists (largely due to confusion over what the implications were for the merging of disciplines – see commentaries in Tracy *et al.* 1982), the inferential benefits of these approaches started to take hold. One prime example is the landmark 1987 publication *New Directions in Ecological Physiology* edited by Martin Feder, Al Bennett, Warren Burggren, and Ray Huey, which emphasized the importance of interindividual and genetic variation in physiological traits as the raw material for evolution – not just a source of noise (Bennett 1987; Koehn 1987; Arnold 1987). At the same time, Richard Sibly and Peter Calow published *Physiological Ecology of Animals: An Evolutionary Approach* (1986) and Peter Calow edited *Evolutionary Physiological Ecology* (1987), which were both early attempts to link physiological mechanisms to classic questions on trade-offs and life history evolution – a topic that has grown into a major area of research (see Flatt & Heyland 2011; Chapters 10 & 13, this volume). More recently, physiologists and ecologists have continued along the path of integration, as a way to understand the mechanisms underlying large-scale ecological patterns (Spicer & Gaston 1999; Chown & Gaston 2008; Lessels 2008; Williams 2008; Gaston *et al.* 2009; Chapter 17, this volume).

A second wave of unification between evolutionary and functional biology occurred throughout the 1990s as new molecular tools appeared. Perhaps the most prominent example was evolutionary developmental biology, or evo-devo (Goodman & Coughlin 2000; Gilbert 2003). Developmental biology was a vibrant but fairly independent field for many years (Gottlieb 2001), but in this period it began to connect with mainstream evolutionary biology, in the process revisiting early attempts at unification by Goldschmidt, Schmalhausen, and Waddington. Its first major contribution was to show that diverse body plans, across phyla, stemmed from conserved patterns of gene expression (Gilbert 2003; Carroll 2005). Since then, the incorporation of evolutionary developmental biology has expanded rapidly (Stern 2010), and can be seen in the emerging fields of ecological development (eco-devo), evolutionary medicine, and mechanistic life history theory (e.g., Gilbert & Epel 2009; Nesse *et al.* 2010; Flatt & Heyland 2011). Still, evolutionary thinking in other branches of functional biology remains limited to a tiny fraction of a large field of researchers.

Today, these subdisciplines, functional and evolutionary alike, are converging around two themes. The first centers on phenotypic plasticity, which attempts to understand how environmental variation interacts with genetic variation to shape the development and function of phenotypes. Plasticity has a long history in evolutionary biology and is increasingly viewed as an interesting problem in organismal biology. The second theme moves beyond reductionism and studying traits in isolation, and instead emphasizes the complexity and integration of traits. The implications of such integration, especially how it manifests mechanistically, warrant attention at multiple levels.

PHENOTYPIC PLASTICITY: THE LINK BETWEEN INDIVIDUALS, ENVIRONMENTS, AND EVOLUTION

A Brief History and Review

That phenotypes emerge from interactions between genotypes and their environments has long been recognized, but it was the work of Woltereck (1909), on parthenogenic lines of *Daphnia* and *Hyalodaphnia*, that first clearly revealed the distinction between genetic and environmental sources of phenotypic variation. Woltereck (1909) introduced the concept of the reaction norm, plotting how individuals from the same genetic background changed predictably in response to environmental cues. Reaction norms are visual representations of phenotypic plasticity, and they typically are depicted as continuous linear or nonlinear functions (Schlichting & Pigliucci 1998; Chapter 3, this volume). While the concept of reaction norms is widely used today, the interactions between genes, phenotypes, and the environment was a major source of confusion and controversy at the turn of the century (Schlichting & Pigliucci 1998). Indeed, the founders of the *Modern Synthesis* struggled with or simply ignored plasticity because they considered environmentally induced variation as nonheritable and thus irrelevant to evolutionary processes (Schlichting & Pigliucci 1998; Pigliucci 2001). Yet, other evolutionary biologists had a different perspective on such environmentally induced variation, as there was no denying that environments could significantly alter patterns of development, physiology, and behavior (e.g., Baldwin 1896; Goldschmidt 1940; Waddington 1942, 1953; Schmalhausen 1949). However, their "alternative synthesis" did not resonate with their contemporaries (Schlichting & Pigliucci 1998).

Today, environmentally induced variation – in the form of developmental plasticity, physiological flexibility, epigenetic inheritance, hidden reaction norms and release of cryptic genetic variation – is increasingly thought to influence adaptive evolution (e.g., Baldwin 1896; Waddington 1942, 1953; Bradshaw 1965; Schlichting & Pigliucci 1998; Pigliucci 2001; West-Eberhard 2003; Jablonka & Lamb 2005; Ghalambor *et al.* 2007; Piersma & van Gils 2011). This renaissance is a reflection both of plasticity's ecological and evolutionary roles, but also its ability to link processes at the individual and subindividual levels (i.e., gene expression) to processes at the population level (i.e., the relative fitness of different phenotypes and the strength of selection). Indeed, developmental, physiological, and behavioral plasticity can facilitate adaptive evolution by helping populations (1) move between adaptive peaks without incurring the cost of selection, and (2) persist in new environments until an opportunity arises for directional selection to act on new mutations (e.g., Robinson & Dukas 1999; Price *et al.* 2003; Ghalambor *et al.* 2007). Conversely, adaptive plasticity can also constrain adaptation by masking heritable genetic variation and weakening the strength of selection (e.g., Huey *et al.* 2003; Borenstein *et al.* 2006; Paenke *et al.* 2007). Environments do not always induce beneficial changes in phenotypes; particularly in cases where the environment is novel or stressful, plastic changes can be nonadaptive (Ghalambor *et al.* 2007). The role of nonadaptive plasticity in evolution has received even less attention but likely results in either extinction or strong directional selection in poor or novel environments (Falconer 1990; Grether 2005; Ghalambor *et al.* 2007; Conover *et al.* 2009).

What role do organismal biologists have in answering evolutionary questions about phenotypic plasticity? An integral one! Studies of development, physiology, and behavior are often mechanistic studies of phenotypic plasticity, whether or not researchers cast their research in that context. Understanding organismal responses to temperature, nutrients, disease, or any other aspect of the environment, external or internal, can be studied within a phenotypic plasticity framework. As such, empirical studies of physiology (e.g., Lee & Peterson 2003; Lee *et al.* 2003: McCairns & Bernatchez 2010; Handelsman *et al.* 2013), morphology (e.g., Chapman *et al.* 2000; Losos *et al.* 2000; Wund *et al.* 2008; Badyaev 2009), development (Badyaev & Landeen 2007; Young & Badyaev 2007), and behavior (Yeh & Price 2004; Losos *et al.* 2006) are already playing an important role in testing the ecological and evolutionary importance of phenotypic plasticity in natural populations.

Plasticity 101: A Q&A Session

Despite straddling the environment-organism interface, phenotypic plasticity remains a source of confusion (Pigliucci 2001), primarily because unlike classical evolutionary theory, the dual roles played by the environment – as the source generating phenotypic variation and the filter through which selection sorts this variation (West-Eberhard 2003). Because plasticity is so integral to the role of the individual in evolutionary biology and as the bridge across scales of biology, we attempt to clarify some of the confusion by answering commonly raised questions.

If I don't study genotypes, can I still study phenotypic plasticity? – Yes. Phenotypic plasticity has been classically defined as the capacity of a given genotype to produce different phenotypes in response to changes in the environment. However, in practice the "genotype" can be any distinct grouping (e.g., a set of clones, recombinant inbred lines, a group of full or half-siblings, different individuals, genetically differentiated populations, or even species) about which some genetic relationships exist. Obviously the more one can control and replicate the genetic background of the groups being compared, the more strongly one can infer how the same set of genes responds to environmental variation. For laboratory-based studies, this is most easily achieved by generating genetic groups through controlled crosses (e.g., by considering siblings to be a genetic group). In natural populations, genetic control can be incorporated into studies via the use of pedigrees, or by repeatedly measuring the same individual (see Chapter 2, this volume). Repeated measurements of flexible traits in the same individual over time or space is also referred to as phenotypic flexibility (Piersma & van Gils 2011), as a means of distinguishing it from nonreversible developmental plasticity.

Is all environmentally induced variation phenotypic plasticity? – No. Phenotypic plasticity represents a repeatable phenotypic response to a given environmental cue by a genetic group. Plasticity can therefore be distinguished from "developmental noise" or other nonrepeatable and unpredictable responses to the environment (Woods 2014).

What do I measure if I'm studying plasticity? – Any phenotypic trait of the same genetic group in more than one environment. At the suborganismal level, plasticity can be measured by quantifying changes in gene expression (e.g., Aubin-Horth & Renn 2009), patterns of alternative splicing (e.g., Marden 2008) or of protein synthesis (e.g., Tomanek & Somero 2000), or at whole-individual changes in metabolic rates, stress responses, growth rates, or immune profiles (e.g., Kingsolver & Huey 1998; Handelsman *et al.* 2013; Chapters 11, 12, & 14, this volume), morphology (e.g., Kingsolver & Huey 1998; Torres-Dowdall *et al.* 2012), life history orientation (Stearns & Koella 1986; Chapter 2, this volume), or behavior (Ghalambor *et al.* 2010; Chapter 4, this volume). Indeed, the list of phenotypes is limited only by our ability to measure them.

Can plasticity evolve in response to natural selection? – Yes. Plasticity is both an attribute of a trait (i.e., how its expression changes in response to different environments), and a trait in and of itself, therefore capable of evolving. Furthermore, selection experiments have shown that plasticity can evolve independently of trait means (Scheiner & Lyman 1991; Murren *et al.* 2014), and that such evolutionary changes can be rapid (e.g., Handelsman *et al.* 2013). Indeed, genetic groups commonly respond to environmental variation in different ways. Such variation is referred to as a genotype × environment interaction (G × E), and it reflects the amount of genetic variation for plasticity.

Collectively, the many facets of phenotypic plasticity give it profound ecological and evolutionary significance. Specifically, when we consider the dual role of the environment – as a source of plasticity within generations but a source of natural selection between generations – we find common ground between functional and evolutionary biology.

Same Wine, New Bottle: Shifting to a Reaction Norm Perspective

Functional biologists often embrace the influence of environmental variation on the traits they measure, or try to eliminate it outright through carefully controlled experiments. For example, physiologists have long been interested in the influence of temperature on traits ranging from metabolism to running speed to rates of enzyme-mediated reactions, and they quantify such changes in ways that are comparable across species or populations (e.g., Q_{10}). However, by considering such responses as thermal reaction norms (Angilletta 2009), comparisons between groups focus on changes in slopes and shapes of functions (see also Chapter 3, this volume). Here we illustrate the usefulness of reaction norms across a range of contexts. We begin with a hypothetical example that considers how the phenotypes of three populations raised in a common garden change as a function of three environmental treatments. In this example the trait of interest could be any phenotype, and the treatment could be any environmental condition. A traditional way of representing these results is shown in Figure 1.1a, where we observe that populations differ in their trait values and that the environment influences the expression of the trait. Now consider the same data represented as reaction norms (Figure 1.1b). In Figure 1.1b, we see that whereas all three populations increase their trait values in response to increases in the environmental treatment, population 3 is most plastic (i.e., has the steepest slope). Divergence in the slope of reaction norms measured under common garden conditions on populations of known relatedness provides evidence that plasticity has evolved. Models for the evolution of plasticity predict that the amount of plasticity should increase when populations are exposed to predictably more variable environments (e.g., Levins 1968; Moran 1992). Thus, if the focal trait was metabolic rate and the treatment was temperature, one testable hypothesis is that population 3 comes from a more variable thermal environment (see also Angilletta 2009). A final approach is to visualize variation within each population (Figure 1.1c). Here, we see a new pattern within the data; population 2 has less genetic variation than the others (Figure 1.1c). The major conclusion would be that population 2 lacks significant genetic variation (G × E), and thus would evolve more slowly in response to selection.

Where will Reaction Norm Thinking Lead?

In the more integrative approach we envision, functional biologists will play key roles in describing the actual mechanisms that underlie plasticity (Cossins *et al.* 2006). Ultimately, integrative organismal biology will likely become the study of individual plasticity in gene expression, physiology, development, endocrinology, neuroscience, behavior, and other organismal properties, but in the context of evolutionary theory. Although that

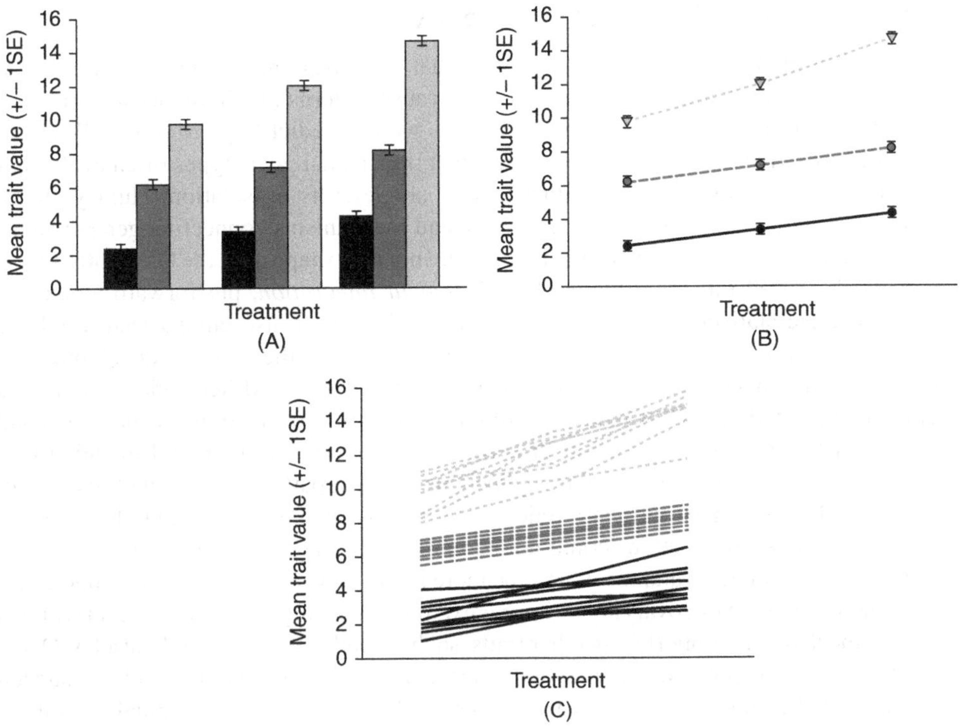

Figure 1.1. Three ways of representing the same data. Shown are three populations under three different environmental treatments. (A) A traditional bar graph showing the mean and variance for each population and each treatment. (B) The mean reaction norm for each population. (C) The individual reaction norms within each population. (See plate section for color version.)

shift is already underway, the real synthesis will come into view when evolutionary theory is extended to incorporate mechanisms at multiple biological levels. For example, understanding mechanism may lead to models that can predict when plasticity should and should not evolve, which traits are more likely to be plastic, or how plasticity in multiple complex interacting traits influences fitness (e.g., Ricklefs & Wikelski 2002; Hau 2007; Marden 2008; McGlothlin & Ketterson 2008; Ketterson *et al.* 2009; Martin *et al.* 2011; Chapter 5, this volume).

THE PROBLEM OF COMPLEXITY

Toward the end of the 17th century, when van Leeuwenhoek developed microscope lenses that allowed him to see microorganisms and other small objects, the apparent complexity of the living world suddenly increased. Since then, our tools and ability to describe and study the building blocks of life have only improved, as has their apparent complexity. Complex biological systems pose a dilemma for evolutionary theory. Although it is obvious that organisms adapt, how variable, viable phenotypes develop from and evolve a diversity of redundant pathways and levels of interacting systems is exceedingly difficult to understand. Nevertheless, organismal and evolutionary biology are converging on a shared theory for mapping genomes to phenotypes and for explaining the evolvability of complex systems.

A Brief History of Biological Complexity

The founders of the *Modern Synthesis* knew that organisms had many genes, and that interactions and correlations among genes could cause patterns of inheritance and trait values to deviate from what simple additive models would predict (e.g., Moore & Williams 2005; Stearns 2010). The founders also knew that organismal phenotypes often evolved as integrated sets of covarying traits rather than as single traits in isolation (Huxley 1932). However, the theory, quantitative descriptions, and mechanisms connecting genetic architecture to integrated "multivariate" phenotypes did not take shape until the late 1950s. Olson and Miller's (1958) influential book, *Morphological Integration,* put forward an explicitly multivariate statistical approach to the study of traits; it also emphasized a holistic view of organismal structure and evolution that considered the genetic, developmental, and functional connections within individuals. They (Olson & Miller 1958) described a method that created groups of traits that were biologically and functionally integrated and could be distinguished using statistical correlations (i.e., pF-groups), and quantitatively compared across species. One of their key insights was that patterns of morphological integration could constrain evolutionary trajectories; because certain traits had to be inherited as groups, organismal functions could not be optimized. They argued that functionally or developmentally related traits would also be more likely to be controlled by the same genes to prevent the breakup of covarying traits. For example, crushing and grinding food requires intimate, mechanistic support from other traits, such as teeth, jawbones, and muscles. Olson and Miller's (1958) point was that evolutionary change in tooth length could not happen without related changes in jaw and muscle traits. Their ideas are conceptual cousins to more modern gene and physiological regulatory networks that coordinate gene expression throughout development (Davidson 2001; Cohen *et al.* 2012) and concepts like symmorphosis (Weibel *et al.* 1991), which seek to understand how multiple parts of a pathway interact to perform a function.

Around the same time, plant biologists were also investigating trait integration. Clausen and Hiesey (1960) were interested in how trait correlations changed among ecotypes, and they conducted breeding experiments to investigate the mechanisms of inheritance underlying what they called "character coherence." They were particularly concerned with how selection acted on constellations of genes and how the tension between character coherence and variation was resolved in natural populations. In a similar vein, Berg (1960) resurrected the concept of "correlation pleiades" to describe patterns of modularity and trait correlations within flowers, and argued for the importance of correlational selection favoring the evolution of integrated floral traits. Today, patterns of trait correlation, and the ways in which selection shapes them, are commonly studied in a quantitative genetic framework (Lande 1979; Hansen 2006).

Collectively, these early ideas laid the foundation for modern views of how interactions of suborganismal factors (e.g., genes, hormones, morphologies) contribute to phenotypic integration and modularity at the individual level and among-individual variation at the population level (e.g., Lande 1979; Cheverud 1982, 1996; Mezey *et al.* 2000; Wagner & Altenberg 1996; Wagner *et al.* 2007; Klingenberg 2008; Martin *et al.* 2011; Wagner & Zhang 2011). This new directive of how genes map to phenotypes, however, poses an evolutionary dilemma: if traits are developmentally and functionally integrated, they will need to evolve as genetically integrated units, or in other words be under strong pleiotropic control. If correlational selection favors integration to maintain function, how do new beneficial mutations and selection act on variation without compromising performance? Breaking up tightly linked groups seems highly unlikely. These issues fall under the umbrella of "evolvability," a concept that comprises how complex traits arise, how selection acts on variation within complex systems, and how constraints and trade-offs bias evolution.

Wagner and Altenberg (1996) partly solved these problems, arguing that genes mapped to phenotypes in highly modular ways. Parcellation of traits into functional modules would offset the negative effects of mutations and pleiotropy on traits belonging to different modules. Thus, the evolution of modularity in otherwise integrated organisms decouples the correlated response to selection. Decoupling also increases evolvability by allowing genetic and phenotypic modules to change independently without compromising function (Wagner & Altenberg 1996). These ideas have been transformative, especially in light of our increasing knowledge about the complex ways genes interact to influence the development of phenotypes (Wagner *et al.* 2007; Wagner & Zhang 2011; Pavličev & Wagner 2012, Chapters 5 & 6, this volume).

The utility of modularity can be seen with a simple caricature of a complex system. In Figure 1.2a, we see a pattern where the symbols could represent any set of characters, genes, molecules, or functional traits. In this caricature of a complex system, traits are able to function and evolve independently. In Figure 1.2b, each character is connected to others by an interaction and correlational selection means that the efficacy/efficiency and evolvability of some traits (and the system generally) is impacted strongly by linkages among elements. Figure 1.2b is therefore a caricature of a highly-integrated complex system, much like the floral and feeding traits described above. Figure 1.2c represents a possible resolution to problems of trait correlations (as well as within generation matching of phenotypes to environments; West-Eberhard 2003; Monaghan 2008; Martin *et al.* 2011); modularity enables traits to evolve as collectives and only those elements with functional connections to others experience correlated responses to selection. In other words, components of modules are highly integrated, but modules are minimally connected to other modules. Such thinking dominates much of systems biology, and the ideas are applicable to all levels of biological investigation.

What role does modularity have in the present volume? A major one. In his 2006 book, *The Music of Life: Biology Beyond the Genome*, Denis Noble claims that the field of systems biology is about

> putting together rather than taking apart, integration rather than reduction. It requires that we develop ways of thinking about integration that are as rigorous as our reductionist programmes, but different.... It means changing our philosophy, in the full sense of the term.

We agree, as it is through the integrated, individual organism that genetic variation is translated into phenotypic variation (Pavličev & Wagner 2012, Chapter 6, this volume).

EMBRACING THE INDIVIDUAL AND THE UNION OF FUNCTIONAL AND EVOLUTIONARY BIOLOGY

This chapter has focused on two emerging themes in organismal biology: phenotypic plasticity and biological complexity. Because these concepts describe properties of individuals, they collectively describe how environmental variation acts on genetic variation to shape phenotypes. We see these and other topics as progress toward a *theory of the individual* (see also Chapter 19, Conclusions to this volume), a theory whereby mechanistic understanding at different levels of biological organization informs, and is guided by, broader

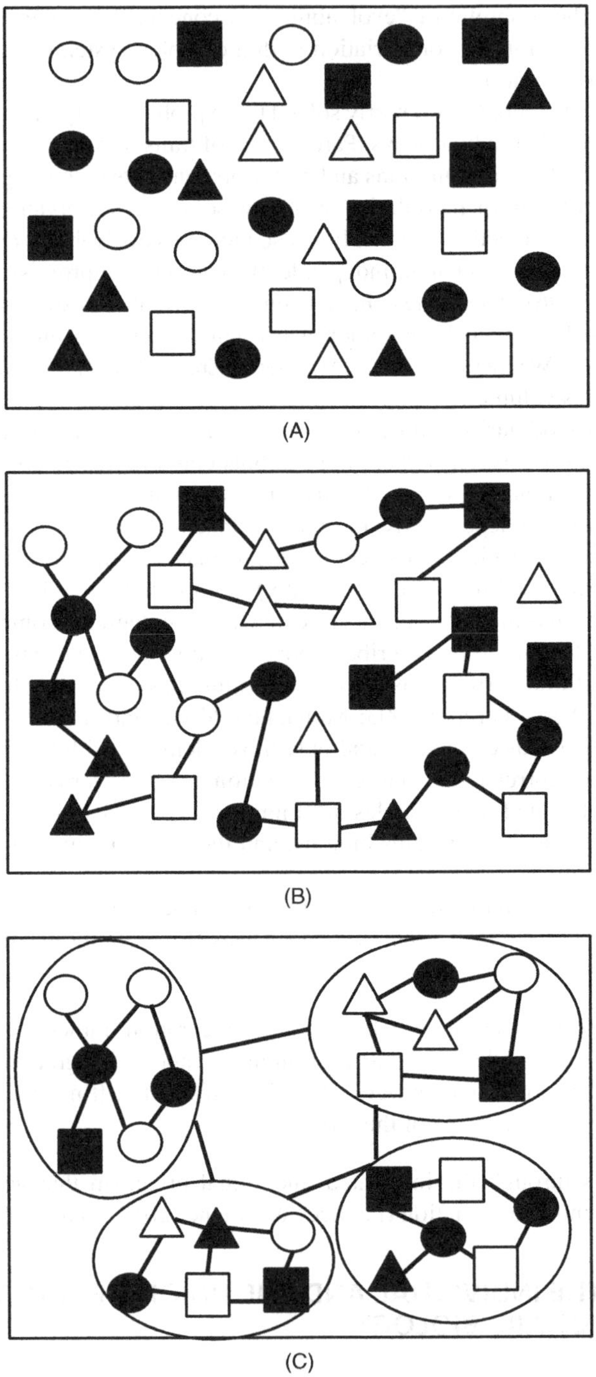

Figure 1.2. A visual caricature of complex systems. (A) A complex system of different symbols, where each symbol could represent genes, traits, or any other set of characters that are independent of each other. (B) An integrated complex system of characters, where the lines connecting the symbols represents correlations and interactions between characters. (C) A modular complex system, where each large circle represents a module containing a highly integrated set of characters.

evolutionary theory. We feel that biologists embracing such a perspective are poised to move beyond statistical descriptions of gene–trait relationships to a more realistic understanding of how environments interface with genetic, developmental, and physiological networks to influence organismal function and performance.

We anticipate skepticism from both functional and evolutionary biologists. For example, in an exchange among Thomas Flatt, Andreas Heyland, and Stephen Stearns (2012), Flatt and Heyland argued that mechanism might force changes in life history theory. Stearns disagreed, claiming that although the study of mechanism has shed some light on trade-offs, it has not fundamentally altered life history theory. On this point, we (and others – e.g., Pigliucci & Muller 2010) side with Flatt and Heyland. To illustrate the importance of mechanism to the development of evolutionary theory, consider the example of organismal responses to heat stress. Functional and evolutionary biologists are interested in the mechanisms of thermal tolerance and adaptation to temperature and climate, but they approach the problem with different motives. The convergence of these interests rests on the highly conserved gene family from bacteria to vertebrates that produce a group of protein chaperones; heat shock proteins (hsps). Initially, as their name implies, the expression of heat shock proteins was viewed in the context of acute thermal stress, but we now know that hsps stabilize and refold unfolded proteins even in benign contexts and are thus involved in diverse processes ranging from gene expression to chromatin remodeling (Lindquist & Craig 1988; Rutherford *et al.* 2007). Not surprisingly, evolutionary biologists and ecologists are often interested in adaptive hsp differences between populations and species in response to different thermal environments, whereas cellular and molecular biologists often seek to understand the mechanisms regulating hsp expression and the resulting functional consequences (Lindquist & Craig 1988; Moseley 1997; Hoffmann *et al.* 2003; Sorenson *et al.* 2003; Fangue *et al.* 2006). As with most phenotypic traits, heat shock proteins are plastic, have the capacity to evolve, and are highly integrated with other systems to perform diverse functions.

What gives hsps to the power to force revisions to evolutionary theory is their ability to act as capacitors, particularly Hsp90. Like other molecular chaperones, Hsp90 assists in proper protein folding and protein degradation in normal cells and during heat stress. However, it also acts on signal transducers to keep unstable signaling proteins ready for activation. Hsp90 thus lies at the interface of many cellular, physiological, and developmental pathways (e.g., Rutherford & Lindquist 1998; Rutherford 2000; Rutherford *et al.* 2007; Taipale *et al.* 2010). In this capacity, Hsp90 acts at the hub of many regulatory circuits and thus masks or ccompensates for mutations that would otherwise have potentially deleterious effects on the phenotype (Taipale *et al.* 2010). This key insight was made when Hsp90 expression was altered across a range of different genetic backgrounds in *Drosophila* and *Arabidopsis*; previously silent mutations were revealed with substantial phenotypic consequences (e.g., Rutherford & Lindquist 1998; Queitsch *et al.* 2002; Milton *et al.* 2003). In other words, although substantial genetic variation existed within different genetic backgrounds, Hsp 90 covered up this variation under typical environments and effectively hid genetic variation from natural selection (e.g., Rutherford *et al.* 2007; Jarosz *et al.* 2010).

The evolutionary implications of these results are significant. In times of environmental stress, when Hsp90's protein clients become destabilized and the buffering capacity of Hsp90 is strained, cryptic genetic and phenotypic variation may be revealed (Rutherford & Lindquist 1998; Queitsch *et al.* 2002; Milton *et al.* 2003; Rutherford 2000; Rutherford *et al.* 2007; Jarosz & Lindquist 2010; Jarosz *et al.* 2010; Rohner *et al.* 2013). This buffering capacity and potential for releasing variation for selection to act on has been referred to as evolutionary capacitance. Importantly from the perspective of this volume, many suborganismal mechanisms may act in a similar way. Indeed, capacitance is a general feature of complex gene networks and a form of epigenetic inheritance (e.g., Bergman & Siegal 2003; Dickins & Rahman 2012) and has been argued to act similarly at physiological (Chapter 9, this volume; Cohen *et al.*, 2012, Martin *et al.*, 2011) and behavioral levels (Duckworth 2009; Ledón-Rettig *et al.* 2013; Snell-Rood 2013). Other capacitors, such as the yeast prion [PSI+] also exist, and both empirical and theoretical work is revealing a role for capacitors in various traits from cancer (Whitesell & Lindquist 2005; Feinberg *et al.* 2006; Feinberg 2007) to the evolvability of populations (e.g., Cowen & Lindquist 2005; Wagner 2008; Masel & Trotter 2010; Jarosz & Lindquist 2010). Thus, capacitors serve as an example of how a mechanistic understanding has led to the development of a new model for evolutionary change. Similar contributions are likely to increase in the future as the barriers between functional and evolutionary biology erode.

CONCLUSION

Integrative Organismal Biology can be envisioned as an exercise in opening, or at least shrinking, the black box in which genetic and environmental variation come together to shape the phenotype. The major impediment to the *Modern Synthesis* was the difficulty of partitioning genetic and environmental variation. Biologists often forget that R.A. Fisher, Sewell Wright, and others laid the foundation for modern statistics mostly because the tools to quantify variation did not exist. In the process, the contributors to the *Modern Synthesis* took the first steps toward a cohesive theory of biology. We are now in the midst of a new synthesis that is using a new set of tools. This new synthesis embraces, rather than simplifies, the nature of plastic, yet integrated, organisms. However, we argue that part of what is holding us back is, ironically enough, the very things that have driven so much progress over the past century: reductionism and theory. Reductionism has driven massive progress in many fields but, at the same time, has conceptually extracted genes from the whole organism and their natural environments. Likewise, theory proliferated to shed light on how different environmental conditions shape the evolvability of different lineages, fluctuations in population sizes, rates of speciation, the distribution of species, the emergence of infectious diseases, and other important biological phenomena, but without an understanding of the mechanisms operating at the individual level. By contrast, an integrative biological philosophy that embraces complex systems might profoundly shift how we approach questions, ultimately catalyzing genuine synthesis across biological disciplines. Such a philosophy would embrace not only the complexity of organisms, but also their complex interactions with environments (i.e., plasticity). Indeed, the causal arrows spanning genes and proteins are now appreciated as multidirectional and inseparable from the environmental context in which they occur (West-Eberhard 2003; Noble 2006, 2013). Again, the environment serves two biological roles: it regulates phenotypic variation through gene function, genetic regulatory networks, and interacting developmental and physiological systems, but it also sorts this variation at the population level through selection. It is therefore only in the natural, ecological context of the organism that integration across levels of biological organization is truly possible. Noble (2013) echoed these ideas in his claim that, "Physiology is rocking the foundations of evolutionary biology." What we think he meant (and what the chapters of much of this volume emphasize) is that we will make great strides in biology by augmenting reductionist approaches with studies of the interactions and integration of genetic, developmental and physiological pathways that comprise whole organisms in real environments.

REFERENCES

Agrawal, A.A. (2001) Phenotypic plasticity in the interactions and evolution of species. *Science*, **294**, 321–6.

Angilletta, M.J. (2009) *Thermal Adaptation: A Theoretical and Empirical Synthesis*, Oxford University Press.

Arnold, S.J. (1983) Morphology, performance and fitness. *American Zoologist*, **23**, 347–61.

Arnold, S.J. (1987) Genetic correlation and the evolution of physiology, in *New Directions in Ecological Physiology* (eds M.E. Feder *et al.*), Cambridge University Press, pp. 189–215.

Aubin-Horth, N. and Renn, S.C. (2009) Genomic reaction norms: using integrative biology to understand molecular mechanisms of phenotypic plasticity. *Molecular Ecology*, **18**, 3763–80.

Badyaev, A.V. (2009) Evolutionary significance of phenotypic accommodation in novel environments: an empirical test of the Baldwin effect. *Philosophical Transactions of the Royal Society B: Biological Sciences*, **364**, 1125–41.

Badyaev, A.V. and Landeen, E.A. (2007) Developmental evolution of sexual ornamentation: model and a test of feather growth and pigmentation. *Integrative and Comparative Biology*, **47**, 221–33.

Baldwin, J.M. (1896) A new factor in evolution. *American Naturalist*, **30**, 441–51.

Bartholomew, G.A. (1958) The role of physiology in the distribution of terrestrial vertebrates. *Zoogeography* (ed. C.L. Hubbs), pp. 81–95, Publ. 51. American Association for the Advancement of Science, Washington, DC.

Bartholomew, G.A. (1964) The roles of physiology and behaviour in the maintenance of homeostasis in the desert environment. *Symposia of the Society for Experimental Biology*, **18**, 7–29.

Bartholomew, G.A. (1966) Interaction of physiology and behavior under natural conditions, in *The Galapagos* (ed R.I. Bowman), University of California Press, Berkeley, pp. 39–45.

Bartholomew, G.A. (1986) The role of natural history in contemporary biology. *BioScience*, **36**, 324–392.

Bell, G. (2009) *Selection: The Mechanism of Evolution*, Oxford University Press.

Bennett, A.F. (1987) Interindividual variability: an underutilized resource, in *New Directions in Ecological Physiology* (eds M.E. Feder *et al.*), University Press, Cambridge, pp. 147–69.

Berg, R.L. (1960) The ecological significance of correlation pleiades. *Evolution*, **14**, 171–80.

Bergman, A. and Siegal, M.L. (2003) Evolutionary capacitance as a general feature of complex gene networks. *Nature*, **424**, 549–52.

Borenstein, E., Meilijson, I. and Ruppin, E. (2006) The effect of phenotypic plasticity on evolution in multipeaked fitness landscapes. *Journal of Evolutionary Biology*, **19**, 1555–70.

Bradshaw, A.D. (1965) Evolutionary significance of phenotypic plasticity in plants. *Advances in Genetics*, **13**, 115–55.

Calow, P. (ed) (1987) *Evolutionary Physiological Ecology*, Cambridge University Press, New York.

Carroll, S.B. (2005) *Endless Forms Most Beautiful: The New Science of Evo Devo and the Making of the Animal Kingdom*, WW Norton & Company.

Carroll, S.P. and Fox, C.W. (eds) (2008) *Conservation Biology: Evolution in Action*, Oxford University Press.

Chapman, L.G., Galis, F. and Shinn, J. (2000) Phenotypic plasticity and the possible role of genetic assimilation: Hypoxia-induced trade-offs in the morphological traits of an African cichlid. *Ecology Letters*, **3**, 387–93.

Cheverud, J.M. (1982) Phenotypic, genetic, and environmental morphological integration in the cranium. *Evolution*, **36**, 499–516.

Cheverud, J.M. (1996) Developmental integration and the evolution of pleiotropy. *American Zoologist*, **36**, 44–50.

Chown, S.L. and Gaston, K.J. (2008) Macrophysiology for a changing world. *Proceedings of the Royal Society B: Biological Sciences*, **275**, 1469–78.

Clausen, J. and Hiesey, W.M. (1960) The balance between coherence and variation in evolution. *Proceedings of the National Academy of Sciences of the United States of America*, **46**, 494.

Cohen, A.A., Martin, L.B., Wingfield, J.C. *et al.* (2012) Physiological regulatory networks: ecological roles and evolutionary constraints. *Trends in Ecology & Evolution*, **27**, 428–35.

Conover, D.O., Duffy, T.A. and Hice, L.A. (2009) The covariance between genetic and environmental influences across ecological gradients. *Annals of the New York Academy of Sciences*, **1168**, 100–29.

Cossins, A., Fraser, J., Hughes, M. and Gracey, A. (2006) Post-genomic approaches to understanding the mechanisms of environmentally induced phenotypic plasticity. *Journal of Experimental Biology*, **209**, 2328–36.

Cowen, L.E. and Lindquist, S. (2005) Hsp90 potentiates the rapid evolution of new traits: drug resistance in diverse fungi. *Science*, **309**, 2185–9.

Darwin, C. (1859) *On the Origins of Species by Means of Natural Selection*, Murray, London.

Davidson, E.H. (2001) *Genomic Regulatory Systems: In Development and Evolution*, Academic Press.

Dickins, T.E. and Rahman, Q. (2012) The extended evolutionary synthesis and the role of soft inheritance in evolution. *Proceedings of the Royal Society B: Biological Sciences*, **279**, 2913–21.

Dobzhansky, T. (1964) Biology, molecular and organismic. *American Zoologist*, **4**, 443–52.

Dobzhansky, T. (1973) Nothing in biology makes sense except in the light of evolution. *American Biology Teacher*, **35**, 125–9.

Duckworth, R.A. (2009) The role of behavior in evolution: a search for mechanism. *Evolutionary Ecology*, **23**, 513–31.

Falconer, D.S. (1990) Selection in different environments: effects on environmental sensitivity (reaction norm) and on mean performance. *Genetical Research*, **56**, 57–70.

Fangue, N.A., Hofmeister, M. and Schulte, P.M. (2006) Intraspecific variation in thermal tolerance and heat shock protein gene expression in common killifish, *Fundulus heteroclitus*. *Journal of Experimental Biology*, **209**, 2859–72.

Feder, M.E., Bennett, A.F., Burggren, W.W. and Huey, R.B. (eds) (1987) *New Directions in Ecological Physiology*, Cambridge University Press.

Feinberg, A.P. (2007) Phenotypic plasticity and the epigenetics of human disease. *Nature*, **447**, 433–40.

Feinberg, A.P., Ohlsson, R. and Henikoff, S. (2006) The epigenetic progenitor origin of human cancer. *Nature Reviews Genetics*, **7**, 21–33.

Flatt, T. and Heyland, A. (eds) (2011) *Mechanisms of Life History Evolution: The Genetics and Physiology of Life History Traits and Trade-offs*, Oxford University Press.

Garland, T. Jr. and Carter, P.A. (1994) Evolutionary physiology. *Annual Review of Physiology*, **56**, 579–621.

Garland, T., Bennett, A.F. and Rezende, E.L. (2005) Phylogenetic approaches in comparative physiology. *Journal of Experimental Biology*, **208**, 3015–35.

Gaston, K.J., Chown, S.L., Calosi, P. *et al.* (2009) Macrophysiology: a conceptual reunification. *The American Naturalist*, **174**, 595–612.

Gerhart, J. and Kirschner, M. (2007) The theory of facilitated variation. *Proceedings of the National Academy of Sciences*, **104** (Suppl 1), 8582–9.

Ghalambor, C.K., McKay, J.K., Carroll, S.P. and Reznick, D.N. (2007) Adaptive versus non-adaptive phenotypic plasticity and the potential for contemporary adaptation in new environments. *Functional Ecology*, **21**, 394–407.

Ghalambor, C.K., Angeloni, L.M. and Carroll, S.P. (2010) Behavior as phenotypic plasticity, in *Evolutionary Behavioral Ecology* (eds C. Fox and D. Westneat), Oxford University Press, pp. 90–108.

Gilbert, S.F. (2003) The morphogenesis of evolutionary developmental biology. *International Journal of Developmental Biology*, **47**, 467–78.

Gilbert, S.F. and Epel, D. (2009) *Ecological Developmental Biology: Integrating Epigenetics, Medicine, and Evolution*, Sinauer Associates, Sunderland.

Goldschmidt, R. (1940) *The Material Basis of Evolution* (Vol. **28**). Yale University Press.

Goodman, C.S. and Coughlin, B.C. (2000) The evolution of evo-devo biology. *Proceedings of the National Academy of Sciences*, **97**, 4424–5.

Gottlieb, G. (2001) *Individual Development and Evolution: The Genesis of Novel Behavior*, Psychology Press.

Grether, G.F. (2005) Environmental change, phenotypic plasticity, and genetic compensation. *The American Naturalist*, **166**, E115–E123.

Hall, B.K. (ed) (2012) *Homology: The Hierarchial Basis of Comparative Biology*, Academic Press.

Handelsman, C.A., Broder, E.D., Dalton, C.M. *et al.* (2013) Predator-induced phenotypic plasticity in metabolism and rate of growth: rapid adaptation to a novel environment. *Integrative and Comparative Biology*, **53**, 975–88.

Hansen, T.F. (2006) The evolution of genetic architecture. *Annual Review of Ecology, Evolution, and Systematics*, **37**, 123–57.

Hansen, T.F. and Houle, D. (2004) Evolvability, stabilizing selection, and the problem of stasis, in *Phenotypic Integration: Studying the Ecology and Evolution of Complex Phenotypes* (eds M. Pigliucci and K. Preston), Oxford University Press, pp. 130–50.

Harrison, J.F., Woods, H.A. and Roberts, S.P. (2012) *Ecological and Environmental Physiology of Insects*, Oxford University Press.

Hau, M. (2007) Regulation of male traits by testosterone: implications for the evolution of vertebrate life histories. *BioEssays*, **29**, 133–44.

Hochachka, P.W. & Somero, G.N. (2002) *Biochemical Adaptation: Mechanism and Process in Physiological Evolution* (Vol. **480**). Oxford University Press.

Hoffmann, A.A., Sørensen, J.G. and Loeschcke, V. (2003) Adaptation of *Drosophila* to temperature extremes: bringing together quantitative and molecular approaches. *Journal of Thermal Biology*, **28**, 175–216.

Huey, R.B. and Bennett, A.F. (1986) A comparative approach to field and laboratory studies in evolutionary biology, in *Predator–Prey Relationships: Perspectives and Approaches from the Study of Lower Vertebrates* (eds M.E. Feder and G.V. Lauder), University of Chicago Press, Chicago, pp. 82–98.

Huey, R.B. and Bennett, A.F. (2008) Bart's familiar quotations: the enduring biological wisdom of George A. Bartholomew. *Physiological and Biochemical Zoology*, **81**, 519–25.

Huey, R.B., Hertz, P.E. and Sinervo, B. (2003) Behavioral drive versus behavioral inertia in evolution: a null model approach. *The American Naturalist*, **161**, 357–66.

Huey, R.B. and Pianka, E.R. (1981) Ecological consequences of foraging mode. *Ecology*, **62**, 991–9.

Huxley, J.H. (1932) *Problems of Relative Growth*, Methuen.

Jablonka, E. and Lamb, M.J. (2005) *Evolution in Four Dimensions: Genetic, Epigenetic, Behavioral, and Symbolic Variation in the History of Life*, MIT Press.

Jarosz, D.F. and Lindquist, S. (2010) Hsp90 and environmental stress transform the adaptive value of natural genetic variation. *Science*, **330**, 1820–4.

Jarosz, D.F., Taipale, M. and Lindquist, S. (2010) Protein homeostasis and the phenotypic manifestation of genetic diversity: principles and mechanisms. *Annual Review of Genetics*, **44**, 189–216.

Kalinka, A.T. and Tomancak, P. (2012) The evolution of early animal embryos: conservation or divergence? *Trends in Ecology & Evolution*, **27**, 385–93.

Ketterson, E.D., Atwell, J.W. and McGlothlin, J.W. (2009) Phenotypic integration and independence: hormones, performance, and response to environmental change. *Integrative and Comparative Biology*, **49**, 365–79.

Kingsolver, J.G. and Huey, R.B. (1998) Evolutionary analyses of morphological and physiological plasticity in thermally variable environments. *American Zoologist*, **38**, 545–60.

Kingsolver, J.G. and Huey, R.B. (2003) Introduction: the evolution of morphology, performance, and fitness. *Integrative and Comparative Biology*, **43**, 361–6.

Kirschner, M. and Gerhart, J. (1998) Evolvability. *Proceedings of the National Academy of Sciences*, **95**, 8420–7.

Klingenberg, C.P. (2008) Morphological integration and developmental modularity. *Annual Review of Ecology, Evolution, and Systematics*, **39**, 115–32.

Koehn, R.K. (1987) The importance of genetics to physiological ecology, in *New Directions in Ecological Physiology* (eds M.E. Feder *et al.*), Cambridge University Press, pp. 170–188.

Krogh, A. (1929) The progress of physiology. *Science*, **70**, 200–4.

Lande, R. (1979) Quantitative genetic analysis of multivariate evolution, applied to brain: body size allometry. *Evolution*, **33**, 402–16.

Ledón-Rettig, C.C., Richards, C.L. and Martin, L.B. (2013) Epigenetics for behavioral ecologists. *Behavioral Ecology*, **24**, 311–24.

Lee, C.E., Remfert, J.L. and Gelembiuk, G.W. (2003) Evolution of physiological tolerance and performance during freshwater invasions. *Integrative and Comparative Biology*, **43**, 439–49.

Lee, C.E. and Petersen, C.H. (2003) Effects of developmental acclimation on adult salinity tolerance in the freshwater-invading copepod *Eurytemora affinis*. *Physiological and Biochemical Zoology*, **76**, 296–301.

Lessells, C.M. (2008) Neuroendocrine control of life histories: what do we need to know to understand the evolution of phenotypic plasticity? *Philosophical Transactions: Biological Sciences*, **363**, 1589–98.

Levins, R. (1968) *Evolution in Changing Environments*, Princeton University Press.

Lindquist, S. and Craig, E.A. (1988) The heat-shock proteins. *Annual Review of Genetics*, **22**, 631–77.

Losos, J.B., Creer, D.A., Glossip, D. *et al.* (2000) Evolutionary implications of phenotypic plasticity in the hindlimb of the lizard *Anolis sagrei*. *Evolution*, **54**, 301–5.

Losos, J.B., Schoener, T.W., Langerhans, R.B. and Spiller, D.A. (2006) Rapid temporal reversal in predator-driven natural selection. *Science*, **314**, 1111.

Losos, J.B., Arnold, S.J., Bejerano, G. *et al.* (2013) Evolutionary biology for the 21st century. *PLoS Biology*, **11**, e1001466.

Marden, J.H. (2008) Quantitative and evolutionary biology of alternative splicing: how changing the mix of alternative transcripts affects phenotypic plasticity and reaction norms. *Heredity*, **100**, 111–20.

Martin, L.B., Liebl, A.L., Trotter, J.H. *et al.* (2011) Integrator networks: illuminating the black box linking genotype and phenotype. *Integrative and Comparative Biology*, **51**, 514–27.

Masel, J. and Trotter, M.V. (2010) Robustness and evolvability. *Trends in Genetics*, **26**, 406–14.

Mayr, E. (1961) Cause and effect in biology. *Science*, **134**, 1501–6.

Mayr, E. and Provine, W.B. (eds) (1998) *The Evolutionary Synthesis: Perspectives on the Unification of Biology*, Harvard University Press.

McCairns, R.J. and Bernatchez, L. (2010) Adaptive divergence between freshwater and marine sticklebacks: insights into the role of phenotypic plasticity from an integrated analysis of candidate gene expression. *Evolution*, **64**, 1029–47.

McGlothlin, J.W. and Ketterson, E.D. (2008) Hormone-mediated suites as adaptations and evolutionary constraints. *Philosophical Transactions of the Royal Society B: Biological Sciences*, **363**, 1611–20.

Mezey, J.G., Cheverud, J.M. and Wagner, G.P. (2000) Is the genotype-phenotype map modular?: a statistical approach using mouse quantitative trait loci data. *Genetics*, **156**, 305–11.

Milton, C.C., Huynh, B., Batterham, P. *et al.* (2003) Quantitative trait symmetry independent of Hsp90 buffering: distinct modes of genetic canalization and developmental stability. *Proceedings of the National Academy of Sciences*, **100**, 13396–13401.

Monaghan, P. (2008) Early growth conditions, phenotypic development and environmental change. *Philosophical Transactions of the Royal Society B: Biological Sciences*, **363**, 1635–45.

Moore, J.H. and Williams, S.M. (2005) Traversing the conceptual divide between biological and statistical epistasis: systems biology and a more modern synthesis. *Bioessays*, **27**, 637–46.

Moran, N.A. (1992) The evolutionary maintenance of alternative phenotypes. *American Naturalist*, **139**, 971–89.

Moran, N.A. (1994) Adaptation and constraint in the complex life cycles of animals. *Annual Review of Ecology and Systematics*, 573–600.

Moseley, P.L. (1997) Heat shock proteins and heat adaptation of the whole organism. *Journal of Applied Physiology*, **83**, 1413–17.

Murren, C., Maclean, H.J., Diamond, S.E., Steiner, U.K., Heskel, M.A., Handelsman, C.A., Ghalambor, C.K., Auld, J.R., Callahan, H.S., Pfennig, D.W., Relyea, R.A., Schlichting, C.D., and J. Kingsolver (2014) Evolutionary change in continuous reaction norms. *American Naturalist*. In press.

Mykles, D.L., Ghalambor, C.K., Stillman, J.H. and Tomanek, L. (2010) Grand challenges in comparative physiology: integration across disciplines and across levels of biological organization. *Integrative and Comparative Biology*, **50**, 6–16.

Nesse, R.M., Bergstrom, C.T., Ellison, P.T. *et al.* (2010) Making evolutionary biology a basic science for medicine. *Proceedings of the National Academy of Sciences*, **107**, 1800–7.

Noble, D. (2006) *The Music of Life*, Oxford University Press.

Noble, D. (2013) Physiology is rocking the foundations of evolutionary biology. *Experimental Physiology*, **98**, 1235–43.

Olson, E.C. and Miller, R.L. (1958) *Morphological Integration*, University of Chicago Press.

Orians, G.H. (1962) Natural selection and ecological theory. *The American Naturalist*, **96**, 257–63.

Paenke, I., Sendhoff, B. and Kawecki, T.J. (2007) Influence of plasticity and learning on evolution under directional selection. *The American Naturalist*, **170**, E47–E58.

Parker, G.A. and Smith, J.M. (1990) Optimality theory in evolutionary biology. *Nature*, **348**, 27–33.

Pavličev, M. and Wagner, G.P. (2012) A model of developmental evolution: selection, pleiotropy and compensation. *Trends in Ecology & Evolution*, **27**, 316–22.

Perlman, R. (2013) *Evolution and Medicine*, Oxford University Press, Oxford.

Piersma, T. and Van Gils, J.A. (2011) *The Flexible Phenotype: A Body-centred Integration of Ecology, Physiology, and Behaviour*, Oxford University Press, Oxford.

Pigliucci, M. (2001) *Phenotypic Plasticity: Beyond Nature and Nurture*, Johns Hopkins University Press.

Pigliucci, M. (2007) Do we need an extended evolutionary synthesis? *Evolution*, **61**, 2743–9.

Pigliucci, M. and Muller, G. (2010) *Evolution – the Extended Synthesis*, MIT Press.

Price, T.D., Qvarnström, A. and Irwin, D.E. (2003) The role of phenotypic plasticity in driving genetic evolution. *Proceedings of the Royal Society of London. Series B: Biological Sciences*, **270**, 1433–40.

Proulx, S.R., Promislow, D.E. and Phillips, P.C. (2005) Network thinking in ecology and evolution. *Trends in Ecology & Evolution*, **20**, 345–53.

Queitsch, C., Sangster, T.A. and Lindquist, S. (2002) Hsp90 as a capacitor of phenotypic variation. *Nature*, **417**, 618–24.

Ricklefs, R.E. and Wikelski, M. (2002) The physiology/life-history nexus. *Trends in Ecology & Evolution*, **17**, 462–8.

Robinson, B.W. and Dukas, R. (1999) The influence of phenotypic modifications on evolution: the Baldwin effect and modern perspectives. *Oikos*, **85**, 582–9.

Rogers, S.O. (2012) *Integrated Molecular Evolution*, CRC Press.

Rohner, N., Jarosz, D.F., Kowalko, J.E. *et al.* (2013) Cryptic variation in morphological evolution: HSP90 as a capacitor for loss of eyes in cavefish. *Science*, **342**, 1372–5.

Rutherford, S.L. (2000) From genotype to phenotype: buffering mechanisms and the storage of genetic information. *BioEssays*, **22**, 1095–1105.

Rutherford, S., Hirate, Y. and Swalla, B.J. (2007) The Hsp90 capacitor, developmental remodeling, and evolution: the robustness of gene networks and the curious evolvability of metamorphosis. *Critical Reviews in Biochemistry and Molecular Biology*, **42**, 355–72.

Rutherford, S.L. and Lindquist, S. (1998) Hsp90 as a capacitor for morphological evolution. *Nature*, **396**, 336–42.

Satterlie, R.A., Pearse, J.S. and Sebens, K.P. (2009) The black box, the creature from the Black Lagoon, August Krogh, and the dominant animal. *Integrative and Comparative Biology*, **49**, 89–92.

Scheiner, S.M. and Lyman, R.F. (1991) The genetics of phenotypic plasticity. II. Response to selection. *Journal of Evolutionary Biology*, **4**, 23–50.

Schlichting, C.D. and Pigliucci, M. (1998) *Phenotypic Evolution: A Reaction Norm Perspective*, Sinauer Associates Incorporated.

Schmalhausen, I.I. (1949) *Factors of Evolution*, Blakiston.

Sharan, R., Suthram, S., Kelley, R.M. *et al.* (2005) Conserved patterns of protein interaction in multiple species. *Proceedings of the National Academy of Sciences of the United States of America*, **102**, 1974–9.

Sibly, R.M. & Calow, P. (1986) *Physiological Ecology of Animals: An Evolutionary Approach* (Vol. **179**). Blackwell Scientific Publications.

Snell-Rood, E.C. (2013) An overview of the evolutionary causes and consequences of behavioural plasticity. *Animal Behaviour*, **85**, 1004–11.

Sørensen, J.G., Kristensen, T.N. and Loeschcke, V. (2003) The evolutionary and ecological role of heat shock proteins. *Ecology Letters*, **6**, 1025–37.

Spicer, J. and Gaston, K. (1999) *Physiological Diversity: Ecological Implications*, Blackwell Science.

Stearns, F.W. (2010) One hundred years of pleiotropy: a retrospective. *Genetics*, **186**, 767–73.

Stearns, S.C. and Koella, J.C. (1986) The evolution of phenotypic plasticity in life-history traits: predictions of reaction norms for age and size at maturity. *Evolution*, **40**, 893–913.

Stern, D.L. (2010) *Evolution, Development and the Predictable Genome*, Roberts & Company.

Strange, K. (2005) The end of "naive reductionism": rise of systems biology or renaissance of physiology? *American Journal of Physiology Cell Physiology*, **288**, C968–974.

Taipale, M., Jarosz, D.F. and Lindquist, S. (2010) HSP90 at the hub of protein homeostasis: emerging mechanistic insights. *Nature Reviews Molecular Cell Biology*, **11**, 515–28.

Tomanek, L. and Somero, G.N. (2000) Time course and magnitude of synthesis of heat-shock proteins in congeneric marine snails (genus *Tegula*) from different tidal heights. *Physiological and Biochemical Zoology*, **73**, 249–56.

Torres-Dowdall, J., Handelsman, C.A., Reznick, D.N. and Ghalambor, C.K. (2012) Local adaptation and the evolution of phenotypic plasticity in Trinidadian guppies (*Poecilia reticulata*). *Evolution*, **66**, 3432–43.

Tracy, C.R., J. Turner, J.S., Bartholomew, G.A. *et al.* (1982) What is physiological ecology? *Bulletin of the Ecological Society of America*, **63**, 340–7.

Waddington, C.H. (1942) Canalization of development and the inheritance of acquired characters. *Nature*, **150**, 563–5.

Waddington, C.H. (1953) Genetic assimilation of an acquired character. *Evolution*, **7**, 118–26.

Wagner, A. (2008) Robustness and evolvability: a paradox resolved. *Proceedings of the Royal Society B: Biological Sciences*, **275**, 91–100.

Wagner, G.P. and Altenberg, L. (1996) Perspective: complex adaptations and the evolution of evolvability. *Evolution*, **50**, 967–76.

Wagner, G.P., Pavličev, M. and Cheverud, J.M. (2007) The road to modularity. *Nature Reviews Genetics*, **8**, 921–31.

Wagner, G.P. and Zhang, J. (2011) The pleiotropic structure of the genotype–phenotype map: the evolvability of complex organisms. *Nature Reviews Genetics*, **12**, 204–13.

Wake, M.H. (2008) Integrative biology: Science for the 21st century. *BioScience*, **58**, 349–53.

Weibel, E.R., Taylor, C.R. and Hoppeler, H. (1991) The concept of symmorphosis: a testable hypothesis of structure-function relationship. *Proceedings of the National Academy of Sciences*, **88**, 10357–61.

West-Eberhard, M.J. (2003) *Developmental Plasticity and Evolution*, Oxford University Press.

Westneat, D., Fox, C. and Fox, C.W. (eds) (2010) *Evolutionary Behavioral Ecology*, Oxford University Press.

Whitesell, L. and Lindquist, S.L. (2005) HSP90 and the chaperoning of cancer. *Nature Reviews Cancer*, **5**, 761–72.

Williams, T.D. (2008) Individual variation in endocrine systems: moving beyond the 'tyranny of the Golden Mean'. *Philosophical Transactions of the Royal Society B: Biological Sciences*, **363**, 1687–98.

Woltereck, R. (1909) Weitere experimentelle Untersuchungen uber Artveranderung, speziell uber das Wesen quantitativer Artunterschiede bei Daphniden. *Verh Dtsch Zool Gesellschaft*, 19.

Woods, H.A. (2014) Mosaic physiology from developmental noise: within-organism physiological diversity as an alternative to phenotypic plasticity and phenotypic flexibility. *Journal of Experimental Biology*, **217**, 35–45.

Wund, M.A., Baker, J.A., Clancy, B. *et al.* (2008) A test of the "flexible stem" model of evolution: ancestral plasticity, genetic accommodation, and morphological divergence in the threespine stickleback radiation. *The American Naturalist*, **172**, 449–62.

Yeh, P.J. and Price, T.D. (2004) Adaptive phenotypic plasticity and the successful colonization of a novel environment. *The American Naturalist*, **164**, 531–42.

Young, R.L. and Badyaev, A.V. (2007) Evolution of ontogeny: linking epigenetic remodeling and genetic adaptation in skeletal structures. *Integrative and Comparative Biology*, **47**, 234–44.

2

THE ECOLOGICAL AND EVOLUTIONARY IMPORTANCE OF VARIATION IN LIFE HISTORY REACTION NORMS

Daniel H. Nussey

Institute of Evolutionary Biology, University of Edinburgh, Edinburgh, United Kingdom

INTRODUCTION

Understanding the causes and consequences of individual differences in phenotypic traits (including life history, behavioral and morphological traits) is of central importance to the field of evolutionary ecology. This variation represents the raw material on which natural selection operates in the wild, as well as underpinning variation in demographic rates which in turn will drive population dynamics. Importantly, a phenotype is not a static property of an individual or genotype. One of the most ubiquitous phenomena in organismal biology is phenotypic plasticity: the capacity of a given genotype (or individual) to change its phenotype in response to a change in the environment (Pigliucci 2001). Plasticity may evolve to allow organisms to maximize fitness in variable environments by matching phenotype to best suit the environmental conditions they find themselves in. However, under this broad definition, it may also reflect nonadaptive or even maladaptive effects of the environment on phenotype. It is now widely appreciated that current efforts to understand how genes link to phenotypes cannot assume additive contributions of genotype and environment, and need to consider plasticity and genotype-by-environment (G × E) interactions (Pigliucci 2001). Furthermore, according to current theoretical models plasticity plays a pivotal role, alongside the potentially much slower process of microevolution, in determining population viability *in situ* in the face of rapid environmental change (Chevin *et al.* 2010; Reed *et al.* 2011). Life history traits are those associated with the amount or rate (e.g., clutch size), timing (e.g., onset of reproduction) or location (e.g., dispersal distance) of events in an organism's life cycle (Roff 1992; Stearns 1992). Life history traits are particularly important

Integrative Organismal Biology, First Edition.
Edited by Lynn B. Martin, Cameron K. Ghalambor, and H. Arthur Woods.

to evolutionary ecologists: they typically have strong effects on demographic rates and fitness, and hence are intimately linked to population dynamics and under strong selection. They are also almost invariably plastic: their expression is influenced by complex interactions between environment and an individual's physiology and neuro-endocrine control systems (Lessells 2008).

Reaction norms have been widely applied in evolutionary biology to characterize phenotypic plasticity in a diversity of traits (Pigliucci 2001; Via *et al.* 1995). A reaction norm is simply a function or series of functions that relates phenotype to the environment. For instance, an extremely simple reaction norm is captured by a linear regression of environment on phenotype, producing a function comprising an intercept (or elevation) and a slope, with the slope reflecting the degree of plasticity (Figure 2.1). The component functions of reaction norms can themselves be considered as properties or traits of an individual or genotype, and evolutionary biologists have long sought to understand how natural selection might maintain genetic variation in plasticity (manifest as a genotype-by-environment interaction or G × E), what the costs and limits of plasticity are, and how variation in reaction norm components map onto fitness (de Witt *et al.* 1998; Pigliucci 2005; Via *et al.* 1995). For very obvious reasons, evolutionary studies of plasticity in animals predominantly involve the study of responses in morphological or life history traits to experimental manipulation of a selectively important environmental cue (e.g., predator presence, levels of intra- or inter-specific competition, light, temperature) under controlled or semi-natural conditions. Such studies have shed important light on the evolutionary processes shaping phenotypic plasticity and maintaining G × E in amenable systems (Pigliucci 2001). The logical next step is to develop complementary laboratory and field systems that will advance a more integrated understanding under natural conditions: linking genes and natural variation in the environment to integrated physiological systems and, ultimately, to behavioral syndromes, life histories and fitness.

Until fairly recently, longitudinal studies of life history plasticity at the individual level were quite rare, and such studies of variation in plasticity in wild animals even rarer (see also Chapter 4, this volume). The drivers of among-individual variation in plasticity in natural populations of animals experiencing complex patterns of environmental variation remain largely obscure, as do their consequences for fitness and demographic rates

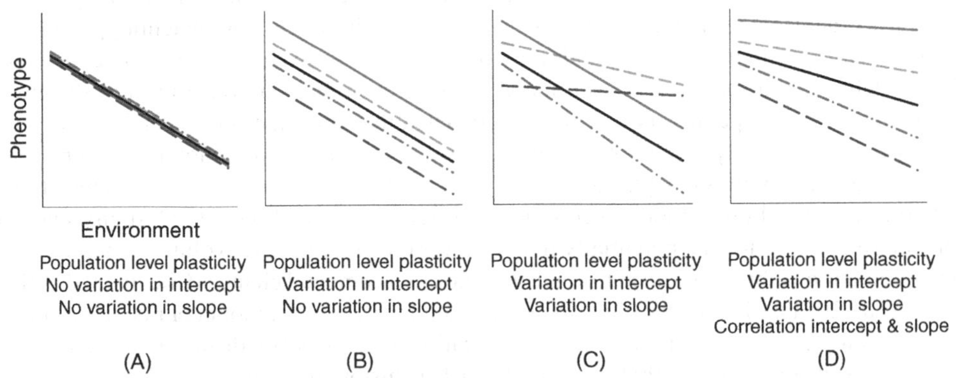

Figure 2.1. Some potential patterns of individual variation in a simple linear reaction norm, illustrating situations in which (A) neither elevations nor slopes vary among individuals, (B) elevations but not slopes vary, (C) both elevations and slopes vary, and (D) elevations and slopes vary and are positively correlated with one another. (See plate section for color version.)

(Nussey *et al.* 2007; Visser & Both 2005). In this chapter, I argue that we need integrative studies involving both laboratory experiments, to dissect the precise environmental drivers of life history plasticity and the genetic and physiological control mechanisms that may underpin individual variation in plasticity, and long-term field studies, to relate ecologically-realistic reaction norms to demographic rates and thus natural selection and population dynamics. I will begin be re-introducing the very general reaction norm framework and the idea of individual by environment interactions (or I × E; Nussey *et al.* 2007). I will then discuss the relative merits and limitations of laboratory and field studies of life history plasticity, and go on to illustrate progress being made from both field and laboratory studies of the phenology of small hole-nesting passerine birds (principally of the great tit, *Parus major*). Finally, I will consider the many challenges ahead if we are to forge a cohesive understanding of the physiological, ecological and evolutionary causes and consequences of variation in life history plasticity.

THE REACTION NORM APPROACH AND INDIVIDUAL BY ENVIRONMENT INTERACTIONS (I × E)

For traits expressed repeatedly or continuously across the lifetime of an organism – so-called "labile" or "flexible" traits, including many of the life history, morphology, behavioral, and physiological traits one might think of – phenotypes can be related to environmental variables at the level of the individual (Pigliucci 2001; Piersma & Drent 2003; Chapter 4, this volume). This perspective gives rise to a key question: do components of a population-level average reaction norm vary among individuals within that population? Most classical experimental work on plasticity did not consider this level of variation, either because it dealt with phenotypes that are expressed only once per individual (e.g., timing of metamorphosis, some morphological induced defences against predators) or because the use of clones or inbred lines allowed researchers to assess variation directly at the genotypic level. At the individual level, variation in reaction norms can be driven by both diverse genetic (e.g., additive, dominance, epistatic and indirect genetic effects) and environmental (e.g., developmental and juvenile food availability, developmental trajectories, persistent differences in habitat quality) factors. Separating the contributions of genetic and environmental causes of variation in reaction norms can be tremendously challenging, but is crucially important for evolutionary ecologists. The additive genetic basis to reaction norms reflects their evolutionary potential and may provide important insights into the selective forces that shaped them in the past. If early-life or persistent environmental differences influence reaction norms then reaction norms themselves become plastic, and can be altered by prevailing ecological conditions (Nussey *et al.* 2007).

If we consider the simplest case of a linear reaction norm, composed simply of an intercept (or elevation) and a slope (illustrated in Figure 2.1), we can see that the former will reflect the phenotype when the environment is at zero (or if the environmental variable is mean-centered, the phenotype in the mean environment) and the latter reflects plasticity (the rate of change in phenotype per unit of the environment). Imagine a situation in which, across all phenotypic data from a population measured in a variable environment, there is a significant negative slope: on average phenotypic values decline as environmental values increase at the population level (Figure 2.1A). As a biological example (which I will return to throughout the chapter), consider the reaction norm relating a bird's first egg laying date to spring temperature. In many temperate passerine systems females start laying eggs earlier when springs are warmer (e.g., Both *et al.* 2004; Visser *et al.* 2002). Now, consider that the individuals in the population breed repeatedly across their lifetimes and experience different environments at each breeding event. How do individual reaction norms vary? It might be, as in Figure 2.1B, that individuals differ in their phenotypic elevation but show consistency in their slope/plasticity. Alternatively, individuals might also vary in their slope, as illustrated in Figure 2.1C, which would constitute evidence for individual variation in plasticity or an individual by environment interaction (I × E). That might seem simple, as all we are asking here is: (a) whether there is on average a change in the phenotype with environment (population level plasticity) and (b) whether there is variation in at the individual level in elevations and slopes. However, there are many complicating factors and challenges to consider, especially if one is using longitudinal field data.

Decomposing population level reaction norms into underlying individual-level variation requires high quality longitudinal data which is time-consuming and difficult to collect in both field and laboratory settings. Insufficient sample sizes either at the population or within-individual level can produce results suggesting no significant I × E is present, when in fact the study simply lacks the power to detect it, potentially leading to a Type II error (Martin *et al.* 2011; van de Pol 2012). Furthermore, it is very important to remember that the different components of a reaction norm (in Figure 2.1, for example, the elevation and the slope) are not necessarily independent. At the individual level, if both intercept and slope vary they may also be positively or negatively correlated. This will have important implications for the pattern of reaction norms we observe and the biology underpinning that pattern. For example, in Figure 2.1D both elevations and slopes vary but the two are positively correlated, resulting in an outward fanning pattern and revealing that individuals with high elevations (e.g., late average egg laying) are less plastic (e.g., respond less strongly to changes in spring temperature). Such correlations among reaction norm components are important as they may reveal ecological and evolutionary constraints on plasticity (e.g., Porlier *et al.* 2012). Another important consideration occurs when individuals have been measured across different ranges of environmental conditions – as may be likely in long-term field studies of short-lived animals or when the environment is changing over time. In such cases associations between phenotype and environment can occur at both among- and within-individual level and it is important to separate these effects (van de Pol & Wright 2009). Finally, it is very important to keep in mind that reaction norms need not be as simple as those illustrated in Figure 2.1: the merits of polynomial and nonlinear functions to describe reaction norms should always be considered and linearity not simply assumed (see Reed *et al.* 2009 for an example of a polynomial reaction norm applied in a wild bird system and Chapter 3, this volume).

The statistical framework of choice for most researchers currently trying to describe and understand variation in individual reaction norms is a form of mixed-effects model known as random regression (Henderson 1982; Kirkpatrick & Heckman 1989). These models estimate and provide confidence intervals around the key parameters that describe variation in a linear reaction norm: variation in intercepts, variation in slopes and the covariation between the two, and they can readily be extended to incorporate polynomial and nonlinear functions (Pinheiro & Bates 2000). They have been widely applied in quantitative genetic and animal breeding contexts for several decades, and have been increasingly applied in the context of the evolutionary ecology of life history plasticity over the last decade or so. They have been used to estimate patterns of variation in reaction norms in both lab and field across a range of taxa (invertebrates: Bonte *et al.* 2007; Hairston *et al.* 2001; fish: Adriaenssens & Johnsson 2011; mammals: Clements *et al.* 2010; Lane *et al.* 2012; birds: Brommer *et al.* 2005; Charmantier *et al.* 2008; reptiles: Schwanz & Janzen 2008). They have also been widely applied in the study of ageing in a complementary manner, simply swapping age for environment on the x-axis of the reaction norm to investigate the causes and consequences of variation in ageing rates in different systems (Nussey *et al.* 2008; Wilson *et al.* 2008). The technical details of these models are beyond the scope of this chapter, but comprehensive consideration and discussion of some of the important issues associated with using random regression in this context can be found in three recent papers: Martin *et al.* 2011; van de Pol 2012; van de Pol & Wright 2009.

WHY DO REACTION NORMS VARY?

The application of statistical techniques like random regression models to estimate the patterns of individual reaction norms, as illustrated in Figure 2.1, is an important but largely descriptive exercise. More important evolutionary, physiological and ecological insights will emerge from studies that attempt to tackle the potentially more challenging question of why reaction norms vary among individuals, and why patterns of individual variation in reaction norms vary among populations and species. Figure 2.2 presents a simplified way of thinking about this problem. We can begin by thinking about the pattern of reaction norms we observe among individuals in a population (e.g., Figure 2.1) in terms of an underlying physiological control system (*sensu* Lessells 2008) which can integrate information from sensory organs about the environment to elicit a life history response. Neuro-endocrine systems regulating timing of breeding in birds are a reasonably well-studied example (Lessells 2008; Dawson 2008). An individual's genes and early-life environmental conditions will interact to influence physiological development and experience which in turn will shape the way the adult control system reacts to the current environment to influence a life history decision (Figure 2.2). Dissecting the contributions of genetic, environmental, developmental and physiological variation to individual reaction norms is a huge challenge, but one of central importance. Even more importantly, the control system concept itself simply places the diverse and complex range of physiological

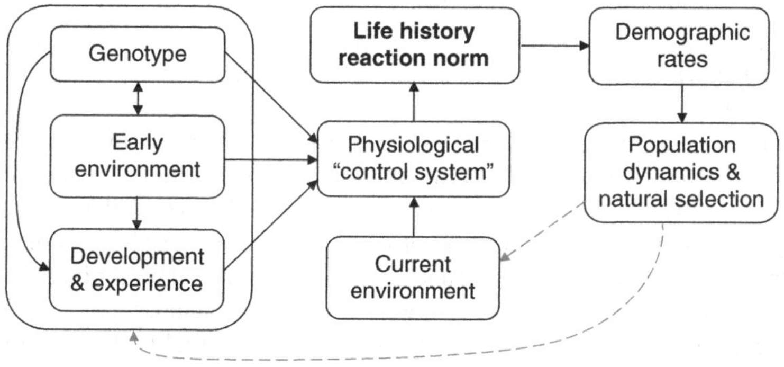

Figure 2.2. A conceptual framework to explore the evolutionary and ecological causes and consequences of individual variation in phenotypic plasticity: genes, environment, development and experience influence the physiological "control system" that detects and processes information from the current environment and uses this to guide life history decisions. The fitness and population dynamics consequences of those decisions are manifest through variation in demographic rates (e.g., survival, fecundity) which result in eco-evolutionary feedbacks (dashed gray arrows) potentially altering the environment and the genetic composition of subsequent years and generations.

processes underpinning life history variation within a black box. In Figure 2.2, this black box sits at the fulcrum of our consideration of life history plasticity. Developing a more complete and integrated understanding that will allow us to unpick this physiological black box and relate genetic and environmental variation through integrated physiological process to organismal life history and fitness represents one of the central topics of this volume, and remains a fundamental goal within evolutionary ecology (Cohen *et al.* 2012).

The ecological and evolutionary consequences of reaction norm variation will depend on how well an individual's reaction norm matches it to the demands of the environments it finds itself in. This will be reflected in the relationship between an individual's reaction norm and their demographic rates (survival rate, fecundity, offspring survival) and, ultimately, in their lifetime reproductive fitness. These relationships can produce ecological and evolutionary feedbacks that will, in turn, influence reaction norms across different time scales (gray arrows in Figure 2.2). When plasticity produces phenotypic changes that maintain fitness by matching life histories to their environment, directional selection on plasticity should be weak and stabilizing selection should predominate and this would be expected to erode genetic variation in plasticity. When plasticity produces phenotypes that are mismatched to the environment, directional selection should predominate and will favor a shift toward a well-matched reaction norm (Chevin *et al.* 2010; Ghalambor *et al.* 2007). If suitable genetic variation exists for reaction norm components, natural selection can alter the distribution of alleles involved, shaping the genetic architecture underpinning plasticity across generations (Visser 2008). From an ecological and demographic perspective, mismatches between life history decisions and environmental demands or resource availability – for instance, where environmental change occurs too fast for current plasticity to track – may result in reduced demographic rates and fitness, ultimately even negatively affecting population growth and viability (Visser & Both 2005; Both *et al.* 2006). Such demographic effects will also feed back to influence environmental conditions via density-dependence (Visser 2008). The outcome of a life history decision may also provide experiential feedback, and individuals may learn across their lifetimes to alter their physiological responses to particular environmental cues (Grieco *et al.* 2002). Ultimately, the persistence of populations *in situ* in the face of rapid environment is likely to hinge on the degree to which plasticity allows individuals to maintain their vital rates and, when it does not, the speed at which natural selection can act to alter the reaction norm (Chevin *et al.* 2010; Visser 2008).

When it comes to dissecting the roles of different components in Figure 2.2, integrative studies combining laboratory experimentation and long-term field work are required. For instance, identifying either the precise environmental drivers of plasticity or the physiological systems underpinning reaction norms is going to be near impossible in the field, where diverse sources of environmental and genetic variation will prohibit controlled experimental study of precise mechanisms involved. On the other hand, laboratory experiments attacking these questions need to be informed by ecological reality. Inbreeding and artificial (and potentially unintentional) selection under lab conditions can result in differences in genetic architecture between laboratory and field populations. Environmental cues and the range of environments within an experiment will need to be ecologically relevant if their outcomes are to generalize back into the wild. Furthermore, the generally benign conditions in the lab may remove many of the constraints and risks associated with behaviors and life history decisions in the wild (e.g., through *ad libitum* feeding, no parasites or predators). Natural selection occurs, by definition, in nature and cannot be removed from its ecological context: it *is* the action of the environment on the phenotype (MacColl 2011). Ecological and evolutionary conclusions drawn from artificial laboratory conditions thus must remain tentative until confirmed under more ecologically-relevant conditions. Conversely, the phenotypic and fitness space covered by reaction norms assessed in the field may already have been strongly limited by past selection and current environmental variation. Laboratory experiments allow researchers to look beyond this and test the costs or constraints associated with plasticity outside the current natural range of environments, which can be important for understanding the longer-term evolution of variation in reaction norms toward the pattern observed currently in the wild (Gienapp *et al.* 2008; Hairston *et al.* 2001).

The literature on phenotypic plasticity in animals is a rich one, and there are a number of study systems that provide important evidence for adaptive plasticity in morphological and life history traits (Pigliucci 2001; Schlichting & Pigliucci 1998; Doughty & Reznick 2004). There are therefore many different systems and life history reaction norms in which researchers consider the sorts of questions and challenges outlined above and are seeking to understand the causes and consequences of plasticity from the kind of integrative perspective sketched in Figure 2.2. Doughty and Reznick (2004) provide discussion of several experimental examples chosen from the literature, including changes in morphological defences in response to predator risk in marine snails, alterations of maternal provisioning of offspring in response to host plant type in seed beetles, and shifts in the initiation of metamorphosis in toad tadpoles in response to pond drying. Similarly, there are a growing number of studies of plasticity emerging from individual-based field studies of animals that have sought to test for variation in reaction norm components and, in some cases, test for genetic or environmental causes and fitness consequences of such variation (reviewed recently in Martin *et al.* 2011; van de Pol 2012). To my knowledge, there is no study system that has been able to comprehensively combine field and laboratory approaches to fully dissect the causes and consequences of reaction norm variation as presented in Figure 2.2. For several decades, the egg laying date – spring temperature reaction norm in birds has been studied both under laboratory and field conditions (McCleery & Perrins 1998; Lessells 2008; Dawson 2008; Visser *et al.* 1998). There is now a good understanding of the physiological processes regulating the onset of breeding in birds (Dawson 2008; Lessells 2008) as well as growing number of very high quality longitudinal field studies (e.g., Brommer *et al.* 2005; Charmantier *et al.* 2008; Reed *et al.* 2006). While there are a great many other animal reaction norms that have been studied in a similar way, I will focus on the examples from the literature on bird phenology in the next two sections for illustrative purposes. First, I will assess progress using long-term field studies of this reaction norm in our understanding variation in I × E within and among populations, and then I will consider some exciting progress attempting to use laboratory experiments on wild-caught animals to inform our understanding of the mechanisms involved.

CAUSES AND CONSEQUENCES OF VARIATION IN AVIAN PHENOLOGICAL PLASTICITY

In field studies where information on relatedness between individuals is available (e.g., through socially- or genetically-determined pedigrees), it is possible to use quantitative genetic models to decompose I × E into constituent additive genetic and nongenetic components of variation in plasticity (Nussey *et al.* 2007). Early attempts to do this utilized a two-step modelling approach involving extracting best-linear unbiased predictors (BLUPs) from random regression models and then estimating their heritability through subsequent quantitative genetic analysis (Brommer *et al.* 2005; Nussey *et al.* 2005b). This approach ignores the uncertainty associated with estimating each BLUP, and has been demonstrated unequivocally to yield highly downwardly-biased P values (Hadfield *et al.* 2010). However, models that combine random regression and quantitative genetic aspects (random regression animal models, RRAM) can be applied to avoid many of these issues. The observation, using the two-step approach, that in a Dutch population there was a significant additive genetic component to I × E (Nussey *et al.* 2005b) was not supported when re-analysed using RRAM (Husby *et al.* 2010). In the latter study, although the proportion of I × E ascribed to genetic sources was quite large (indeed, larger than that ascribed to nongenetic sources), it was not statistically significant (Husby *et al.* 2010). Another study applying RRAM to laying date – spring temperature in common gulls, found similarly that while the majority of that variance in plasticity was attributed to additive genetic sources in the model, there was no statistical support for G × E (Brommer *et al.* 2008). It is currently unclear to what degree field studies like this actually have the power to disentangle genetic and nongenetic components of I × E: the two studies described both suggest substantial proportions of I × E may be genetic but the errors on the estimates of G × E are very large (Brommer *et al.* 2008; Husby *et al.* 2010). There is risk of emerging publication bias here: other researchers may have tried to use RRAM in similar contexts and then failed to publish because of similarly equivocal findings. Both meta-analyses that collate estimates from RRAMs across populations and species and simulation-based power analyses to elucidate the limitations on detecting G × E in such settings could be useful ways forward here.

An important pattern emerging from avian field studies is that while many studies have now documented I × E in phenology in wild populations, a reasonable number of others have not, despite evidently having the statistical power to do so. For example, a study of laying date – North Atlantic Oscillation reaction norms in a guillemot population in the UK did not find statistical support for I × E (Reed *et al.* 2006) and a subsequent power analysis with a similar data structure and population level plasticity parameters suggested that power to detect I × E was not limited (Martin *et al.* 2011). However, another study of laying date in an American population of the same species detected a rather complex pattern of I × E across a quadratic reaction norm, although a different environmental variable was used, incorporating a wider set of candidate environmental drivers of phenology (Reed *et al.* 2009). The importance of standardized direct comparisons among populations is quite clear here: they allow direct statistical comparison of variation in reaction norms, and require the use of comparable environmental variables across populations. Recent studies doing just this also reveal notable differences in levels of I × E among populations (Husby *et al.* 2010; Porlier *et al.* 2012). A study comparing laying date with spring temperature reaction norms between a UK and Dutch population of great tits found differences in both average plasticity and I × E: the Dutch population showed lower average plasticity but significantly greater I × E (Husby *et al.* 2010). A study of four populations of blue tits (one in Southern France, the others in Corsica) also found differences: two of the four populations did not show I × E and these were also the only two in which significant directional selection on laying date was observed (Porlier *et al.* 2012).

Differences in habitat and heterogeneity in timing of key food sources during the breeding season have been suggested to explain these differences (Husby *et al.* 2010; Porlier *et al.* 2012). In both the above studies, the populations that did not show I × E had, broadly speaking, more homogenous habitats – and possibly also tighter peaks in food availability and more a more limited range of prey types – which may result in selection favoring synchronization of plasticity and a loss or reduction of I × E (Charmantier *et al.* 2008; Porlier *et al.* 2012). Alternatively, in heterogeneous habitats, environmental cues (such as spring temperatures) may have much weaker associations with the timing of food availability, resulting in limited selection on the slope of the laying date – spring temperature reaction norm (Porlier *et al.* 2012). Both scenarios assume historical genetic variation in I × E, which selection may then erode in some populations. However, there is also evidence that the ecological experiences of individuals can influence their phenological plasticity and generate I × E (Brommer *et al.* 2005; Nussey *et al.*, 2005a). Remarkably, in the blue tit study, even though two of the Corsican populations were only 6 km apart, with little evidence of genetic structure between them, they differed markedly in both their mean plasticity and in levels of I × E (Porlier *et al.*, 2012). It is unclear whether natural selection could drive and maintain genetically differentiated patterns of plasticity under such conditions. Instead population-level differences in habitat and climate could explain differences in average plasticity, while the degree of fine-scale habitat or microclimate heterogeneity could either itself drive I × E or generate I × E as individuals gain different patterns of breeding experience and learn to alter their breeding phenology in different ways depending on their precise circumstances.

Spring temperatures have increased rapidly over the last few decades in many regions. Long-term field studies of birds have been at the forefront of work to understand the demographic consequences of these changes in climate (Visser *et al.* 2004; Visser & Both 2005). There is now very strong evidence that egg laying dates are under directional selection in many wild birds: individuals that start laying earlier tend to have a greater proportion of their offspring fledge or recruit back into the population as adults (Visser *et al.* 2004). In some systems, the strength of selection on phenology is increasing through time potentially as a result of changing climate conditions (e.g., Visser *et al.* 1998). Such environment-dependent selection would be expected to lead to selection for increased plasticity, which could lead to a micro-evolutionary change if variation in reaction norm components is heritable (Nussey *et al.* 2005b). This situation will also mean many individuals mismatch their breeding phenology with the optimal timing with respect to food availability, resulting in depressed reproductive performance and declines in population size (Gienapp *et al.* 2013). As discussed above, the genetic component of I × E has proved hard to determine with confidence in wild birds, and the push of natural selection toward a reaction norm associated with higher fitness may be slow relative to the demographic load imposed by rapid environmental changes. Some recent studies do suggest that weaker species- and population-level plasticity is associated with reduced growth rates (Both *et al.* 2006; Moller *et al.* 2008). Interestingly, a recent study of Dutch great tits has suggested that even apparently strong individual-level directional selection on relative egg laying dates may have quite limited demographic consequences at the population level (Reed *et al.* 2013).

It is notable that very few studies have directly measured selection on individual reaction norm components. It is debatable at what scale selection should be measured: selection on reproductive life history traits in annual breeders occurs annually and most studies measure it at this time scale (e.g., Gienapp & Bregnballe 2012; Husby *et al.* 2011). However, the reaction norm perspective considers components of the reaction norm as individually varying traits under genetic and physiological control. As such, it could be argued it is the reaction norm components (e.g., individual elevation and slope) that are the targets for selection (Schlichting & Pigliucci 1998). Measuring natural selection on reaction norms in this way could be achieved by estimating the covariation among the reaction norm components and a measure of lifetime fitness. Previous attempts to do just this have suffered the same issue as efforts to measure G × E, discussed above: they extracted BLUPs for reaction norm components and these were regressed against lifetime reproductive fitness (Brommer *et al.* 2005; Nussey *et al.* 2005a, 2005b). A recent study has shown how reaction norm – fitness associations may be estimated within a multivariate random regression framework (Brommer *et al.* 2012), avoiding many of the issues highlighted by Hadfield *et al.* (2010). Wider application of Brommer *et al.*'s (2012) multivariate approach may help us better understand whether and how strongly selection can act on reaction norms. Ultimately, understanding the links between reaction norms and fitness at the individual level, and demographic rates and population growth rates at the population level, are vitally important if we want to understand the ecological consequences of variation in plasticity and the potential for evolutionary rescue when populations find themselves out of synch with changing environmental conditions (Gienapp *et al.* 2013; Reed *et al.* 2013). A combination of theoretical models and high quality field data look likely to provide useful insights, but these endeavors remain at a very early stage (Gienapp *et al.* 2008, 2013).

INTEGRATING FIELD AND LABORATORY STUDIES OF LIFE HISTORY PLASTICITY

Long-term field studies are providing important insights into the patterns of among-individual and among-population variation in life history reaction norms, although the underlying ecological and evolutionary causes of this variation remain very largely undetermined. Laboratory experiments have long been the mainstay of research into phenotypic plasticity: the control they offer of environment and genotype is essential for precise dissection of the way environment impacts on physiology and life history (Pigliucci 2001; Pigliucci 2005). That said, without grounding in ecological reality such experiments represent valuable tests of predictions from general evolutionary theory – rather than necessarily providing insight into the complex feedbacks shaping and in turn being shaped by organismal plasticity in nature. Excitingly, there are growing efforts to integrate an experimental, common garden approach with field research and ecological context in the study of responses to recent climate warming (e.g., Bradshaw & Holzapfel 2001; Pelini *et al.* 2009; Franks *et al.* 2007). Recently, this kind of approach has been used to shed new light on the relationship between spring temperatures and breeding phenology in hole-nesting passerines: this work both illuminates the power of such integrated lab-field study systems to help understand variation in life history reaction norms and also the challenges ahead, particularly for work on longer-lived species including many vertebrates.

A recent study examined variation in free-running period, a measure of an individual's circadian periodicity, in wild-caught great tits (Helm & Visser 2010). Because these birds came from a study site that had been the subject of long-term individual-based monitoring, estimates of the effects of parent's identity could be obtained. There was evidence that under controlled laboratory conditions, variation in free-running period existed among families and, remarkably, was also influenced by whether an individual was the product of an extra-pair mating or not (Helm & Visser 2010). Recent studies have also identified possible links between allelic variants at Clock genes involved in the intrinsic control of circadian cycles and timing of breeding in wild birds, with further evidence emerging that such associations may be sex-specific (Caprioli *et al.* 2012; Liedvogel *et al.* 2009). This research suggests important genetic, early-life environment and/or epigenetic influences on an individual's circadian rhythms, which could be an important source of any physiological variation underpinning phenological reaction norms (Helm & Visser 2010). However, exactly how this might scale up to influence responses to photoperiod and other environmental cues associated with timing of breeding and other life history traits remains to be determined.

Photoperiod has been clearly established as the primary environmental cue determining the onset of reproductive development in birds, and the physiological responses to photoperiodic cues triggering the onset of reproduction are quite well understood (Dawson 2008). However, the role of other environmental cues associated with the onset and timing of breeding – including temperature, rainfall, and visual cues associated with the phenology of other species across the food web – remain much less well understood (Schaper *et al.* 2011). Recent experiments involving wild-caught great tits have begun to elucidate both the way such supplementary cues influence phenology and how and why the sensitivity to these cues might vary in the wild. Providing visual cues associated with the onset of spring, in the form of budding vegetation or availability of caterpillar prey, did not influence the onset of reproduction under controlled conditions, suggesting such cues play little if any role (Schaper *et al.* 2011). However, similar experiments have begun to elucidate an important causal role for temperature in the onset of egg laying whilst also documenting marked among-population and among-individual variation in the sensitivity to temperature cues. A cross-population population study brought male great tits from populations at differing latitudes in Europe into the laboratory and compared the response of onset of gonadal development to different absolute temperature conditions (4 °C and 20 °C) under day lengths that are known to stimulate gonads. They found that birds from the most southerly population, where temperatures were thought to provide more predictable cues for onset of laying, showed a stronger response in both luteinizing hormone (LH) and testes growth to temperature compared to northern populations (Silverin *et al.* 2008). This suggests a genetic basis in reproductive plasticity, which may reflect selective pressures that vary across latitudes depending on the reliability and importance of temperature as a cue for conditions during the breeding period.

A study of Dutch great tits found that wild-caught birds exposed to temperature conditions mimicking those in a warm spring year (in which wild birds laid early) tended to start egg laying earlier in the laboratory compared to individuals in a group experiencing conditions mimicking a colder spring year in the wild (Visser *et al.* 2009). Since photoperiod and food availability (*ad libitum*) were controlled, the researchers were able to conclude that temperature did not affect the onset of breeding through changes in food availability, but rather that is was used as a cue for food availability by the birds. The data further suggested that it was temperatures just prior to the onset of breeding that best predicted a bird's reproductive timing. Importantly, because the experimental birds were part of a long-term field study, it was possible to demonstrate that phenological variation generalized between lab to field: individuals that started laying relatively early in a laboratory experimental group, also tended to lay relatively early under natural conditions (Visser *et al.* 2009). A follow-up experiment exposed groups of wild-caught birds to consistently higher or lower temperatures from December through the breeding season – maintaining a 4 °C difference in conditions – but this did not affect the onset of gonadal growth or laying, although warmer conditions did result in an earlier termination of the breeding cycle (Visser *et al.* 2011). This study used entire broods taken in from the wild as nestlings in two separate years, and the pedigree of the birds and the phenology of their relatives in the wild were known. Despite the puzzling lack of a temperature effect – which was interpreted as indicating that variation or changes in temperature across spring, rather than absolute temperature differences, might be the important cue – the authors showed substantial variation associated with family and year on the onset of laying in the lab. In particular, the laying date of an individual's mother in the wild predicted an offspring's laying date in the laboratory (Visser *et al.* 2011). This again suggests persistent differences in life history decision-making carry from field to lab, presumably due to genetic or early-life causes. More recent work, which explored the effects of different forms of temperature variation in the run-up to the breeding season, revealed little evidence for effects of cold periods or variation in temperature per se, or indeed increases in temperature in early spring (potentially mimicking early spring onset), or the onset of egg laying (Schaper *et al.* 2012). Rather, it was a progressive increase in temperature in the period just prior to the onset of breeding that resulted in advanced egg laying dates (Schaper *et al.* 2012). The generally limited experimental effects on LH profiles and gonadal development led the authors to speculate that a potentially different neuro-endocrine pathway could be responsible for reproductive responses to these late temperature cues (Schaper *et al.* 2012). However, the experimental effects interacted with maternal phenotype – how a treatment influenced the onset of laying differed depending on whether the individual had a mother who tended to lay relatively early or late in the wild. Although complex, these patterns reveal that the identity and phenotype of an individual's parents impacts the way they respond to temperature cues under controlled laboratory conditions, and supports the idea that I × E evident in egg laying – spring temperature reaction norms in wild populations is due to either (or potentially both) genetic or parental effects.

THE CHALLENGES AHEAD

In the above, I have mainly focused on just one particular life history reaction norm: the avian first egg laying date – spring temperature reaction norm. I have argued that a full understanding of the evolutionary ecology of life history plasticity demands an integration of studies in both field and laboratory, and the laying date – spring temperature reaction norm of passerine birds is a good example of emerging efforts to apply such an approach. Researchers have used long-term field data to map the patterns of individual reaction norms in the wild (Husby *et al.* 2010; Porlier *et al.* 2012), brought wild-caught birds of known natural phenotype and pedigree into the laboratory to experimentally test which aspects of environmental variation really matter and why individuals differ in their life history reaction norms (Schaper *et al.* 2012; Visser *et al.* 2009). Although other systems may not be so amenable to such a complete integration of field and laboratory research, the synthesis of experimental work to understand causal mechanisms of plasticity and its variation with field-based work to discern the patterns in nature and relate these to fitness and ecological dynamics is going to be increasingly important (Figure 2.2). Such work will provide important observations with which to test evolutionary theories of plasticity, especially for our understanding of the costs and limits of plasticity (Lessells 2008; de Witt *et al.* 1998), while providing empirical data to feed into demographic models and theory to predict population-level consequences of environmental change (Chevin *et al.* 2010; Gienapp *et al.* 2013; Reed *et al.* 2011). While there is clear progress here, even in the best-studied systems we are still a long way from a full understanding of how current environment influences physiological control systems to influence life history decisions, and the precise causes and consequences of individual differences in sensitivities to those environmental factors remain largely obscure.

A major challenge in the study of life history plasticity lies in understanding how different aspects of the environment (e.g., photoperiod, temperature, rainfall, food availability) interact to influence life history. Most studies to date in both laboratory and field have tended to focus on a single aspect of the environment, although clearly in reality different environmental factors could influence the phenotype independently or through interactions.

Several studies have begun to address this issue of multidimensional plasticity (e.g., Westneat *et al.* 2009). It is also important to widen focus wherever possible to understand reaction norms of different life history traits to determine what degree plastic responses are consistent or whether plasticity in one trait might come at the expense of plasticity or elevation in another. There has been considerable work on so-called "phenotypic integration" on plants (Pigliucci 2003) and field studies of life history traits have begun to consider both multitrait environmental reaction norms (e.g., laying date and clutch size responses to spring temperatures in great tits: Husby *et al.* 2010) and trait-trait reaction norms (e.g., laying date – clutch size reaction norms in Ural owls: Brommer *et al.* 2012). Ultimately, only a multitrait and multienvironment perspective is likely to reflect the complexities of ecological reality and may reveal important constraints on variation in reaction norms. The role of intrinsic changes in the organism associated with development and senescence must also be considered, as must experiential feedback. It is well known in population ecology that the demographic rates of younger and older animals tend to be more sensitive to environmental conditions (e.g., Coulson *et al.* 2001; Gaillard *et al.* 2000), and if not properly controlled for such age-related heterogeneity in plasticity could be mistaken for I × E. The use of "developmental reaction norms" in which phenotypes vary as a function of both age and environment (Schlichting and Pigliucci 1998) have been proposed as way of incorporating within-individual intrinsic changes over time into the more classical phenotype-environment reaction norm framework (Wilson *et al.* 2008), but have yet to be widely applied. Finally, understanding the way that variation in life history plasticity relates to fitness (and hence is shaped by natural selection) and impacts on demographic rates (and hence influences population dynamics and viability under environmental change) crucially depends on long-term studies in natural contexts. Laboratory studies can provide the invaluable dissection the precise effects of environment on physiology and life history, and tests of general evolutionary theory but may not realistically reflect the constraints and complexities of plasticity in the wild. We need to couple and integrate mechanism-oriented laboratory research with long-term studies of individuals in natural settings to place our understanding of the causes and consequences of variation in life history reaction norms in ecological and evolutionary context.

REFERENCES

Adriaenssens, B. and Johnsson, J.I. (2011) Shy trout grow faster: exploring links between personality and fitness-related traits in the wild. *Behavioral Ecology*, **22**, 135–43.

Bonte, D., Bossuyt, B. and Lens, L. (2007) Aerial dispersal plasticity under different wind velocities in a salt marsh wolf spider. *Behavioral Ecology*, **18**, 438–43.

Both, C., Artemyev, A.V., Blaauw, B. *et al.* (2004) Large-scale geographical variation confirms that climate change causes birds to lay earlier. *Proceedings of the Royal Society of London Series B-Biological Sciences*, **271**, 1657–62.

Both, C., Bouwhuis, S., Lessells, C.M. and Visser, M.E. (2006) Climate change and population declines in a long-distance migratory bird. *Nature*, **441**, 81–3.

Bradshaw, W.E. and Holzapfel, C.M. (2001) Genetic shift in photoperiodic response correlated with global warming. *Proceedings of the National Academy of Sciences of the United States of America*, **98**, 14509–11.

Brommer, J.E., Kontiainen, P. and Pietiainen, H. (2012) Selection on plasticity of seasonal life-history traits using random regression mixed model analysis. *Ecology and Evolution*, **2**, 695–704.

Brommer, J.E., Merila, J., Sheldon, B.C. and Gustafsson, L. (2005) Natural selection and genetic variation for reproductive reaction norms in a wild bird population. *Evolution*, **59**, 1362–71.

Brommer, J.E., Rattiste, K. and Wilson, A.J. (2008) Exploring plasticity in the wild: laying date-temperature reaction norms in the common gull Larus canus. *Proceedings Of The Royal Society B-Biological Sciences*, **275**, 687–93.

Caprioli, M., Ambrosini, R., Boncoraglio, G. *et al.* (2012) Clock gene variation is associated with breeding phenology and maybe under directional selection in the migratory barn swallow. *PLoS ONE*, **7**, e35140.

Charmantier, A., Mccleery, R.H., Cole, L.R. *et al.* (2008) Adaptive phenotypic plasticity in response to climate change in a wild bird population. *Science*, **320**, 800–3.

Chevin, L.-M., Lande, R. and Mace, G.M. (2010) Adaptation, plasticity, and extinction in a changing environment: towards a predictive theory. *Plos Biology*, **8**, e1000357.

Clements, M.N., Clutton-Brock, T.H., Albon, S.D. *et al.* (2010) Getting the timing right: antler growth phenology and sexual selection in a wild red deer population. *Oecologia*, **164**, 357–68.

Cohen, A.A., Martin, L.B., Wingfield, J.C. *et al.* (2012) Physiological regulatory networks: ecological roles and evolutionary constraints. *Trends In Ecology & Evolution*, **27**, 428–35.

Coulson, T., Catchpole, E.A., Albon, S.D. *et al.* (2001) Age, sex, density, winter weather, and population crashes in Soay sheep. *Science*, **292**, 1528–31.

Dawson, A. (2008) Control of the annual cycle in birds: endocrine constraints and plasticity in response to ecological variability. *Philosophical Transactions of the Royal Society B: Biological Sciences*, **363**, 1621–33.

De Witt, T.J., Sih, A. and Wilson, D.S. (1998) Costs and limits of phenotypic plasticity. *Trends in Ecology and Evolution*, **13**, 77–91.

Doughty, P. and Reznick, D.N. (2004) Patterns and analysis of adaptive phenotypic plasticity in animals, in *Phenotypic Plasticity: Functional and Conceptual Approaches* (eds T.J. Dewitt and S.M. Schieiner), Oxford University Press, Oxford, pp. 126–50.

Franks, S.J., Sim, S. and Weis, A.E. (2007) Rapid evolution of flowering time by an annual plant in response to a climate fluctuation. *Proceedings of the National Academy of Sciences of the United States of America*, **104**, 1278–82.

Gaillard, J.M., Festa-Bianchet, M., Yoccoz, N.G. *et al.* (2000) Temporal variation in fitness components and population dynamics of large herbivores. *Annual Review of Ecology And Systematics*, **31**, 367–93.

Ghalambor, C.K., Mckay, J.K., Carroll, S.P. and Reznick, D.N. (2007) Adaptive versus non-adaptive phenotypic plasticity and the potential for contemporary adaptation in new environments. *Functional Ecology*, **21**, 394–407.

Gienapp, P. and Bregnballe, T. (2012) Fitness consequences of timing of migration and breeding in cormorants. *PLoS ONE*, **7**, e46165.

Gienapp, P., Lof, M., Reed, T.E. *et al.* (2013) Predicting demographically sustainable rates of adaptation: can great tit breeding time keep pace with climate change? *Philosophical Transactions of the Royal Society B-Biological Sciences*, **368**, 20120289.

Gienapp, P., Teplitsky, C., Alho, J.S. *et al.* (2008) Climate change and evolution: disentangling environmental and genetic responses. *Molecular Ecology*, **17**, 167–78.

Grieco, F., Van Noordwijk, A.J. and Visser, M.E. (2002) Evidence for the effect of learning on timing of reproduction in blue tits. *Science*, **296**, 136–8.

Hadfield, J.D., Wilson, A.J., Garant, D. *et al.* (2010) The misuse of BLUP in ecology and evolution. *American Naturalist*, **175**, 116–25.

Hairston, N.G., Holtmeier, C.L., Lampert, W. *et al.* (2001) Natural selection for grazer resistance to toxic cyanobacteria: Evolution of phenotypic plasticity? *Evolution*, **55**, 2203–14.

Helm, B. and Visser, M.E. (2010) Heritable circadian period length in a wild bird population. *Proceedings of the Royal Society B-Biological Sciences*, **277**, 3335–42.

Henderson, C.R. (1982) Analysis of covariance in the mixed model – higher-level, non-homogeneous, and random regressions. *Biometrics*, **38**, 623–40.

Husby, A., Nussey, D.H., Visser, M.E. *et al.* (2010) Contrasting patterns of phenotypic plasticity in reproductive traits in two great tit populations. *Evolution*, **64**, 2221–37.

Husby, A., Visser, M.E. and Kruuk, L.E.B. (2011) Speeding up microevolution: The effects of increasing temperature on selection and genetic variance in a wild bird population. *Plos Biology*, **9**, e1000585.

Kirkpatrick, M. and Heckman, N. (1989) A quantitative genetic model for growth, shape, reaction norms and other infinite-dimensional characters. *Journal of Mathematical Biology*, **27**, 429–50.

Lane, J.E., Kruuk, L.E.B., Charmantier, A. *et al.* (2012) Delayed phenology and reduced fitness associated with climate change in a wild hibernator. *Nature*, **489**, 554–7.

Lessells, C.M. (2008) Neuorendocrine control of life histories: what do we need to know to understand the evolution of phenotypic plasticity? *Philosophical Transactions of the Royal Society B: Biological Sciences*, **363**, 1589–98.

Liedvogel, M., Szulkin, M., Knowles, S.C.L. *et al.* (2009) Phenotypic correlates of Clock gene variation in a wild blue tit population: evidence for a role in seasonal timing of reproduction. *Molecular Ecology*, **18**, 2444–56.

Maccoll, A.D.C. (2011) The ecological causes of evolution. *Trends in Ecology & Evolution*, **26**, 514–22.

Martin, J.G.A., Nussey, D.H., Wilson, A.J. and Réale, D. (2011) Measuring individual differences in reaction norms in field and experimental studies: a power analysis of random regression models. *Methods in Ecology and Evolution*, **2**, 362–74.

McCleery, R.H. and Perrins, C.M. (1998) Temperature and egg-laying trends. *Nature*, **391**, 30–1.

Moller, A.P., Rubolini, D. and Lehikoinen, E. (2008) Populations of migratory bird species that did not show a phenological response to climate change are declining. *Proceedings of the National Academy of Sciences of the United States of America*, **105**, 16195–16200.

Nussey, D.H., Clutton-Brock, T.H., Elston, D.A. *et al.* (2005a) Phenotypic plasticity in a maternal trait in red deer. *Journal of Animal Ecology*, **74**, 387–96.

Nussey, D.H., Coulson, T., Festa-Bianchet, M. and Gaillard, J.M. (2008) Measuring senescence in wild animal populations: towards a longitudinal approach. *Functional Ecology*, **22**, 393–406.

Nussey, D.H., Postma, E., Gienapp, P. and Visser, M.E. (2005b) Selection on heritable phenotypic plasticity in a wild bird population. *Science*, **310**, 304–6.

Nussey, D.H., Wilson, A.J. and Brommer, J.E. (2007) The evolutionary ecology of individual phenotypic plasticity in wild populations. *Journal of Evolutionary Biology*, **20**, 831–44.

Pelini, S.L., Dzurisin, J.D.K., Prior, K.M. *et al.* (2009) Translocation experiments with butterflies reveal limits to enhancement of poleward populations under climate change. *Proceedings of the National Academy of Sciences of the United States of America*, **106**, 11160–5.

Piersma, T. and Drent, J. (2003) Phenotypic flexibility and the evolution of organismal design. *Trends in Ecology & Evolution*, **18**, 228–33.

Pigliucci, M. (2001) *Phenotypic Plasticity*, John Hopkins University Press, Baltimore.

Pigliucci, M. (2003) Phenotypic integration: studying the ecology and evolution of complex phenotypes. *Ecology Letters*, **6**, 265–72.

Pigliucci, M. (2005) Evolution of phenotypic plasticity: where are we going now? *Trends in Ecology & Evolution*, **20**, 481–6.

Pinheiro, J.C. and Bates, D.M. (2000) *Mixed-Effects Models in S and S-Plus*, Springer-Verlag, New York.

Porlier, M., Charmantier, A., Bourgault, P. *et al.* (2012) Variation in phenotypic plasticity and selection patterns in blue tit breeding time: between- and within-population comparisons. *Journal of Animal Ecology*, **81**, 1041–51.

Reed, T.E., Jenouvrier, S. and Visser, M.E. (2013) Phenological mismatch strongly affects individual fitness but not population demography in a woodland passerine. *Journal of Animal Ecology*, **82**, 131–44.

Reed, T.E., Schindler, D.E. and Waples, R.S. (2011) Interacting effects of phenotypic plasticity and evolution on population persistence in a changing climate. *Conservation Biology*, **25**, 56–63.

Reed, T.E., Wanless, S., Harris, M.P. *et al.* (2006) Responding to environmental change: plastic responses vary little in a synchronous breeder. *Proceedings of the Royal Society B-Biological Sciences*, **273**, 2713–19.

Reed, T.E., Warzybok, P., Wilson, A.J. *et al.* (2009) Timing is everything: flexible phenology and shifting selection in a colonial seabird. *Journal of Animal Ecology*, **78**, 376–87.

Roff, D.A. (1992) *The Evolution of Life Histories: Theory and Analysis*, Chapman & Hall, London.

Schaper, S.V., Dawson, A., Sharp, P.J. *et al.* (2012) Increasing temperature, not mean temperature, is a cue for avian timing of reproduction. *American Naturalist*, **179**, E55–E69.

Schaper, S.V., Rueda, C., Sharp, P.J. *et al.* (2011) Spring phenology does not affect timing of reproduction in the great tit (Parus major). *Journal of Experimental Biology*, **214**, 3664–71.

Schlichting, C. and Pigliucci, M. (1998) *Phenotypic Evolution: A Reaction Norm Perspective*, Sinauer Associates, Sunderland.

Schwanz, L.E. and Janzen, F.J. (2008) Climate change and temperature-dependent sex determination: can individual plasticity in nesting phenology prevent extreme sex ratios? *Physiological and Biochemical Zoology*, **81**, 826–34.

Silverin, B., Wingfield, J., Stokkan, K.-A. *et al.* (2008) Ambient temperature effects on photo induced gonadal cycles and lehormonal secretion patterns in Great Tits from three different breeding latitudes. *Hormones and Behavior*, **54**, 60–8.

Stearns, S.C. (1992) *The Evolution of Life Histories*, Oxford University Press, Oxford.

Van De Pol, M. (2012) Quantifying individual variation in reaction norms: how study design affects the accuracy, precision and power of random regression models. *MEthods in Ecology and Evolution*, **3**, 268–80.

Van De Pol, M.V. and Wright, J. (2009) A simple method for distinguishing within- versus between-subject effects using mixed models. *Animal Behaviour*, **77**, 753–8.

Via, S., Gomulkiewicz, R., De Jong, G. *et al.* (1995) Adaptive phenotypic plasticity: consensus and controversy. *Trends in Ecology and Evolution*, **10**, 212–17.

Visser, M.E. (2008) Keeping up with a warming world: assessing the rate of adaptation to climate change. *Proceedings of the Royal Society of London, Series B: Biological Sciences*, **275**, 649–60.

Visser, M.E., Adriaensen, F., Van Balen, J.H. *et al.* (2002) Variable responses to large-scale climate change in European *Parus* populations. *Proceedings of the Royal Society London, Series B*, **270**, 367–72.

Visser, M.E. and Both, C. (2005) Shifts in phenology due to global climate change: the need for a yardstick. *Proceedings of the Royal Society B-Biological Sciences*, **272**, 2561–9.

Visser, M.E., Both, C. and Lambrechts, M.M. (2004) Global climate change leads to mistimed avian reproduction. *Advances in Ecological Research*, **35**, 89–110.

Visser, M.E., Holleman, L.J.M. and Caro, S.P. (2009) Temperature has a causal effect on avian timing of reproduction. *Proceedings of the Royal Society B-Biological Sciences*, **276**, 2323–31.

Visser, M.E., Schaper, S.V., Holleman, L.J.M. *et al.* (2011) Genetic variation in cue sensitivity involved in avian timing of reproduction. *Functional Ecology*, **25**, 868–77.

Visser, M.E., Van Noordwijk, A.J., Tinbergen, J.M. and Lessells, C.M. (1998) Warmer springs lead to mistimed reproduction in great tits (Parus major). *Proceedings of the Royal Society of London Series B-Biological Sciences*, **265**, 1867–70.

Westneat, D.F., Stewart, I.R. and Hatch, M.I. (2009) Complex interactions among temporal variables affect the plasticity of clutch size in a multi-brooded bird. *Ecology*, **90**, 1162–74.

Wilson, A.J., Charmantier, A. and Hadfield, J.D. (2008) Evolutionary genetics of ageing in the wild: empirical patterns and future perspectives. *Functional Ecology*, **22**, 431–42.

3

CURVE-THINKING: UNDERSTANDING REACTION NORMS AND DEVELOPMENTAL TRAJECTORIES AS TRAITS

Joel Kingsolver[1], Sarah Diamond[2,3], and Richard Gomulkiewicz[4]

[1]*Department of Biology, University of North Carolina, Chapel Hill, NC, USA*
[2]*Department of Biology, North Carolina State University, Raleigh NC, USA*
[3]*Department of Biology, Case Western Reserve University, Cleveland, OH, USA*
[4]*School of Biological Sciences, Washington State University, Pullman, WA, USA*

INTRODUCTION

A central challenge in organismal biology is to understand how the phenotypic traits of an individual organism determine its performance and fitness in specific environmental conditions. The long traditions of research in functional morphology, environmental physiology, and animal behavior have addressed this issue for different types of traits and different environments in diverse organisms. In combination with phylogenetic information, these functional studies help us understand patterns of evolutionary adaptation in nature. More recently, many researchers have explored how phenotypic variation among individuals within populations determines variation in performance and fitness – that is, phenotypic selection – under different environmental conditions. In combination with genetic and population information, these studies help us understand the microevolutionary processes that generate adaptation, integrating individual- and population-level perspectives. Both means and variances in functionally important traits are central to analyzing patterns and processes of adaptive evolution.

Many important phenotypes are not single traits or sets of traits, but curves – traits that are functions of some continuous index variable. In general, such traits are called function-valued traits (FVTs). (A curve is a function of a single index, and a function of two or more index variables is often called a surface.) For example, phenotypic plasticity

Integrative Organismal Biology, First Edition.
Edited by Lynn B. Martin, Cameron K. Ghalambor, and H. Arthur Woods.

involves traits whose values vary with one or more environmental variables, such as temperature, light intensity, nutrient concentration or density; in this case the index is an environmental variable. The relationship between trait value and environment for a genotype is often called the reaction norm. Developmental reaction norms indicate plastic traits in which environmental conditions during development determine trait values later in development. For example, in many animals, higher temperatures during development reduce adult body size; in many plants, greater plant density or reduced light intensity results in longer stems. In these cases, an individual organism only expresses a single fixed trait value that is determined by earlier environmental conditions. By contrast, performance curves indicate biological rates (and other aspects of performance) that vary with current or recent environmental conditions. For example locomotory rates of ectothermic animals vary with current body temperatures; photosynthetic rates of plants vary with light intensity. In these cases an individual organism may express a flexible range of different trait values depending on environmental variation. Similarly, many behavioral traits may vary flexibly with the environmental conditions experienced by the individual organism (Chapter 4, this volume). Combinations of fixed and flexible plasticity can also occur (e.g., reversible physiological acclimation). Epigenetic variation may also influence patterns of plasticity (Chapter 7, this volume).

A second important class of function-valued traits is phenotypes that are functions of the age of an individual organism. For example a growth curve or developmental trajectory represents the size of an individual as a function of its age throughout development. Similarly, most life history traits vary with age. By definition an individual organism expresses a range of sizes and other traits throughout its development. Of course, some phenotypes may vary as functions of both age and environmental variables (Chapter 2, this volume).

Developmental reaction norms, performance curves, growth curves and other FVTs can have a variety of different shapes. However, most theoretical analyses and empirical studies of phenotypic plasticity consider only two environmental levels, and quantify plasticity in terms of reaction norm slopes (or their equivalent). For example, random regression and similar statistical models may be used to estimate the variance in plasticity in terms of the variation in reaction norm slopes among individuals or genotypes (Chapter 2, this volume). While powerful, this approach assumes that reaction norms are strictly linear – that reaction norms and performance curves aren't curved. By contrast, a recent meta-analysis of continuous reaction norms suggests that variation in curvature and higher-order aspects of curve shape is greater than variation in slope for most organisms and traits (Murren *et al.* 2014).

Understanding organismal traits that are curves poses important methodological, statistical and conceptual challenges (Kirkpatrick & Heckman 1989; Kirkpatrick & Lande 1989; Kirkpatrick *et al.* 1990; Kingsolver *et al.* 2001; Stinchcombe *et al.* 2012). For example, when measuring growth rates of tadpoles at different rearing temperatures, or body mass of mice at different ages, how do we best quantify the phenotypic variation or mean differences in these curves? How can we partition genetic and environmental components of total variation in curves? If selection acts over a limited range of temperatures or ages, how does this determine evolution of the entire curve (Gomulkiewicz & Kirkpatrick 1992)? An important feature of curves is that, as the difference in environmental levels or ages decreases, the trait values at those levels or ages must necessarily become more correlated (with a perfect correlation when the difference is zero). As a result, variation and differences in the shapes of curves may be more strongly constrained than for other trait types, with implications for their evolution. How do we characterize such constraints, and assess their importance for evolutionary divergence in reaction norms and developmental trajectories (Kirkpatrick & Lofsvold 1992)?

This chapter is an exercise in curve-thinking for integrative biologists, as applied to phenotypic plasticity and developmental trajectories. The ideas and methods are quite general, but we use two main case studies to illustrate curve-thinking: thermal performance curves of insects, and growth trajectories of plants. First, we discuss various approaches to quantifying variation in curves, and their advantages and disadvantages. A key point is to distinguish between the shape of a curve, and the pattern of variation among curves. Second, we review ways of characterizing and visualizing phenotypic and genetic variation in curves using principal components analysis. We emphasize how different biological hypotheses about curve variation can generate distinctive patterns in the principal components. An important and familiar challenge is how to provide biological interpretations of PCA results for curves. Third, we will describe and illustrate a recent statistical method that identifies simple, interpretable patterns of phenotypic and genetic variation in curves. The key insight here is that we can decompose this variation in terms of different "directions" of simplicity in the curves. Finally, we briefly describe how patterns of genetic variation in curves may affect and constrain evolutionary changes in the mean curve for a population. A key idea here is that some patterns of selection on curves may lead to little or no evolutionary change, even when there is substantial genetic variation in the trait. Our overall goal is to provide conceptual and statistical tools for integrative and evolutionary biologists to develop and assess biological hypotheses about variation in curves and other function-valued traits. We conclude by discussing some challenges and promising directions for curve-thinking in integrative biology.

CHARACTERIZING CURVES

There are three general approaches to characterizing organismal curves. One approach is to choose a landmark value that represents a key feature of the curve: the body size at adulthood; the critical thermal maximum, indicating the maximum temperature for survival or activity; the saturation light intensity for photosynthetic rate; the LD50 for a toxin. These can provide valuable information that allows straightforward comparisons among organisms and systems, but this essentially avoids the issue by reducing a continuous curve to a single point – not curve-thinking. The second approach is to choose a specific functional form for the curve, and estimate the parameters of the function. In this case, the parameters become the traits of interest. This approach has been widely used for characterizing mean curves for both developmental trajectories (e.g., logistic and Gompertz functions for growth) and reaction norms (e.g., Michaelis-Menten model for physiological rates, modified Gaussian functions for thermal performance curves). This makes efficient use of limited data – in most organismic studies, traits are only measured at 3–6 index values – and the parameters often have simple biological interpretations. An important limitation of this approach is that for many (probably most) reaction norms and developmental trajectories, the appropriate parametric model is unknown. More generally, choosing a specific parametric model assumes that all individuals and genotypes follow the same model; differing only their parameter values. This assumption can be quite limiting when considering patterns of phenotypic and genetic variation and potential evolutionary changes in reaction norms and developmental trajectories. In addition, model parameters are estimated with some associated error, making it difficult to consider parameter values as traits of an individual organism for further analysis (but see Hadfield (2010) for one approach to this problem). As a consequence, parametric models are a powerful way to estimate mean curves for a sample of individuals, but are less useful for characterizing variation among individuals (see next page).

A third, nonparametric way of characterizing curves is by using a set of orthogonal polynomials, Fourier series, splines or other sets of functions (known as basis functions). An important advantage of this approach is that one can use the data to choose the number of basis functions (called the order) that best characterize mean and variation in the curves: the greater the order, the more complex (wiggly) the curves that can be represented. This approach can also be readily extended to quantify patterns of phenotypic and genetic variation in curves.

It is important to recognize that, as for any variable object, the mean of a sample of curves can differ greatly from the individual curves. Among other things, this implies that the mean curve need not represent the shape of the individual curves (see Figures 3.1 and 3.2 for examples). Our point here is that individual variation can affect mean curves estimated for populations or species; this may be particularly important when populations or species differ in the amount of individual variation.

VARIATION AMONG CURVES

For many reaction norms and developmental trajectories, the qualitative shape of the curve for a given trait may be similar for different individuals and genotypes within a population or species. As a result there may be important constraints on the patterns of phenotypic and genetic variation in these curves. This can lead to simple biological hypotheses about variation among curves (Huey & Kingsolver 1989; Izem & Kingsolver 2005). For example, imagine a sample of thermal performance curves (TPCs) for 5 individuals or genotypes (Figure 3.1, left panels). One pattern or "direction" of variation, called *vertical shift* (Figure 3.1, first row), describes variation in overall performance across all temperatures: relative to the mean curve, an individual with high performance at one temperature has high performance at all temperatures. *Maximum shift* (Figure 3.1, second row) describes variation in maximal performance at the optimal temperature: individuals have similar, low performance at very low and high temperatures, with greater variation among individuals at temperatures closer to the optimum. *Horizontal shift* (Figure 3.1, third row) describes variation in the location (e.g., the optimal temperature) of the curve: in this case individuals with relatively high performance at low temperatures have relatively low performance at high temperatures (and vice versa). A fourth pattern of variation involves tradeoffs between thermal breadth and maximal performance (aka *specialist-generalist* variation): here individuals with higher maximal performance at intermediate (optimal) temperatures have relatively low performance at both low and high temperatures (Figure 3.1, fourth row). These different hypotheses about variation have been widely explored by evolutionary physiologists interested in thermal adaptation (Huey and Kingsolver 1989; Angilletta 2009).

These different patterns of correlation in relative performance across different temperatures can be quantified using principal components analysis. Figure 3.1 (right panels) shows the loadings for the first principal component (explaining the greatest proportion of total variance) for each of the four hypotheses. The loadings describe the contributions from each temperature to differences among individuals in their performance across temperatures. For the *vertical shift* example, the loadings all have the same sign (i.e., the line is above zero) across all temperatures (Figure 3.1, first row): individuals have relatively high (or low) performance at all temperatures. (Recall that a principal component with all positive loadings is equivalent to one with all negative loadings – it is the change in sign across temperatures that is relevant.) For *maximum shift*, the loadings all have the same sign across

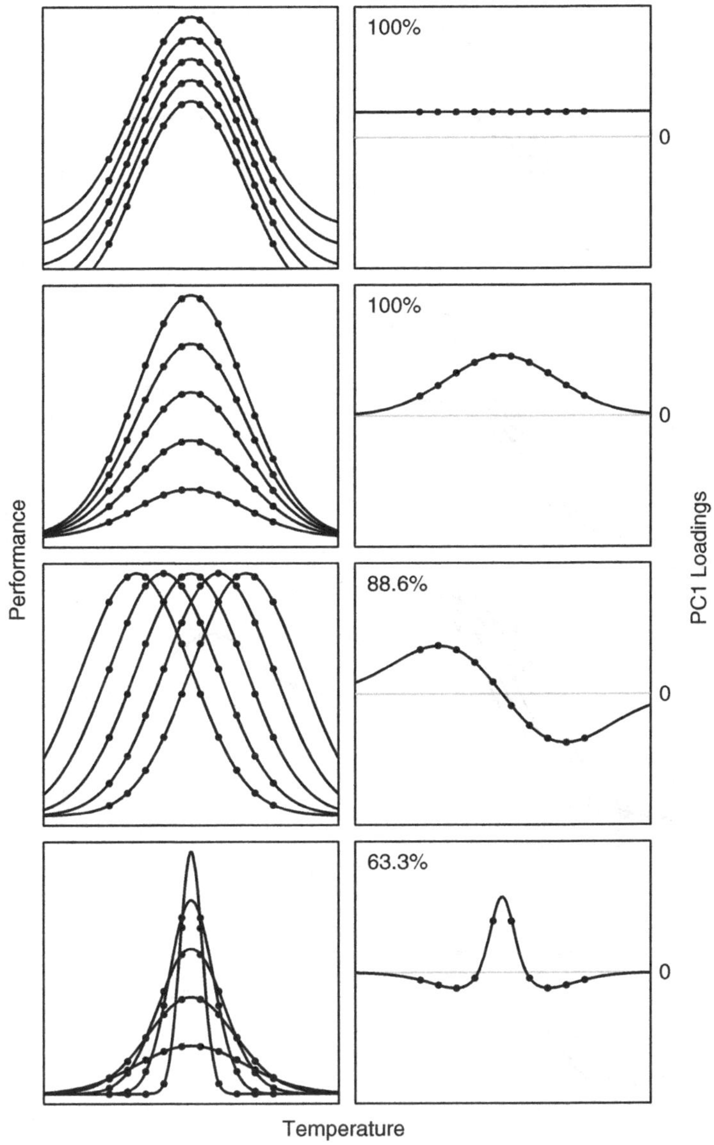

Figure 3.1. Toy thermal performance curves based on normal functions (left panels) and associated eigenfunction loadings for the first principal component (PC1) (right panels). Row 1: *vertical shift*; row 2: *maximum shift*; row 3: *horizontal shift*; row 4: *specialist-generalist tradeoff*. The points along each thermal performance curve correspond with the points along the eigenfunction. Percent variance explained by PC1 is indicated in the upper left of each panel.

all temperatures, but the magnitude of the loadings increases at intermediate temperatures (Figure 3.1, second row), which indicates that some individuals achieve higher performance than others at intermediate temperatures. For *horizontal shift*, the first principal component has positive loadings at low temperatures and negative loadings at high temperatures (or vice versa) (Figure 3.1, third row), which indicates a reversal of relative performance at

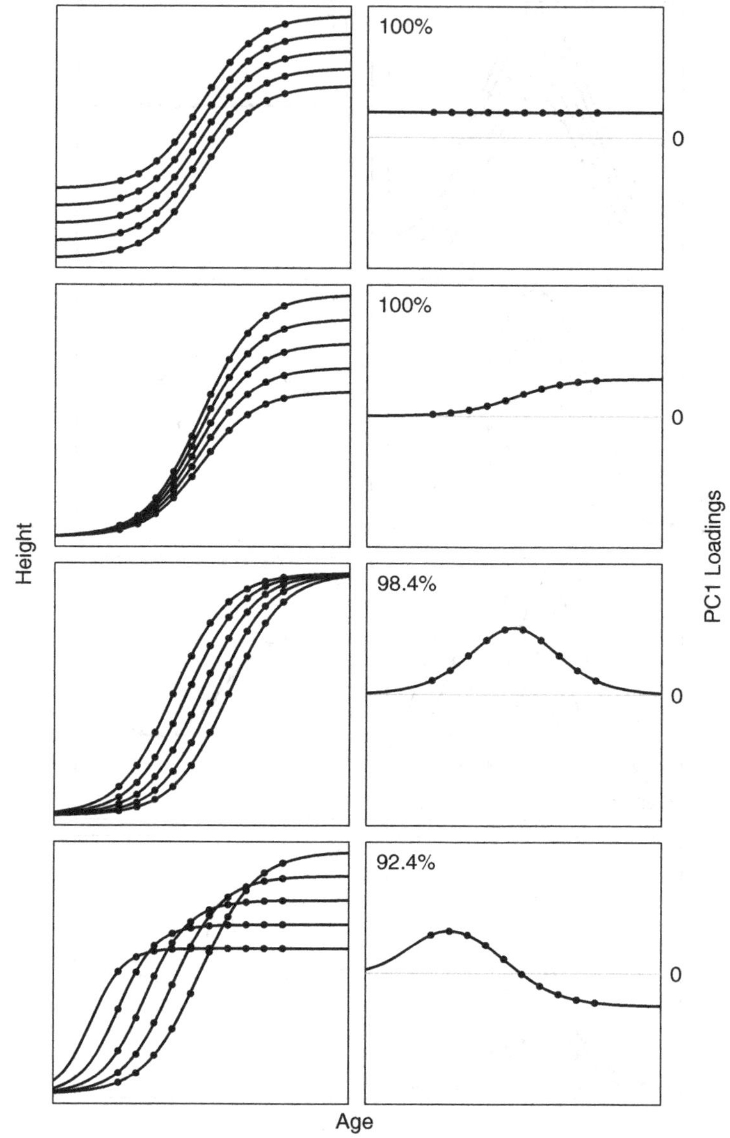

Figure 3.2. Toy growth curves based on logistic functions (left panels) and associated eigenfunction loadings for the first principal component (PC1) (right panels). Row 1: *vertical shift*; row 2: *maximum shift*; row 3: *horizontal shift*; row 4: *early growth-final size tradeoff*. The points along each growth curve correspond with the points along the eigenfunction. Percent variance explained by PC1 is indicated in the upper left of each panel.

low and high temperatures. In this sense there are "cool" and "warm" adapted individuals or genotypes. For the *specialist-generalist* case, the combination of positive loadings at intermediate temperatures with negative loadings at both low and high temperatures (or vice versa) indicates that specialist individuals with high relative performance at intermediate temperatures have low performance at low *and* high temperatures (Figure 3.1, fourth row).

The key insight here is that biological patterns of variation generate principal components with distinctive, characteristic shapes.

The principal components described above, and represented as the lines or curves in the right panels of Figure 3.1, are called eigenfunctions. There is a simple and direct relationship between eigenfunctions and more familiar eigenvectors. Imagine that we measure performance at 10 different temperatures for each individual (points in Figure 3.1, left panels). We can consider performance at each temperature as a separate, but potentially correlated, trait, so there are 10 traits for each individual: in this sense we represent the curve for each individual as a vector of 10 traits. We can estimate the variance-covariance matrix for the 10 traits, yielding the variance in performance at each temperature, and the covariance in performance across each pair of temperatures. For example, element [2,3] in this matrix indicates the covariance between performance at temperature 2 and performance at temperature 3. We can then use PCA to compute the eigenvectors and associated eigenvalues for the variance-covariance matrix. The loadings for each trait (i.e., performance at temperature 1, at temperature 2, etc.) on the first principal component (the first eigenvector) are the points along the principal component (Figure 3.1, right panels). In this sense, an eigenfunction is the natural extension of an eigenvector as we move from a discrete set of points to a continuous function. As a result we can approximate continuous curves as discrete vectors, and use the familiar method of principal components analysis (PCA) to decompose variation in the curves into a set of orthogonal vectors (the eigenvectors), based on the amount of variance explained. As illustrated above (Figure 3.1), plotting the loadings of the eigenvectors (e.g., the first PC) allows us to visualize patterns of variation in the curves.

Similar hypotheses about patterns of variation may occur for other types of reaction norms and for developmental trajectories, even if the shape of the mean curve is different. For example, for logistic growth curves we can consider *vertical shifts* that reflect variation in overall size at all ages (Figure 3.2, first row, left panel). The resulting principal component (Figure 3.2, first row, right panel) is identical to that for vertical shift of the TPCs (Figure 3.1, first row, right panel). *Maximum shift* represents variation in final size (Figure 3.2 second row), resulting in a first principal component with loadings that are near zero at early ages but that increase with age until reaching final (maximum) size. For growth curves, *horizontal shift* represents variation in the age at the inflection (midway) point (where growth rate is most rapid (Figure 3.2, third row). Loadings for the resulting principal component are near zero at initial and later ages, and are largest at intermediate ages. Note the shape of the principal component in this case is similar to that for the maximum shift case for TPCs (Figure 3.1, second row). We can also consider tradeoffs between the timing of growth (age at the inflection point) and final size (*early growth-final size tradeoff*), such that individuals who grow to larger size at early ages have relatively small final sizes (Figure 3.2, fourth row). Loadings for the resulting principal component have positive values at earlier ages and negative values at later ages, illustrating the tradeoff between early growth and final size. Note that the shape of the principal component for this case is qualitatively similar to that for the horizontal shift case in TPCs (Figure 3.1, third row). The key to this similarity in both cases is the changes in relative ranks of the curves: relative performance at low vs high temperatures, and relative height at early vs late ages.

Note that for both the TPC and growth curve examples, the *vertical shift* and *maximum shift* patterns (Figures 3.1 and 3.2, top two rows) are fully explained by a single principal component: PC1 explains 100% of the variance. This is because these patterns represent linear changes or deformations of the curves (and principal components analysis is a linear method of decomposition). In contrast the other two patterns of variation (Figures 3.1

and 3.2, bottom two rows) represent nonlinear deformations in the curves, which cannot be fully described by a single principal component: in these cases, the variance explained by PC1 is less than 100%.

There are two important insights from these toy examples. First, there is often a direct relationship between simple biological patterns of variation in curves and the shape of their associated principal component (eigenvector or eigenfunction), for reaction norms, developmental trajectories, and other biological curves. The key to understanding this association is that changes in the relative rankings of curves across (e.g.) ages or temperatures produce changes in the sign of the associated principal component with age or temperature. Second, the shape of the principal components may have different biological meanings for curves of different shapes (e.g., normal vs logistic). This is a key insight as we develop biological interpretations about important patterns of phenotypic and genetic variation in curves.

Armed with these conceptual and statistical tools, let us now move from toy examples to data for real biological curves. The methods can be applied to both phenotypic and genetic components of variance. For example, phenotypic data for a sample of individuals from a population can be used to estimate the phenotypic variance-covariance matrix, P. Quantitative genetic studies frequently decompose this total phenotypic variation into genetic and environmental components, to estimate the genetic variance-covariance matrix, G (Chapter 4, this volume). Estimation of G is useful for predicting evolutionary responses of quantitative traits, including curves, to selection (Lande & Arnold 1983; Via & Lande 1985). In our two case studies we will consider G matrices that approximate patterns of genetic variation for curves.

The first example considers thermal performance curves for short-term growth rates of *Pieris rapae* (Imported Cabbageworm) larvae, in which each 4th instar larva was measured at a series of 6 different temperatures (for details and data see (Kingsolver *et al.* 2004). These data were used to estimate the (broad-sense) genetic variance-covariance matrix, G. Principal components analysis of G shows that the first 4 principal components (eigenvectors) explain over 99% of the total genetic variance (Gaydos *et al.* 2013) (Figure 3.3). The shape of the curve for the first PC (explaining 60% of the variance) is quite complex, but is dominated by strong positive loadings at the highest temperature (40 °C) and strong negative loadings at 35 °C (and vice versa). This pattern indicates that most differences among individual curves involve higher growth rates near the optimal temperature (around 35 °C for most individuals and families) coupled with lower growth rates at high temperature (and vice versa; see Izem & Kingsolver (2005) for additional analysis and discussion). The second PC, explaining 19% of the variance, has positive loadings for all temperatures between 17 and 40 °C, and loadings near zero at 11 °C (Figure 3.3). Similar to the *vertical shift* or *maximum shift* patterns (Figure 3.1, rows 1–2), this secondary axis of variation suggests among-genotype variation that corresponds to curves with relatively high (or low) performance across a wide range of temperatures. The other two PCs are difficult to interpret.

The second example considers growth curves (height as a function of age) for *Impatiens capensis* (Jewelweed) plants raised in full-sun conditions at high density, in which each plant was measured at 6 different ages (see Stinchcombe *et al.* (2010) for details and data). Forty nine clonal genotypes were used to estimate the G matrix. PCA of G shows that the first two principal components explain more than 99% of the total genetic variance (Stinchcombe *et al.* 2010; Gaydos *et al.* 2013) (Figure 3.4). The first PC, explaining 99.8% of the variation, has positive loadings at all ages with the magnitude increasing with age, a pattern similar to that for *maximum shift* (see Figure 3.2, second row). This indicates variation in overall growth rate among clones, such that genotypes that are tall (or short) at one age are tall (or short) at all ages, and with increasing variation with age. The second

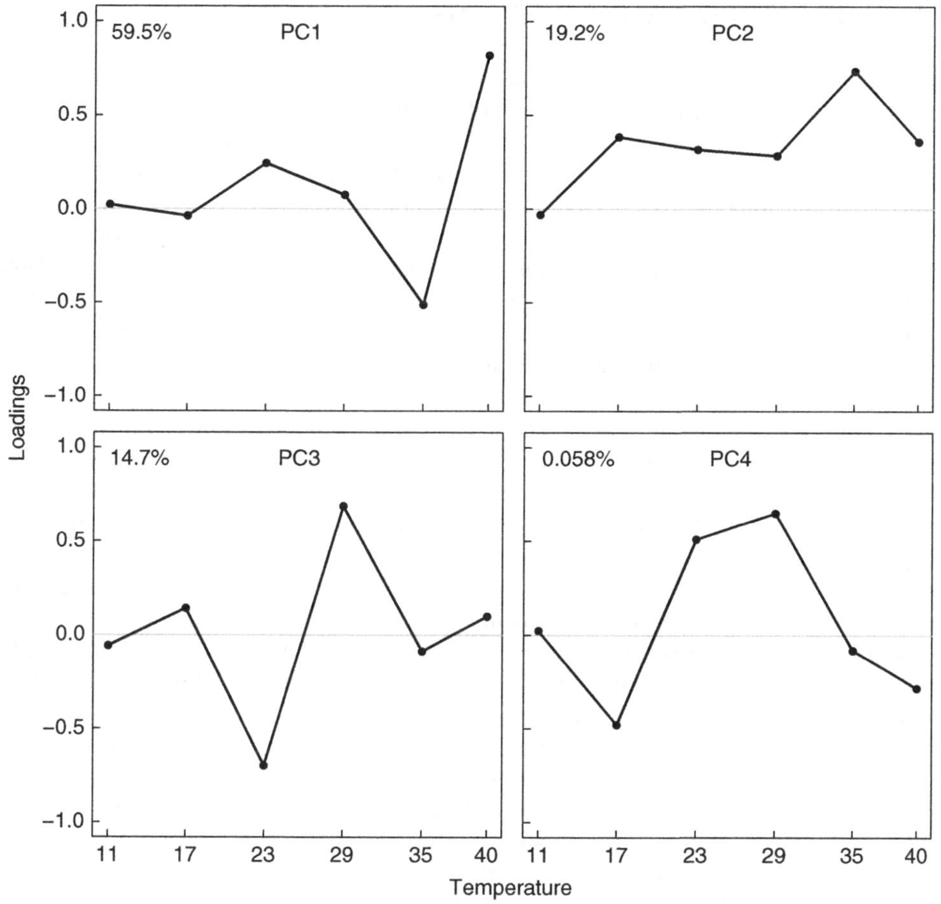

Figure 3.3. The first four principal components of the G-matrix for thermal performance curves of growth rates for Imported Cabbageworms. Percent variance explained by each component is indicated in the upper left of each panel.

PC, explaining only 0.004% of the genetic variation, is qualitatively similar to the *early growth-final size tradeoff* pattern (Figure 3.2, fourth row), with negative loadings at early ages and positive loadings at later ages (or vice versa). This secondary axis of variation suggests that (once overall variance is accounted for by PC1) genotypes that are relatively short at early ages are also relatively tall at later ages (and vice versa).

The structure of genetic covariation in these two examples is quite different. Only two PCs explain nearly all of the genetic variance in the Jewelweed growth trajectories, whereas four PCs are required for the Cabbageworm TPCs; in this sense the patterns of genetic variance are simpler (of lower dimension) in Jewelweed (Stinchcombe *et al.* 2010). In addition, much of the genetic variation in Jewelweed growth can be interpreted in terms of a single biological hypothesis (*maximum shift*: Figure 3.2, second row), but this is not the case for the Cabbageworm TPCs. Whether these differences are specific to these particular examples, or reflect more general differences between growth trajectories and TPCs, is unknown. PC analyses of P and G matrices for a variety of datasets for TPCs, developmental reaction norms, and growth trajectories would be most valuable in addressing this issue.

SIMPLICITY AND BIOLOGICAL HYPOTHESES

Our examples illustrate the important and well-known limitation of PCA that the principal components are often difficult to interpret biologically. The principal components estimated from real datasets (Figures 3.3 and 3.4) may rarely match those predicted by biological hypotheses (Figures 3.1 and 3.2). One solution is to develop statistical methods that directly evaluate specific hypotheses. For example, the "template mode of variation" (TMV) method was developed to decompose and quantify variation in thermal performance curves in terms of the easily interpreted vertical shift, horizontal shift, and generalist-specialist tradeoffs (Izem & Kingsolver 2005). The method places certain constraints on the qualitative shape of the curves (e.g., the curve is continuous, differentiable, and has a single maximum value), but does not assume a specific parametric form for the curve. This nonlinear approach has been used to analyze TPCs in a number of different study systems (Stinchcombe *et al.* 2012). However, the method has not been generalized to other types of function-valued traits, and it involves fancy mathematics like non-Euclidean manifolds and Dirichlet variances (Izem 2004).

An alternative, more general approach is to analyze variation in curves in terms of the simplicity of their eigenfunctions or eigenvectors. For current purposes we can consider simplicity to describe how wiggly or curvy a curve is: a simpler curve is less wiggly. The rationale for this approach is that, for many biological hypotheses, the qualitative shape of the associated eigenfunction is quite simple (Figures 3.1 and 3.2). For example, a flat (nonzero) eigenfunction indicates a vertical shift pattern of variation (Figures 3.1–3.2, top row). An eigenfunction that crosses zero once (e.g., is negative at low index values and positive at high index values, or vice versa) indicates a tradeoff in performance or size across temperatures (Figure 3.1, 3rd row) or ages (Figure 3.2, 3rd row). An eigenfunction that crosses zero twice indicates a specialist-generalist tradeoff: relative performance or size at intermediate temperatures or ages is inversely related to relative performance or size at low and high temperatures (Figure 3.1, 4th row).

Thus, eigenfunctions or eigenvectors with simple shapes can be readily interpreted in biological terms. We have seen the principal components analyses typically do not result in

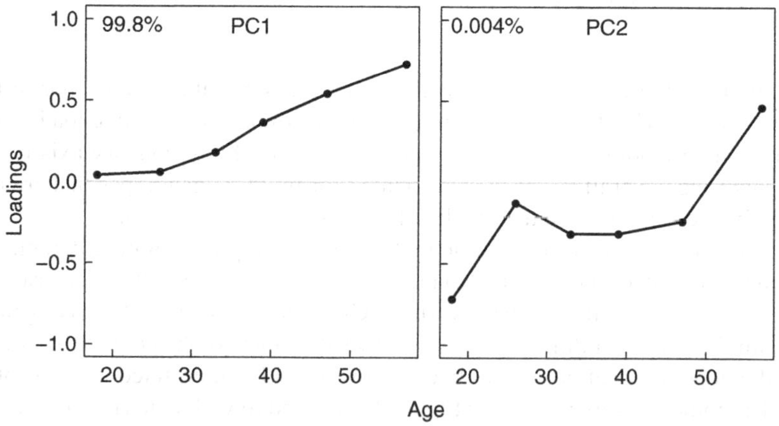

Figure 3.4. The first two principal components of the G-matrix for Jewelweed growth trajectories (height as a function of age). Percent variance explained by each component is indicated in the upper left of each panel.

simple eigenfunctions or eigenvectors. Recall that PCA decomposes a variance-covariance matrix into a set of orthogonal components (the eigenvectors) based on the amount of variance explained by each component. However, other sets of orthogonal vectors (called the basis vectors for the space of variation) can be used to quantify and visualize variation. Gaydos and colleagues (Gaydos *et al.* 2013) have recently developed a method that defines a set of orthogonal basis vectors based on a quadratic metric of simplicity, and quantifies the variation associated with each basis vector.

The Simplicity metric S takes on values between 0 (least simple, most wiggly) and 4 (most simple, a flat line). These basis vectors are ordered from highest to lowest simplicity (Figure 3.5, left panel). For the sake of discussion, suppose performance is measured at 6 temperatures. The loadings for the simplest basis vector (vector 1) give a flat line of constant (nonzero) magnitude (Figure 3.5), indicating *vertical shift*, with variation in overall performance across all temperatures (see Figure 3.1, first row). The Simplicity value for this vector is S = 4. The loadings for the 2nd basis vector (S = 3.7) give a monotonic line with negative loadings at low temperatures and positive loadings at high temperatures (or vice versa), indicating a tradeoff between high performance at low temperatures and low performance at high temperatures (and vice versa). The loadings for the 3rd basis vector (S = 3) represent a *specialist–generalist tradeoff*, in which relatively high maximal performance at intermediate temperatures is correlated with relatively low performance at both low and high temperatures. The 4th (S = 2), 5th (S = 1) and 6th (S = 0.27) basis vectors indicate increasingly complex (wiggly) patterns of variation across temperature, that are more difficult to interpret biologically.

In addition, we can compute the fraction of total variance associated with each Simplicity basis vector (Gaydos *et al.* 2013). This allows us to quantify the variation in a sample of reaction norms, developmental trajectories or other biological curves in terms of simple, biologically interpretable directions. An R library (*prinsimp*) was recently developed to implement this method for analyzing variance-covariance matrices (Cubranic *et al.* 2013).

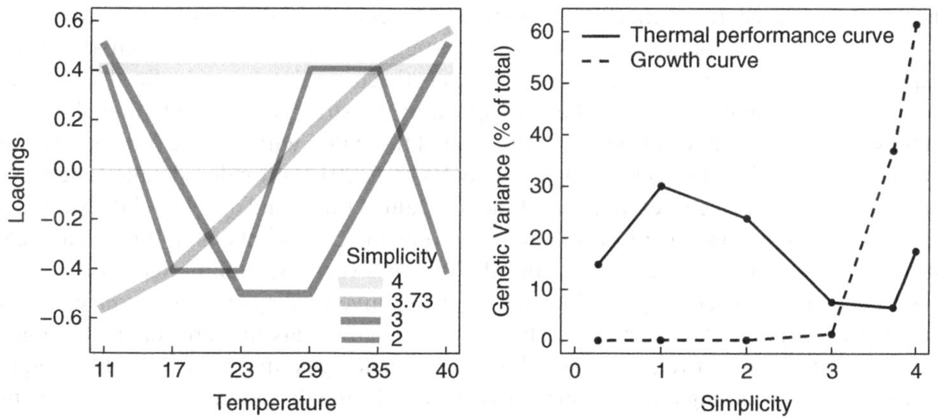

Figure 3.5. Left panel: The first four simplest basis vectors for thermal performance curves measured at 6 different temperatures. Simplicity values are indicated with different shading (4 = light grey to 2 = dark grey) and line thicknesses (higher values of simplicity have thicker line widths). Right panel: Percent variance explained as a function of simplicity based on thermal performance curves of Imported Cabbageworm growth rates (solid line) and Jewelweed growth trajectories (dashed line).

We can apply this approach to our G matrices for TPCs of Imported Cabbageworms and for growth curves of Jewelweed, by representing the matrices in terms of the Simplicity basis, and quantifying the percent variance as a function of the Simplicity value for each vector (Figure 3.5, right panel). For the TPCs, the simpler vectors (e.g., S = 3–4) explain a relatively small fraction of the total genetic variance. For example, the three simplest vectors (see Figure 3.5, left panel) account for only 32% of the variance, with the remainder accounted for by more complex patterns of variation. Returning to the PC analysis for these data (Figure 3.3), this complexity may be the result of the large loadings of opposite sign between performance at 35 °C and 40 °C – a very wiggly feature of the variation.

The Jewelweed growth curves show a very different pattern (Figure 3.5, right panel). The simplest vector (S = 4), representing overall variation in size across all ages (*vertical shift*), explains more than 61% of the total genetic variance. The 2nd vector (3.7), representing a tradeoff between relatively small size at early ages and large size at later ages (and vice versa), explains an additional 37% of the variance. Together these two simple and interpretable patterns explain over 98% of the total genetic variance in these curves. Conversely, the other 4, more complex vectors explain less than 2% of the variance (see below).

These results for Simplicity have some important implications for the evolution of these curves, because genetic variation directly determines the evolutionary responses to selection on function-valued traits (Gomulkiewicz & Kirkpatrick 1992; Kirkpatrick & Lofsvold 1992; Stinchcombe *et al.* 2012). Imagine there is directional selection that favors increased performance at all temperatures, or increased height at all ages. We would expect a strong evolutionary response to such selection in Jewelweed growth curves, because more than 61% of the genetic variance is in this "vertical shift" direction. In contrast we would expect a weaker evolutionary response to selection in Cabbageworm TPCs, as only 17% of the genetic variance is in the vertical shift direction. Whether these are general differences between growth curves and TPCs, or specific to these particular study systems, is unknown.

These results also provide some interesting insights about genetic constraints (Kirkpatrick & Lofsvold 1992). In quantitative genetics, genetic constraints are present if strong phenotypic selection results in little or no evolutionary response due to the lack of appropriate genetic variance. In practice, PCA can be used to detect and describe genetic constraints on adaptive evolution of curves: constraints are suggested by the principal components that explain the least amount of overall variation. Genetic constraints are visualized as the loadings corresponding to these PCs; the corresponding constraints can be interpreted in terms of these loadings since selection in the direction of these PCs will produce little or no evolutionary response. Gaydos and colleagues (Gaydos *et al.* 2013) provide a detailed analysis of genetic constraints on curves and other function-valued traits, and developed the simplicity basis approach as a means of visualizing these genetic constraints using these same case studies. For the Jewelweed example, consider the basis vector with Simplicity S = 3, which has positive loadings at early and late ages with negative loadings at intermediate ages (or vice versa) (Figure 3.5, left panel). This indicates individuals that are relatively large at early and late ages, but relatively small at intermediate ages: this pattern could emerge if relatively large seedlings delay their period of rapid growth, but end up at a larger final size (not shown). The simplicity analysis (Figure 3.5, right panel) shows that this pattern of variation represents less than 1% of the total genetic variance in the Jewelweed growth curves (Gaydos *et al.* 2013). If there was phenotypic selection favoring this pattern of variation (larger size at early and late ages, and smaller size at intermediate ages), there would be little or no evolutionary response to selection due to the lack of genetic variation in this "direction" – even though there is genetic variation in height at all ages. In this way, simplicity analyses can help quantify and visualize patterns of both genetic variation and genetic constraints in curves and other function-valued traits (Gaydos *et al.* 2013).

SUMMARY AND FUTURE DIRECTIONS

The goal of this chapter has been to explore curve-thinking: the conceptual and statistical tools for characterizing, quantifying and visualizing variation in biological curves such as reaction norms and developmental trajectories. Integrative biologists are familiar with the "parameters as traits" approach, in which a specific parametric model is chosen to describe the curve of interest, and the model parameters are estimated for a sample of individuals. This approach can be useful for describing mean curves, but is too restrictive for characterizing phenotypic and genetic variation for most biological curves. Here we have emphasized a familiar method, principal components, as a tool for visualizing and quantifying variation in curves.

A major theme of our discussion is developing biological hypotheses about curve variation among individuals or genotypes. Using toy examples we illustrated how different patterns of variation generate principal components with different shapes that can be readily visualized and interpreted. Our comparison between normal TPCs and logistic growth curves shows that the mapping of a hypothesis to PC shape can sometimes depend on the shape of the curve itself. This approach also provides a natural way to visualize tradeoffs: a tradeoff involves changes in the relative rankings of traits values across ages or environmental levels, and is reflected in a change in the sign of the loadings for the associated PC across ages or environmental levels. It would be useful to explore additional hypotheses about variation in thermal performance curves and growth curves beyond those considered here, and to explore hypotheses for other types and shapes of biological curves.

These hypotheses can be used to interpret the results of principal components analyses on real biological curves. We considered two case studies of genetic variation, in thermal performance curves of Imported Cabbageworms and growth curves of Jewelweeds. One striking result is that most of the genetic variance in Jewelweed growth curves is consistent with a single biological hypothesis (variation in *maximum shift*), whereas none of the hypotheses we considered explained most of the genetic variance in Cabbageworm TPCs. It would be valuable to have similar analyses of TPCs, growth curves and developmental reaction norms for other traits and study systems to explore possible generalities about these patterns. We emphasize that these methods apply to both phenotypic and genetic variation. At a minimum, trait measurements at 4 or more index levels (ages or environments) for at least 15–20 individuals or genotypes (clones, sibs, etc.) will likely be required for the analyses to be informative (Griswold *et al.* 2008).

Unfortunately, the genetic and physiological bases for natural variation in performance curves, reaction norms and growth trajectories are poorly understood. Genetic analyses with bacteriophage show that individual base substitutions (point mutations) can affect multiple aspects of performance curve shape such as optimal temperature and thermal breadth (Knies *et al.* 2006). This suggests that the different "directions" of variation described by our hypotheses may not be independent at the genetic level (Knies *et al.* 2009). The physiological bases of thermal reaction norms for adult body size have been identified recently for some holometabolous insects (Chapter 13, this volume). Despite this, how variation in the underlying physiological parameters determines variation in reaction norm shape has not been explored. Connecting physiological mechanisms to patterns of genetic and phenotypic variation in biological curves remains an important challenge for integrative and evolutionary physiologists.

We have emphasized the use of discrete, matrix-based methods for computing eigenvectors and eigenvalues using PCA because of its familiarity to most integrative and population biologists. However, this discrete approach to curves has important limitations. For example, it requires that all individuals be measured at the same index values;

it does not take into account the order or spacing of index values; it is less efficient and less powerful in hypothesis testing (Griswold *et al.* 2008). True curve-thinking requires consideration of eigenfunctions and variance-covariance functions, rather than their more familiar discrete counterparts. These can be estimated using random regression and related models (see Stinchcombe *et al.* (2012) for a recent overview). However for cases where traits are consistently measured at a fixed and limited set of index values (e.g., 3–8), the discrete approximations can be very useful.

Another general theme we have explored is Simplicity. PCA decomposes total variation onto different orthogonal axes (principal components) based on the amount of variance explained, but the principal components are often difficult to interpret. We describe an alternative approach that decomposes variance onto orthogonal axes based on the simplicity of the basis vectors (curves) that may have simple biological interpretations. One interesting result is that more than 98% of the variance in Jewelweed growth curves is explained by two, simple vectors that can be interpreted in terms of two hypotheses (*vertical shift* and *early growth-final size tradeoff*). Applying this new approach to other traits, study systems and types of curves will be interesting to see whether there are general patterns of simplicity in the variation of biological curves.

ACKNOWLEDGMENTS

We thank Travis Gaydos, Nancy Heckman and Steve Marron for useful discussion; Davor Cubranic and Nancy Heckman for help with *prinsimp*; and Cameron Ghalambor, Marty Martin, and Art Woods for suggestions on an earlier version of the paper. Work supported in part by NSF IOS-1120500 to JGK and NSF DEB-1354264 to RG.

REFERENCES

Angilletta, M.J. (2009) *Thermal Adaptation: A Theoretical and Empirical Synthesis*, Oxford University Press, Oxford.

Cubranic, D., Zhang, J., Heckman, N.E., Gaydos, T.L. & Marron, J.S. (2013) prinsimp: Finding and plotting simple basis vectors for multivariate data. Comprehensive R Archive Network, URL: http://CRAN.R-project.org/package=prinsimp.

Gaydos, T.L., Heckman, N.E., Kirkpatrick, M. *et al.* (2013) Visualizing genetic constraints. *Annals of Applied Statistics*, **7**, 860–82.

Gomulkiewicz, R. and Kirkpatrick, M. (1992) Quantitative genetics and the evolution of reaction norms. *Evolution*, **46**, 390–411.

Griswold, C., Gomulkiewicz, R. and Heckman, N. (2008) Hypothesis testing in comparative and experimental studies of function-valued traits. *Evolution*, **62**, 1229–42.

Hadfield, J. (2010) MCMC methods for multi-response generalized linear mixed models: the MCMCglmm R package. *Journal of Statistical Software*, **33**, 1–22.

Huey, R.B. and Kingsolver, J.G. (1989) Evolution of thermal sensitivity of ectotherm performance. *Trends in Ecology and Evolution*, **4**, 131–5.

Izem, R. (2004) Analyzing variation in curves of common shape. PhD, University of North Carolina.

Izem, R. and Kingsolver, J.G. (2005) Variation in continuous reaction norms: Quantifying directions of biological interest. *The American Naturalist*, **166**, 277–89.

Kingsolver, J.G., Gomulkiewicz, R. and Carter, P.A. (2001) Variation, selection and evolution of function-valued traits. *Genetica*, **112/113**, 87–104.

Kingsolver, J.G., Ragland, G.J. and Shlichta, J.G. (2004) Quantitative genetics of continuous reaction norms: Thermal sensitivity of caterpillar growth rates. *Evolution*, **58**, 1521–9.

Kirkpatrick, M. and Heckman, N. (1989) A quantitative genetic model for growth, shape, reaction norms, and other infinite-dimensional characters. *Journal of Mathematical Biology*, **27**, 429–50.

Kirkpatrick, M. and Lande, R. (1989) The evolution of maternal characters. *Evolution*, **43**, 485–503.

Kirkpatrick, M. and Lofsvold, D. (1992) Measuring selection and constraint in the evolution of growth. *Evolution*, **46**, 954–71.

Kirkpatrick, M., Lofsvold, D. and Bulmer, M. (1990) Analysis of the inheritance, selection and evolution of growth trajectories. *Genetics*, **124**, 979–93.

Knies, J.L., Izem, R., Supler, K.L. *et al.* (2006) The genetic basis of thermal reaction norm evolution in lab and natural phage populations. *PLoS Biology*, **4**, 1–8.

Knies, J.L., Kingsolver, J.G. and Burch, C.L. (2009) Hotter Is better and broader: Thermal sensitivity of fitness in a population of bacteriophages. *The American Naturalist*, **173**, 419–30.

Lande, R. and Arnold, S.J. (1983) The measurement of selection on correlated characters. *Evolution*, **37**, 1210–26.

Murren, C.J., MacLean, H.J., Diamond, S.E., Steiner, U.K., Keskel, M.A., Handelsman, C.A., Ghalambor, C.K., Auld, J.R., Callahan, H.S., Pfennig, D.W., Relyea, R.A., Schlichting, C.D. & Kingsolver, J.G. (2014) Evolutionary change in continuous reaction norms. *American Naturalist*. In press.

Stinchcombe, J.R., Group, F.-V.T.W. and Kirkpatrick, M. (2012) Genetics and evolution of function-valued traits: understanding environmentally responsive phenotypes. *Trends in Ecology and Evolution*, **27**, 637–47.

Stinchcombe, J.R., Izem, R., Herschel, M.S. *et al.* (2010) Across-environment genetic correlations and the frequency of selective environments shape the evolutionary dynamics of growth rate in *Impatiens capensis*. *Evolution*, **64**, 2887–2903.

Via, S. and Lande, R. (1985) Genotype-environment interaction and the evolution of phenotypic plasticity. *Evolution*, **39**, 505–22.

4

PLASTICITY AND PERSONALITY

Kimberley J. Mathot and Niels J. Dingemanse

Max Planck Institute for Ornithology, Seewiesen, Germany

ANIMAL PERSONALITY: MOVING BEYOND OPTIMALITY AND EMBRACING INDIVIDUAL DIFFERENCES

Behavioral ecology has traditionally used what is known as the optimality approach to study animal behavior. This approach is typically used to predict the optimal expression of a behavioral trait for the average individual in a population given any constraints that may limit the expression of that trait (Davies *et al.* 2012). For example, in order to minimize the probability of being depredated, individuals can invest in anti-predator vigilance. However, individuals are limited in how much time they can invest in vigilance, because time is a limited resource and individuals must also engage in other time consuming behaviors such as foraging. The trade-off between the costs and benefits of increased time invested in vigilance leads to individuals accruing the maximum net-benefit from vigilance at some intermediate level of expression, and this level of expression is considered optimal.

The optimality approach has been successfully used to predict mean behaviors in populations but not variation among individuals. However, individuals often show consistent differences in behavior over time or across contexts, a phenomenon referred to as animal personality. This is represented statistically by a significant proportion (i.e., repeatability, *r*) of the total variation in behavior occurring between individuals [i.e., $V_{\text{between individuals}}/(V_{\text{between individuals}} + V_{\text{within individuals}})$], and suggests that individuals may differ from the optimum or mean behavior at the population level. For example, female Ural owls (*Strix uralensis*) aggressively defend their nests, and while the intensity of nest defense varies depending on brood size or environmental factors like food availability, an

Integrative Organismal Biology, First Edition.
Edited by Lynn B. Martin, Cameron K. Ghalambor, and H. Arthur Woods.

individual's aggressiveness in one year predicts its aggressiveness in the subsequent year (Kontiainen *et al.* 2009). Consistent individual differences in behavior are widespread, with an average of 37% of the total variation in behavior occurring between individuals (Bell *et al.* 2009). Although individual differences in behavior have traditionally been viewed as nonadaptive noise around the true adaptive mean (Dall *et al.* 2004), there is a growing awareness that individual differences sometimes also reflect adaptive variation, and this view has given rise to a new field of animal behavior research known as adaptive animal personality (Réale *et al.* 2010).

Most adaptive models for animal personality are based on state-dependent behavior (Dingemanse & Wolf 2010). When the costs and benefits associated with particular behaviors depend on an individual's state, then stable state differences between individuals will give rise to consistent individual differences in behavior. For example, individual differences in physiology, such as metabolic rates, are often stable over relatively long time periods (Speakman *et al.* 2004), and may influence the expression of any behavior associated with energy acquisition or expenditure (Biro & Stamps 2010). In Atlantic salmon (*Salmo salar*) for example, higher metabolic rates in juveniles predict future expression of aggressive behavior and subsequent dominance status (Cutts *et al.* 1998). When individual differences in state are not inherently stable, state-dependent behavior combined with positive feedbacks between state and behavior can generate animal personality (Luttbeg & Sih 2010). For example, if larger individuals are better able to aggressively defend resources, the act of defending resources will allow them to remain large, thereby reinforcing their tendency to be aggressive.

One common misconception is that the presence of consistent individual differences in behavior (animal personality) necessarily implies that individuals are not behaviorally plastic. However, under changing environmental conditions, individuals can adjust their behavior while remaining consistently different from each other (Dingemanse *et al.* 2010). In this chapter, we will discuss how animal personality and behavioral plasticity can be studied together and provide an overview of the statistical methods used to achieve this aim (Box 4.1). We will highlight a few empirical examples of correlations between personality and plasticity and briefly discuss adaptive explanations for such correlations. Next, we will discuss the ecological and evolutionary consequences of personality-related differences in plasticity, and review the existing evidence for environmental versus genetic underpinnings for this phenomenon. Finally, we will conclude with a discussion of experimental designs for the study of personality and plasticity and identify some promising avenues for future research.

Box 4.1. The Behavioral Reaction Norm Approach

The "behavioral reaction norm" framework provides a means of studying personality and plasticity together (Smiseth *et al.* 2008; Dingemanse *et al.* 2010). In this approach, each individual is characterized by a function that describes how its behavior changes across a mean-centered environmental gradient (see also Chapter 2, Figure 2.1). Individuals may differ both in their behavioral "intercept" (i.e., their average behavior is the position where the mean-centered environmental gradient has the value zero) and "slope" (i.e., units change in behavior as a function of units change in environment). For simplicity, we assume that such reaction norms are linear, though we note that such nonlinearity might not always be a valid assumption (Brommer *et al.* 2010). Research on animal personality meets research on phenotypic plasticity through the application of behavioral reaction norm (BRN) approaches when considering that personality variation represents between-individual variance in average behavior (i.e., variance among intercepts) and plasticity is captured in reaction norm slopes (Martin & Réale 2008; Dingemanse *et al.* 2010; Mathot *et al.* 2012).

Variation within a population of animals in reaction norm components can be captured by a single "phenotypic equation" that is commonly applied by evolutionary ecologists to study labile traits (cf. Nussey *et al.* 2007; Dingemanse & Dochtermann 2013) and is known as a "random regression" (Equation (4.1a)):

$$y_{ij} = (\beta_0 + ind_{0j}) + (\beta_1 + ind_{1j})x_{ij} + e_{0ij} \tag{4.1a}$$

Here, a single phenotypic response (y_{ij}), such as the level of activity by individual j exhibited at instance i is modeled as a function of x_{ij} (a covariate representing for example the perceived risk of predation experienced by individual j at instance i). This phenotypic response (y_{ij}) is decomposed into five distinct elements: (1) the population-mean reaction norm intercept (β_0; the grand mean value of average individual responses), (2) the population-mean reaction slope (β_1; the coefficient relating x_{ij} to y_{ij}), (3) the individual's deviation in reaction norm intercept (ind_{0j}) from the population-mean intercept (β_0), (4) the individual's deviation in reaction norm slope (ind_{1j}) from the population-mean slope (β_1), and (5) the instance's deviation from the individual's reaction norm (e_{0ij}). This model is called a "random" regression (Schaeffer 2004) because the individual-specific deviations from the population-mean value with respect to intercepts (ind_{0j}) and slopes (ind_{1j}) are assumed to be "drawn" from (i.e., follow) a bivariate normal distribution (MVN) distribution with a mean of zero and covariance matrix (Ω_{ind}) to be estimated from the data (though other error distributions may also be used). The variances and covariances for this distribution are defined by the variance in intercepts among individuals (V_{ind_0}; "personality"), between-individual variance in slopes (V_{ind1}; "individual plasticity"), and the covariance between intercepts and slopes (Cov_{ind_0,ind_1}; "personality-related plasticity"); the variance in deviations from individual reaction norms for each instance (e_{0ij}) are modeled as normally distributed with a mean of zero and an estimated within-individual variance (V_{e_0}) (Equation (4.1b)):

$$\begin{aligned} \begin{bmatrix} ind_{0_k} \\ ind_{1_k} \end{bmatrix} \sim MVN(0, \Omega_{ind}) \quad &: \quad \Omega_{ind} = \begin{bmatrix} V_{ind_0} & Cov_{ind_0,ind_1} \\ Cov_{ind_0,ind_1} & V_{ind_1} \end{bmatrix} \\ [e_{0ik}] \sim N(0, \Omega_e) \quad &: \quad \Omega_e = [V_{e_0}] \end{aligned} \tag{4.1b}$$

Ω_{ind} is a symmetrical matrix: the elements below the diagonal are mirrored above the diagonal. The intercept-slope covariance can, notably, be expressed as a correlation (r_{ind_0,ind_1}), where $r_{ind_0,ind_1} = Cov_{ind_0,ind_1} / \sqrt{V_{ind_0} V_{ind_1}}$. Personality variation is present in the data in cases where $V_{ind_0} > 0$, individuals differ in behavioral plasticity when $V_{ind_0} > 1$, and personality-related differences in plasticity exists when $r_{ind_0,ind_1} \neq 0$.

This behavioral reaction norm approach provides a powerful means to study patterns of phenotypic variation at different hierarchical levels (Westneat *et al.* 2011). For example, the phenotypic equation (Equation (4.1)) can easily be expanded to include variation in population-average reaction norms. Eqn. 1 provides us with information about the population-average norm of reaction, estimated as the population-average reaction norm intercept β_0 and slope β_1. Since selection pressures acting on labile phenotypes are often population specific, population average reaction norms likely differ between populations depending on population-specific predictability of environmental change, costs, limits, and benefits of plasticity (Gabriel *et al.* 2005). Indeed, population-average behavioral reaction norms are known to vary among populations (Briffa *et al.* 2008), although

(*continued*)

this is not always the case (Dingemanse *et al.* 2012b). Variation in reaction norms at this level of the hierarchy of phenotypic variation might be captured by expanding the phenotypic equation into (Equation (4.2)):

$$y_{ijk} = (\beta_0 + ind_{0jk} + pop_{0k}) + (\beta_1 + ind_{1jk} + pop_{1k})x_{ij} + e_{0ijk} \quad (4.2)$$

In this expanded phenotypic equation, a single phenotypic response (y_{ijk}) now varies across populations (k), individuals within populations (j) and instances within individuals (i). As before, individuals can deviate in both their reaction norm intercept ($+ind_{0jk}$) and slope ($+ind_{1jk}$) but now we also obtain information about among-population variation in intercepts ($+pop_{0k}$) and slopes ($+pop_{1k}$). For brevity, the associated random effect structure is not detailed as done previously for Equation (4.1b), since this structure would simply be expanded with another hierarchal level. In short, the expanded phenotypic equation would provide information about the amount of variance among populations in reaction intercepts (V_{pop_0}) and slopes (V_{pop_1}) and the covariance between these two reaction norm components (Cov_{pop_0,pop_1}). Such information is of great interest because it would provide clues as to whether the same proximate mechanism might generate variation in reaction norms at different hierarchical levels. For example, if the covariance between intercepts and slopes was positive within-populations (i.e., $Cov_{ind_0,ind_1} > 0$) but negative among populations (i.e., $Cov_{pop_0,pop_1} < 0$), one would have to conclude that more than one mechanism is involved in shaping variation in reaction norms in nature (cf. Armbruster 1991; Armbruster & Schwaegerle 1996).

INDIVIDUAL DIFFERENCES IN BEHAVIORAL PLASTICITY

Consistent differences in behavior between individuals (animal personality) suggest that any given individual shows only a subset of the behavioral diversity present in the whole population, and therefore does not exhibit the range of plasticity observed at the population level (Sih *et al.* 2004). Because being able to adjust one's behavior to current conditions has obvious advantages, the occurrence of limited plasticity is somewhat surprising and various explanations for limited plasticity have been proposed (DeWitt *et al.* 1998). The behavioral reaction norm (BRN) approach provides a tool for studying personality and plasticity simultaneously (see Chapters 1 & 2, this volume; and Box 4.1). A BRN describes changes in the phenotype (behavior) of an individual across a mean-centered environmental gradient where personality is characterized by individual differences in intercept, and plasticity is characterized by the presence of a significant slope (Dingemanse *et al.* 2010; Ghalambor *et al.* 2010). Application of the BRN method has revealed that individuals often differ in both their average behavior (i.e., intercept of a BRN, personality) and in their degree of plasticity (Dingemanse *et al.* 2010). For example, even though vigilance generally increases with increasing predation danger at the population level, detailed observations of vigilance behavior in red knots (*Calidris canutus*) shows that individuals differ significantly in how strongly they adjust the proportion of time spent vigilant with increasing predation danger (Mathot *et al.* 2011).

Given the obvious potential advantage of being able to adjust one's behavior to current conditions, understanding the proximate underpinnings and ultimate consequences of individual differences in plasticity are a major interest in behavioral ecology and evolutionary biology (Dingemanse & Wolf 2013). As with animal personality, individual differences in the degree of plasticity may arise if individuals differ in the benefits they can derive from behavioral plasticity, and these differences can be either genetic or environmental in origin (Wolf & Weissing 2010). For example, small fish are more vulnerable to gape-limited predators, and so smaller fish may be expected to show stronger responses to predator cues (i.e., greater plasticity). If differences in body size or growth rate are genetic in origin (e.g., Serbezov *et al.* 2010), then differences in plasticity will have genetic underpinnings and will be expected to be consistent over the long term. However, individuals may show consistent differences in behavioral reaction norms across gradients of predation danger even when differences in body size are due to environmental effects. For example, variation in body size could initially arise because of stochastic between individual differences in foraging success. However, once these differences arise, if small individuals show stronger responses to predators, such as hiding in a refuge, while larger individuals show relatively small reactions and so are able to continue feeding, then over the long term, positive feedbacks between state (body size) and behavior (willingness to forage when confronted with a predator), will tend to reinforce the between-individual differences in state and behavior, leading to differences in plasticity that are consistent over time (Luttbeg & Sih 2010). Note that when individual differences in plasticity arise because individuals differ in the benefits they can obtain by behaving plastically, individuals may not achieve equal fitness even while behaving in the way that is optimal for their phenotype. In the example outlined above, large fish may obtain higher fitness because they are both less vulnerable to predators and better able to maintain high foraging rates even in the presence of predators.

The payoffs for behavioral plasticity may depend not only on an individual's state, but also on the behavior expressed by other individuals in the population. The behavior expressed by other individuals in a population represents an environmental effect that can lead to predictable changes in the degree of plasticity for any given individual. For example, animals should generally invest more in anti-predator vigilance as predation danger increases. However, if an individual can be alerted to the presence of a predator by any other group member that detects the predator, how much any given individual will benefit from increasing its vigilance as predation danger increases will depend on how much other group members increase their vigilance. In this case, the payoffs for plasticity are negatively frequency-dependent because with each individual that exhibits plasticity, the benefit for subsequent individuals to be behave plastically decreases (Wolf *et al.* 2008). When plasticity is less costly for individuals that have previously exhibited plasticity (a form of positive feedback), then individual differences in plasticity can be consistent over time and across contexts (Wolf *et al.* 2008). The importance of negative frequency-dependent payoffs in generating individual differences in plasticity was tested explicitly in red knots (*Calidris canutus*), where it was found that individual differences in plasticity were present for vigilance behavior, a trait with negatively-frequency dependent payoffs, but not anti-predator escape flights, a trait with positively frequency-dependent payoffs (Mathot *et al.* 2011). Similarly, consistent individual differences in behavioral plasticity (measured as the time required to switch to a new foraging patch) were observed in threespined sticklebacks (*Gasterosteaus aculeatus*) when alternative patches were available simultaneously (and so the benefits of switching patches depended on the behavioral decisions taken by other group members), but not when patches were made available sequentially (Laskowski & Bell 2013).

The BRN approach has shown that individuals often differ consistently in their degree of behavioral plasticity, and important theoretical advances have been made toward understanding how such variation can arise (e.g., frequency-dependent payoffs: Wolf *et al.* 2008; e.g., state-dependent payoffs: Wolf & Weissing 2010). Empirical studies are now needed that test predictions based on state- and frequency-dependent explanations for consistent individual differences in plasticity.

PERSONALITY-RELATED DIFFERENCES IN PLASTICITY

Application of the reaction norm approach to the study of animal behavior has revealed that personality (I) and plasticity (I × E) often covary (Mathot *et al.* 2012), referred to as personality-related differences in behavioral plasticity. Personality-related differences in behavioral plasticity can take at least two forms. First, the average level of behavior A can correlate with the degree of plasticity for behavior A (i.e., elevation-slope covariance), illustrated in Figure 4.1a. For example, house sparrows (*Passer domesticus*) exhibit consistent individual differences in provisioning behavior, with some individuals consistently provisioning their young at a higher rate than others, even after controlling for factors that affect brood demand, such as brood size, nestling age and season (Westneat *et al.* 2011). The slope of the reaction norms to these factors also differs between individuals, with individuals that have higher provisioning rates also increasing their provisioning rate more strongly as a function of increasing nestling age (Westneat *et al.* 2011). Nest defense behavior in female Ural owls provides another example of this type of personality-related differences in plasticity. Females increase the intensity of their nest defense in relation to changes in prey abundance, presumably because this increases the potential value of their brood. However, females that defend their nests more aggressively on average also show greater plasticity in nest defense in response to changes in prey abundance (Kontiainen *et al.* 2009).

The second type of personality-related differences in behavioral plasticity occurs when the average level of behavior A is correlated with the level of plasticity for behavior B, illustrated in Figure 4.1b. For example, how quickly Iberian wall lizards (*Podarcis hispanica*)

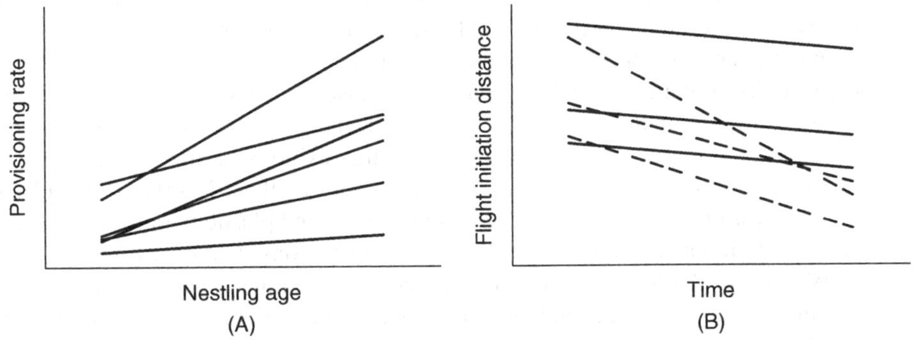

Figure 4.1. Personality-related differences in behavioral plasticity illustrated using a behavioral reaction norm approach. In (A) individuals differ in elevation (personality) and slope (plasticity), with elevation-slope covariance. House sparrows that provision at higher rates also increase their provisioning rate more strongly in response to changes in brood demand, such as nestling age. In (B) individuals differ in elevation (personality) for behavior A, as indicated by the type of line (solid or dashed). The lines indicate individual behavioral plasticity for behavior B. Iberian wall lizards scored fast for exploration (dashed lines) decrease their flight initiation distance to human observers more strongly across repeated encounters compared to individuals scored slow for exploration (solid lines). *Sources*: A: Adapted from Westneat, D.F., Hatch, M.I., Wetzel, D.P. & Ensminger, A.L. (2011) Individual variation in parental care reaction norms: integration of personality and plasticity. *American Naturalist*, **178**, 652–67. B: Adapted from Rodriguez-Prieto, I., Martín, J. & Fernández-Juricic, E. (2011) Individual variation in behavioral plasticity: direct and indirect effects of boldness, exploration and sociability on habituation to predators in lizards. *Proceedings of the Royal Society B*, **278**, 266–73.

habituate to a threatening stimulus (approach of a human observer) can be predicted by their behavior in a novel environment (Rodriguez-Prieto *et al.* 2011), termed exploration behavior. Individuals that move through the novel environment more quickly (fast explorers) are also faster to habituate to human disturbances, reflected by a steeper decrease in flight initiation distance across repeated human disturbance trials. Another example of this form of personality-related differences in plasticity comes from the "coping style" literature. Animals can adopt either proactive or reactive strategies for coping with environmental stressors (termed coping styles), and the coping style that an individual adopts is associated with particular levels of various behavioral characteristics, including aggression (Koolhaas *et al.* 1999). The behavior of proactive individuals is not guided by environmental stimuli, and these individuals readily develop rigid behavioral routines (i.e., low behavioral plasticity). In contrast, the behavior of reactive individuals is guided by environmental stimuli, and so they show more plastic adjustment of behavior. As such, higher levels of aggression are often associated with lower behavioral plasticity (Koolhaas *et al.* 1999). For example, when house mice (*Mus musculus domesticus*) are left to solve a maze task, more aggressive individuals develop rigid routines and do not adjust their attempted route through the maze in response to changes in the environment (Benus *et al.* 1991).

Thus, personality-related differences in plasticity can take at least two forms (illustrated in Figure 4.1). An emerging challenge in animal personality research is to understand at both the proximate and ultimate level why personality and plasticity should covary.

ADAPTIVE EXPLANATIONS FOR PERSONALITY-RELATED DIFFERENCES IN PLASTICITY

Application of the reaction norm approach to behavioral data is beginning to reveal that personality-related differences in plasticity are widespread. Understanding why this should be is a major interest for behavioral and evolutionary biologists because it can have important ecological and evolutionary consequences. Links between personality and plasticity may reflect evolutionary tradeoffs between selection acting on the elevation and slope of behavioral reaction norms (DeWitt *et al.* 1998; Dingemanse *et al.* 2010). However, in some cases, links between personality and plasticity may also be adaptive in the sense that they are the outcome of decision rules that have been shaped by natural selection (sensu Houston & McNamara 1999). Here we discuss two broad classes of adaptive explanations for personality-related differences in plasticity: state-dependence and alternative strategies for managing uncertainty.

State-dependence has been invoked as an explanation for both animal personality and individual differences in plasticity, and could therefore also generate personality-related differences in plasticity (Figure 4.2a). For example, individuals with higher metabolic rates may be expected to have higher foraging rates than individuals with low metabolic rates (Biro & Stamps 2008). They may also be less behaviorally plastic if their higher metabolic rates mean they are more energetically constrained to continue foraging and so do not adjust investment in vigilance across a gradient of predation danger. Additionally, positive feedbacks between behavior and state may help to maintain individual differences over extended time periods (Luttbeg & Sih 2010; Wolf & Weissing 2010). For example, having high feeding rates even when faced with increased predation danger may allow high metabolic rate individuals to achieve the energy intake required to maintain their expensive metabolic machinery.

Several other explanations for personality-related differences in plasticity revolve around the notion that individuals differ in how they value information or in how they detect

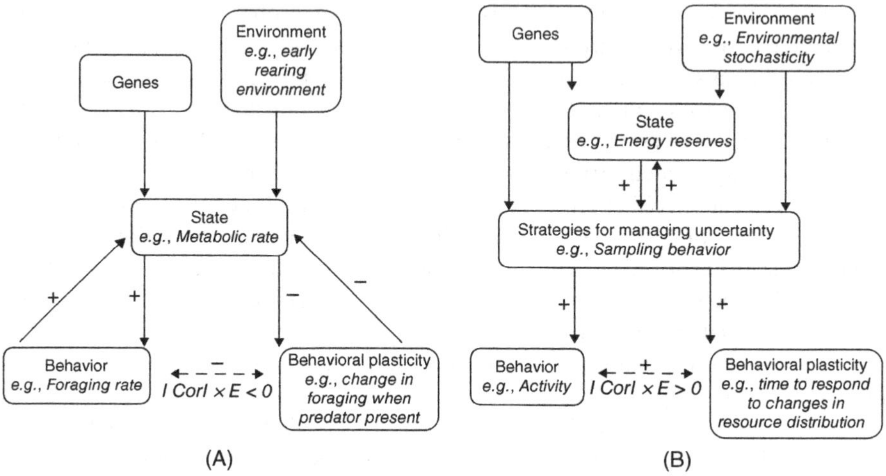

Figure 4.2. Schematic of potential causal pathways (solid arrows) for personality related-differences in plasticity (dashed arrows indicate that relationships arise indirectly). A) Illustration of state-dependent personality and plasticity, with positive feedbacks between behavior and state. Note that since the effect of state (in this case metabolic rate) on behavioral plasticity is negative, the negative effect of behavioral plasticity on state constitutes a positive feedback loop. In the example illustrated here, there is a negative correlation between personality and plasticity (I Corr I × E < 0). B) Illustration of how alternative strategies for managing uncertainty (such as sampling behavior) can generate personality-related differences in plasticity. Differences in sampling behavior may be state-dependent, or may be directly shaped by genetic and/or environmental effects. Where they are state-dependent, positive feedbacks between behavior and state are also possible, for example if sampling helps individuals exploit resources more effectively, allowing them to build energy reserves, which in turn favors higher levels of sampling (Dall & Johnstone 2002). In both A and B, differences in state (and differences in strategies for managing uncertainty in B) can arise due to genetic and/or environmental effects.

changes in the environment (reviewed in Mathot *et al.* 2012; Sih & Del Giudice 2012). Animals face uncertainty in virtually every decision they face, and one means of reducing uncertainty is by investing (time or energy) in gathering information (i.e., sampling) (Dall & Johnstone 2002). If animals differ in the way that they invest in sampling behavior, this will affect their level of expression of a range of behaviors (Figure 4.2). For example, if cues for the resources being sampled are easy to detect, higher sampling may be associated with more rapid movement through the environment (i.e., higher activity). In contrast, when cues for the resources being sampled are difficult to detect, individuals that invest more in sampling may be expected to move through the environment more slowly (i.e., lower activity). Because sampling allows animals to track changes in the environment, animals that invest more in sampling may recognize changes in the environment sooner, and consequently, respond more quickly when conditions change (i.e., show greater behavioral plasticity).

Several empirical studies report patterns of personality-related differences in behavioral plasticity that are consistent with the notion that variation in sampling may underlie both personality and plasticity. For example, great tits (*Parus major*) scored slow for exploration (activity in a novel environment) respond more quickly to changes in the location of food (Verbeek *et al.* 1994), and convict cichlids (*Amatitlania nigrofasciata*) scored slow for exploration respond more quickly to the presence of predators (Jones & Godin 2010). Even more compelling evidence comes from a study in guppies (*Poecilia reticulata*) which found that more accurate decision making was associated with both slower exploration and larger brain size (Burns & Rodd 2008).

ECOLOGICAL AND EVOLUTIONARY IMPLICATIONS OF PERSONALITY AND PLASTICITY

Animal personality can have important implications for a variety of ecological processes, including (but not limited to) species interactions, population dynamics and disease transmission (Sih *et al.* 2012; Wolf & Weissing 2012). For example, the behavioral types of pollinators can influence the traits of flower communities, with low-nectar (i.e., unrewarding) flowers potentially being able to persist only when fast inaccurate bumble bees are present (Sih *et al.* 2012). Behavioral diversity among group members can also increase group success when different behavioral types play complementary roles. For example, colony success in the social spider *Anelosimus studiosus* is greater when there is more behavioral variation among colony members because different behavioral types provide different colony services in terms of prey capture, predator deterrence and reduction of colony agonism (Pruitt *et al.* 2011; Pruitt & Riechert 2011). The behavioral types of individuals can also affect species interactions, including parasite and disease transmission if for example more active individuals are more likely to encounter and become infected by parasites (Barber & Dingemanse 2010; Kortet *et al.* 2010). Animal personality can also have important evolutionary consequences. When differences in behavioral types reflect genetic differences, then populations with higher behavioral diversity will be expected to be at an advantage when faced with a new selective regime (Wolf & Weissing 2012). At the same time, genetic correlations between multiple behavioral traits can constrain adaptive evolution (meta-analysis: Dochtermann & Dingemanse 2013).

Between individual differences in behavioral plasticity can also have important ecological and evolutionary consequences. When an individual's plastic response is in the same direction as that favored by directional selection, then plasticity is said to be adaptive. In contrast, a plastic response that is in the opposite direction to that favored by directional selection is termed nonadaptive. Generally, adaptive plasticity will allow individuals to establish and persist in novel environments, but also shields them from strong selection if their phenotype already provides a close match to the new optimum (Ghalambor *et al.* 2007). In contrast, the consequences of nonadaptive plasticity are less predictable. Individuals (or populations) expressing nonadaptive plasticity will be expected to go extinct when nonadaptive plasticity affects the mean phenotype of individuals, but may provide opportunities for adaptation to new environments when it affects the variance in expressed phenotype (Ghalambor *et al.* 2007). Thus, the mix of individuals expressing different degrees and directions of behavioral plasticity will have critical implications for how populations cope with rapid environmental change. Generally, the greater the diversity of behavioral plasticity within a population, the greater the likelihood that the population includes some individuals that exhibit adaptive plasticity in the novel conditions (Dingemanse & Wolf 2013). Consequently, individual differences in behavioral plasticity can lead to greater population stability and persistence. Individual differences in behavioral plasticity also implies that the repeatability (the proportion of phenotypic variation due to differences between-individuals) and possibly also the heritability (the proportion of phenotypic variance caused by additive genetic effects) are environment specific (e.g., Dingemanse *et al.* 2009; Dingemanse *et al.* 2012a). This implies that the rate of evolutionary change in response to selection will depend on the environment (Dingemanse & Wolf 2013).

Novel ecological and evolutionary consequences emerge when personality and plasticity are correlated (Dingemanse & Wolf 2013). For example, conflicting selection pressures acting on personality and plasticity can have important consequences for species interactions. In Iberian wall lizard, faster explorers show greater plasticity in flight initiation distance (FID) from predators (Rodriguez-Prieto *et al.* 2011), but fast exploration may also be associated with greater exposure to parasites and pathogens (Barber & Dingemanse 2010; Kortet *et al.* 2010). Thus on an ecological timescale, the strength of past selection on anti-predator responsiveness will have important consequences for parasite transmission, and the strength of past selection on anti-predator responsiveness will have important consequences for predator–prey dynamics. Genetic correlations between personality and plasticity may constrain the ability of animals to express optimal combinations of traits (Dingemanse & Wolf 2013).

GENETIC AND ENVIRONMENTAL UNDERPINNINGS

Whether constraints on the expression of the optimal combination of elevation and slope of reactions norms act on ecological versus evolutionary time scales, and the ability to overcome such constraints, will depend in part on whether personality-related differences in plasticity are genetic or environmental in origin. As described above in Section 4.2, personality-related differences in plasticity can be underpinned by genetic variation, including genetic variation in state variables that influence the optimal expression of behavior and behavioral plasticity. They may also be environmental in origin, such as when the early environment experienced by an individual produces lasting effects on its phenotype (e.g., maternal effects or developmental plasticity). Even small, stochastic environmental effects can produce lasting correlations between personality and plasticity if positive feedbacks between state and behavior act to maintain differences (Luttbeg & Sih 2010).

Individual differences in sampling behavior have been suggested as a proximate cause of personality-related differences in plasticity (Mathot *et al.* 2012; Sih & Del Giudice 2012). Genetic variation in alternative strategies (e.g., fast/inaccurate versus slow/accurate decision makers) could be maintained in populations as part of mixed evolutionarily stable strategy because they perform equally well on average (Sih & Del Giudice 2012). There is evidence for genetic variation in cognitive styles in *Drosophila melanogaster*, where variation in the foraging gene ("*rover*" and "*sitter*" alleles) is associated with differences in both spatial movement (presumably related to sampling behavior), and the ability to recall old information (Reaume *et al.* 2011). Individual differences in the use of sampling may also come about via environmental effects or state-dependence. For example, early rearing conditions may affect the ability of animals to develop costly sensory apparatus and cognitive functions needed for effecting sampling (Dingemanse & Wolf 2013), which in turn will have lasting effects on both behavior and behavioral plasticity.

Personality-related differences in plasticity that arise via state-dependence may also have genetic or environmental underpinnings. For example, body size may affect the cost and benefits of boldness in the face of a gape-limited predator, with larger individuals being less vulnerable and so being both more active, and changing their activity less across gradients in predation risk (Luttbeg & Sih 2010). Variation in body size can have genetic underpinnings (e.g., Serbezov *et al.* 2010), but can also come about due to chance differences in foraging success that are in turn enforced by positive feedbacks between behavior and state (Rands *et al.* 2003; Luttbeg & Sih 2010).

Quantitative genetic studies have confirmed the importance of both genetic and environmental effects in shaping personality-related differences in plasticity. Studies in great tits (*Parus major*) suggest potentially strong genetic correlations between elevation and slope for plasticity in lay date as a function of spring temperature (discussed in Chapter 2, this volume). However, another study found that two Corsican populations of blue tits (*Cyanistes caruleus*) that show no genetic differentiation nonetheless exhibit different patterns of plasticity in lay date, suggesting an important role of environmental conditions in shaping individual reaction norms (Porlier *et al.* 2012). In stickleback (*Gasterosteus aculeatus*) significant genetic correlations were found between exploration and plasticity for exploration, but the sign of the correlation differed between populations (Dingemanse *et al.* 2012a).

In summary, personality-related differences in plasticity are shaped by both genetic and environmental effects. However, it is not clear under what circumstances genetic versus environmental effects should be more important, or why genetic correlations should exist for the same traits in some populations but not others (Nussey *et al.* 2007). More studies quantifying the genetic and environmental underpinnings of personality-related differences in plasticity are needed if we hope to address this question.

DESIGNS TO STUDY PERSONALITY-RELATED DIFFERENCES IN PLASTICITY

There is great contemporary interest in studying personality-related differences in plasticity given its potential ecological and evolutionary implications, but what study design is needed to estimate variation within populations in personality and behavioral plasticity? The simple answer is that one requires repeated observations of the same behavioral phenotype collected at different positions of an environmental gradient for a set of individuals (Dingemanse & Dochtermann 2013). The specific details in terms of optimal designs with respect to sample size in terms of individuals and repeated observations per individual per environmental context matter and depend on the researcher's objective, specifically whether statistical power, precision, or accuracy is mostly of interest (Martin *et al.* 2011; Van de Pol 2012).

Both observational and experimental approaches can be applied. Observational field studies have the advantage that reaction norms are assayed under natural conditions, and consequently the environmental conditions across which plasticity is observed are necessarily ecologically relevant (cf. Martin & Réale 2008). However, one important consideration for observational studies is the extent to which individuals are in control over the environmental conditions in which they are assayed. When individuals have no control over the environmental conditions under which they are assayed and there is no repeatable variation in the conditions under which individuals are observed, then observational studies will yield unbiased samples of individuals across the environmental gradient. For example, great tits have no control over the spring temperatures they experience, and so plasticity in lay date as a function of spring temperature can readily be studied using nonmanipulative approaches. However, individuals are often able to exert control over the environmental conditions they experience, and therefore may differ in both the mean and range of the focal environmental gradient over which their reaction norm is assayed. For example, in house sparrows, both average brood size and variance in brood size differ between females (Westneat *et al.* 2011), and consequently the gradient in brood size across which plasticity in provisioning behavior can be observed differs between females. This means that slopes of reaction norms may not be estimated with the same confidence for all individuals, and special statistical approaches are required (van de Pol & Wright 2009; Dingemanse *et al.* 2010).

Alternatively, one can apply an experimental approach to the study of behavioral reaction norms, where the researcher exposes the same set of subjects to a range of environmental conditions. This offers the advantage of ensuring that all subjects experience the same range of environmental conditions, and allows for balanced observations across the environmental gradient. While experimental approaches are more powerful, they may be practically impossible to implement in some field studies. Experimental approaches where subjects are either raised in or taken to the laboratory and assayed repeatedly under a range of conditions can provide a standardized means to quantify variation among individuals in their behavioral reaction norms (Biro *et al.* 2010; Mathot *et al.* 2011; Dingemanse *et al.* 2012a; Dingemanse *et al.* 2012b). This approach is powerful, though at the same time partly precludes the study of environmentally-induced variation in behavioral reaction norms (Nicolaus *et al.* 2012; Dingemanse & Wolf 2013).

CONCLUSIONS AND FUTURE DIRECTIONS

Traditional approaches in behavioral ecology have typically shown a lack of interest in quantifying and explaining different levels of variation (Araya-Ajoy & Dingemanse 2013; Westneat *et al.* 2013), such as variation between individuals. However, individuals often differ in their average behavior (animal personality), and in how they adjust their behavior across an environmental gradient (behavioral plasticity). The behavioral reaction norm approach provides a means of studying personality and plasticity together, and application of this method has revealed that personality and plasticity often covary, referred to as personality-related differences in behavioral plasticity. Understanding how personality-related differences in plasticity can come about is a major interest for behavioral and evolutionary biologists, but adaptive explanations are only beginning to be developed. Personality-related differences in behavioral plasticity can have important ecological and evolutionary consequences, most notably because it may constrain the ability of individuals to express the optimal level of both traits. More studies quantifying the environmental and genetic underpinnings of personality-related differences in plasticity are needed to begin to address questions regarding how selection will act on these integrated aspects of individuals' behavioral phenotype.

REFERENCES

Araya-Ajoy, Y.G. and Dingemanse, N.J. (2013) Characterizing behavioural "characters": a conceptual and statistical framework. *Proceedings of the Royal Society B*, **281**, 2013–645.

Armbruster, W.S. (1991) Multilevel analysis of morphometric data from natural plant populations – insights into ontogenic, genetic, and selective correlations in *Dalechampia scandens*. *Evolution*, **45**, 1229–44.

Armbruster, W.S. and Schwaegerle, K.E. (1996) Causes of covariation of phenotypic traits among populations. *Journal of Evolutionary Biology*, **9**, 261–76.

Barber, I. and Dingemanse, N.J. (2010) Parasitism and the evolutionary ecology of animal personality. *Philosophical Transactions of the Royal Society B*, **365**, 4077–88.

Bell, A.M., Hankison, S.J. and Laskowski, K.L. (2009) The repeatability of behaviour: a meta-analysis. *Animal Behaviour*, **77**, 771–83.

Benus, R.F., Bohus, B., Koolhaas, J.M. and van Oortmerssen, G.A. (1991) Heritable variation for aggression as a reflection of individual coping strategies. *Experientia*, **47**, 1008–19.

Biro, P.A., Beckmann, C. and Stamps, J.A. (2010) Small within-day increases in temperature affects boldness and alters personality in coral reef fish. *Proceedings of the Royal Society B*, **277**, 71–7.

Biro, P.A. and Stamps, J.A. (2008) Are animal personality traits linked to life-history productivity? *Trends in Ecology & Evolution*, **23**, 361–8.

Biro, P.A. and Stamps, J.A. (2010) Do consistent individual differences in metabolic rate promote consistent individual differences in behavior? *Trends in Ecology & Evolution*, **25**, 653–9.

Briffa, M., Rundle, S.D. and Fryer, A. (2008) Comparing the strength of behavioural plasticity and consistency across situations: animal personalities in the hermit crab *Pagurus bernhardus*. *Proceedings of the Royal Society B*, **275**, 1305–11.

Brommer, J., Rattiste, K. and Wilson, A. (2010) The rate of ageing in a long-lived bird is not heritable. *Heredity*, **104**, 363–70.

Burns, J.G. and Rodd, F.H. (2008) Hastiness, brain size and predation regime affect the performance of wild guppies in a spatial memory task. *Animal Behaviour*, **76**, 911–22.

Cutts, C.J., Metcalfe, N.B. and Caylor, A.C. (1998) Aggression and growth depression in juvenile Atlantic salmon: the consequences of individual variation in standard metabolic rate. *Journal of Fish Biology*, **52**, 1026–37.

Dall, S.R.X., Houston, A.I. and McNamara, J.M. (2004) The behavioural ecology of personality: consistent individual differences from an adaptive perspective. *Ecology Letters*, **7**, 734–9.

Dall, S.R.X. and Johnstone, R.A. (2002) Managing uncertainty: information and insurance under the risk of starvation. *Philosophical Transactions of the Royal Society B*, **357**, 1519–26.

Davies, N.B., Kreb, J.R. and West, S.A. (2012) *An Introduction to Behavioural Ecology*, John Wiley & Sons, Oxford.

DeWitt, T.J., Sih, A. and Wilson, D.S. (1998) Costs and limits of phenotypic plasticity. *Trends in Ecology & Evolution*, **13**, 77–81.

Dingemanse, N.J., Barber, I., Wright, J. and Brommer, J.E. (2012a) Quantitative genetics of behavioural reaction norms: genetic correlations between personality and behavioural plasticity vary across stickleback populations. *Journal of Evolutionary Biology*, **25**, 485–96.

Dingemanse, N.J., Bouwman, K.M., van de Pol, M. *et al.* (2012b) Variation in personality and behavioural plasticity across four populations of the great tit *Parus major*. *Journal of Animal Ecology*, **81**, 116–26.

Dingemanse, N.J. and Dochtermann, N.A. (2013) Quantifying individual variation in behaviour: mixed-effect modelling approaches. *Journal of Animal Ecology*, **82**, 39–54.

Dingemanse, N.J., Kazem, A.J.N., Réale, D. and Wright, J. (2010) Behavioural reaction norms: animal personality meets individual plasticity. *Trends in Ecology & Evolution*, **25**, 81–9.

Dingemanse, N.J., Van der Plas, F., Wright, J., Réale, D., Schrama, M., Roff, D.A., Van der Zee, E. & Barber, L. (2009) Individual experience and evolutionary history of predation affect expression of heritable variation in fish personality and morphology. *Proceedings of the Royal Society B*. **276**, 1285–93.

Dingemanse, N.J. and Wolf, M. (2010) Recent models for adaptive personality differences: a review. *Philosophical Transactions of the Royal Society B*, **365**, 3947–58.

Dingemanse, N.J. and Wolf, M. (2013) Between-individual differences in behavioural plasticity within populations: causes and consequences. *Animal Behaviour*, **85**, 1031–9.

Dochtermann, N. and Dingemanse, N.J. (2013) Behavioral syndromes as evolutionary constraints. *Behavioral Ecology*, **24**, 806–11.

Gabriel, W., Luttbeg, B., Sih, A. and Tollrian, R. (2005) Environmental tolerance, heterogeneity, and the evolution of reversible plastic responses. *American Naturalist*, **166**, 339–53.

Ghalambor, C.K., Angeloni, L.M. and Carroll, S.P. (2010) Behavior as phenotypic plasticity, in *Evolutionary Behavioral Ecology* (eds D.F. Westneat and C.W. Fox), Oxford University Press, Oxford, pp. 90–107.

Ghalambor, C.K., McKay, J.K., Carroll, S.P. and Reznick, D.N. (2007) Adaptive versus non-adaptive phenotypic plasticity and the potential for contemporary adaptation in new environments. *Functional Ecology*, **21**, 394–407.

Houston, A.I. and McNamara, J.M. (1999) *Models of Adaptive Behaviour: An Approach Based on State*, Cambridge University Press, Cambridge.

Jones, K.A. and Godin, J.-G.J. (2010) Are fast explorers slow reactors? Linking personality type and anti-predator behaviour. *Proceedings of the Royal Society B*, **277**, 625–32.

Kontiainen, P., Pietiainen, H., Huttunen, K. *et al.* (2009) Aggressive Ural owl mothers recruit more offspring. *Behavioral Ecology*, **20**, 789–96.

Koolhaas, J.M., Korte, S.M., De Boer, S.F. *et al.* (1999) Coping styles in animals: current status in behavior and stress-physiology. *Neuroscience & Biobehavioral Reviews*, **23**, 925–35.

Kortet, R., Hedrick, A.V. and Vainikka, A. (2010) Parasitism, predation and the evolution of animal personalities. *Ecology Letters*, **13**, 1449–58.

Laskowski, K.L. and Bell, A.M. (2013) Competition avoidance drives individual differences in response to a changing food resource in sticklebacks. *Ecology Letters*, **16**, 746–53.

Luttbeg, B. and Sih, A. (2010) Risk, resources and state-dependent adaptive behavioural syndromes. *Philosophical Transactions of the Royal Society B*, **365**, 3977–90.

Martin, J.G.A., Nussey, D.H., Wilson, A.J. and Réale, D. (2011) Measuring individual differences in reaction norms in field and experimental studies: a power analysis of random regression models. *Methods in Ecology and Evolution*, **2**, 362–74.

Martin, J.G.A. and Réale, D. (2008) Temperament, risk assessment and habituation to novelty in easter chipmunks, *Tamias striatus*. *Animal Behaviour*, **75**, 309–18.

Mathot, K.J., van den Hout, P.J., Piersma, T. *et al.* (2011) Disentangling the roles of frequency-vs. state-dependence in generating individual differences in behavioural plasticity. *Ecology Letters*, **14**, 1254–62.

Mathot, K.J., Wright, J., Kempenaers, B. and Dingemanse, N.J. (2012) Adaptive strategies for managing uncertainty may explain personality-related differences in behavioural plasticity. *Oikos*, **121**, 1009–20.

Nicolaus, M., Tinbergen, J.M., Bouwman, K.M. *et al.* (2012) Experimental evidence for adaptive personalities in a wild passerine bird. *Proceedings of the Royal Society B*, **279**, 4885–92.

Nussey, D.H., Wilson, A.J. and Brommer, J.E. (2007) The evolutionary ecology of individual phenotypic plasticity in wild populations. *Journal of Evolutionary Biology*, **20**, 831–44.

Porlier, M., Charmantier, A., Bourgault, P. *et al.* (2012) Variation in phenotypic plasticity and selection patterns in blue tit breeding time: between- and within-population comparisons. *Journal of Animal Ecology*, **81**, 1041–51.

Pruitt, J., Riechert, S. and Harris, D. (2011) Reproductive consequences of male body mass and aggressiveness depend on females' behavioral types. *Behavioral Ecology and Sociobiology*, **65**, 1957–66.

Pruitt, J.N. and Riechert, S.E. (2011) How within-group behavioural variation and task efficiency enhance fitness in a social group. *Proceedings of the Royal Society B*, **278**, 1209–15.

Rands, S.A., Cowlishaw, G., Pettifor, R.A. *et al.* (2003) Spontaneous emergence of leaders and followers in foraging pairs. *Nature*, **423**, 432–4.

Réale, D., Dingemanse, N.J., Kazem, A.J.N. and Wright, J. (2010) Evolutionary and ecological approaches to the study of personality. *Philosophical Transactions of the Royal Society B*, **365**, 3937–46.

Reaume, C.J., Sokolowski, M.B. and Mery, F. (2011) A natural genetic polymorphism affects retroactive interference in Drosophila melanogaster. *Proceedings of the Royal Society B*, **278**, 91–8.

Rodriguez-Prieto, I., Martín, J. and Fernández-Juricic, E. (2011) Individual variation in behavioral plasticity: direct and indirect effects of boldness, exploration and sociability on habituation to predators in lizards. *Proceedings of the Royal Society B*, **278**, 266–73.

Schaeffer, L.R. (2004) Application of random regression models in animal breeding. *Livestock Production Science*, **86**, 35–45.

Serbezov, D., Bernatchez, L., Olsen, E.M. and Vøllestad, L.A. (2010) Quantitative genetic parameters for wild stream-living brown trout: heritability and parental effects. *Journal of Evolutionary Biology*, **23**, 1631–41.

Sih, A., Bell, A. and Johnson, J.C. (2004) Behavioral syndromes: an ecological and evolutionary overview. *Trends in Ecology & Evolution*, **19**, 372–78.

Sih, A., Cote, J., Evans, M. *et al.* (2012) Ecological implications of behavioural syndromes. *Ecology Letters*, **15**, 278–89.

Sih, A. and Del Giudice, M. (2012) Linking behavioural syndromes and cognition: a behavioural ecology perspective. *Philosophical Transactions of the Royal Society B*, **367**, 2762–72.

Smiseth, P.T., Wright, J. and Kolliker, M. (2008) Parent-offspring conflict and co-adaptation: behavioural ecology meets quantitative genetics. *Proceedings of the Royal Society B*, **275**, 1823–30.

Speakman, J.R., Król, E. and Johnson, M.S. (2004) The functional significance of individual variation in basal metabolic rate. *Physiological and Biochemical Zoology*, **77**, 900–15.

Van de Pol, M. (2012) Quantifying individual variation in reaction norms: how study design affects the accuracy, precision and power of random regression models. *Methods in Ecology and Evolution*, **3**, 268–80.

van de Pol, M. and Wright, J. (2009) A simple method for distinguishing within-verus between-subject effects using mixed models. *Animal Behaviour*, **77**, 753–8.

Verbeek, M.E.M., Drent, P.J. and Wiepkema, P.R. (1994) Consistent individual differences in early exploratory behaviour of male great tits. *Animal Behaviour*, **48**, 1113–21.

Westneat, D.F., Hatch, M.I., Wetzel, D.P. and Ensminger, A.L. (2011) Individual variation in parental care reaction norms: integration of personality and plasticity. *American Naturalist*, **178**, 652–67.

Westneat, D.F., Schofield, M. and Wright, J. (2013) Parental behavior exhibits among-individual variance, plasticity, and heterogeneous residual variance. *Behavioral Ecology*, **24**, 598–604.

Wolf, M., van Doorn, G.S. and Weissing, F.J. (2008) Evolutionary emergence of responsive and unresponsive personalities. *Proceedings of the National Academy of Sciences of the United States of America*, **105**, 15825–30.

Wolf, M. and Weissing, F.J. (2010) An explanatory framework for adaptive personality differences. *Philosophical Transactions of the Royal Society B*, **365**, 3959–68.

Wolf, M. and Weissing, F.J. (2012) Animal personalities: consequences for ecology and evolution. *Trends in Ecology & Evolution*, **27**, 452–61.

5

INDUCTION AND FUNCTION OF POLYPHENIC MORPHS: PROXIMATE REGULATORY MECHANISMS AND EVOLUTIONARY IMPLICATIONS

Anthony J. Zera[1] and Jennifer A. Brisson[2]

[1]*School of Biological Sciences, University of Nebraska, Lincoln, NE, USA*
[2]*Department of Biology, University of Rochester, Rochester, NY, USA*

INTRODUCTION

Background

Phenotypic plasticity, the ability of a single genotype to produce different phenotypes in different environments, has become a central topic of evolutionary study during the past few decades. The topic of plasticity has many facets such as the costs and benefits of plasticity, the proximate mechanisms underlying plasticity, the role of plasticity in adaptation, the evolution of plasticity itself, and evolutionary consequences of plasticity (Schlichting & Pigliucci 1998; West-Eberhard 2003; DeWitt & Scheiner 2004; Ghalambor *et al.* 2007: Whitman & Ananthakrishnan 2009; Piersma & van Gils 2011; Ledón-Rettig & Pfennig 2011; see also Chapters 1, 2, 4, 8, & 9, this volume).

A long-standing issue in phenotypic plasticity, which is the main focus of this chapter, concerns the physiological mechanisms underlying the production and functioning of multiple phenotypes. These mechanisms have been particularly well studied in polyphenisms, that is, plasticity involving a few discontinuous phenotypes, rather than a continuous array of phenotypes (Zera & Denno 1997; Nijhout 1999, 2003; Whitman & Ananthakrishnan 2009; Hartfelder & Emlen 2012). Classic examples of polyphenism include reproductive, worker, and soldier castes of social insects, dispersing/low fecundity versus flightless/high-fecundity morphs of many insects, wing-pattern variants in moths and butterflies, horned versus hornless beetles, and carnivorous versus omnivorous morphs of frogs (Figure 5.1). The hallmark of these adaptive polyphenisms is that morphs differ

Integrative Organismal Biology, First Edition.
Edited by Lynn B. Martin, Cameron K. Ghalambor, and H. Arthur Woods.

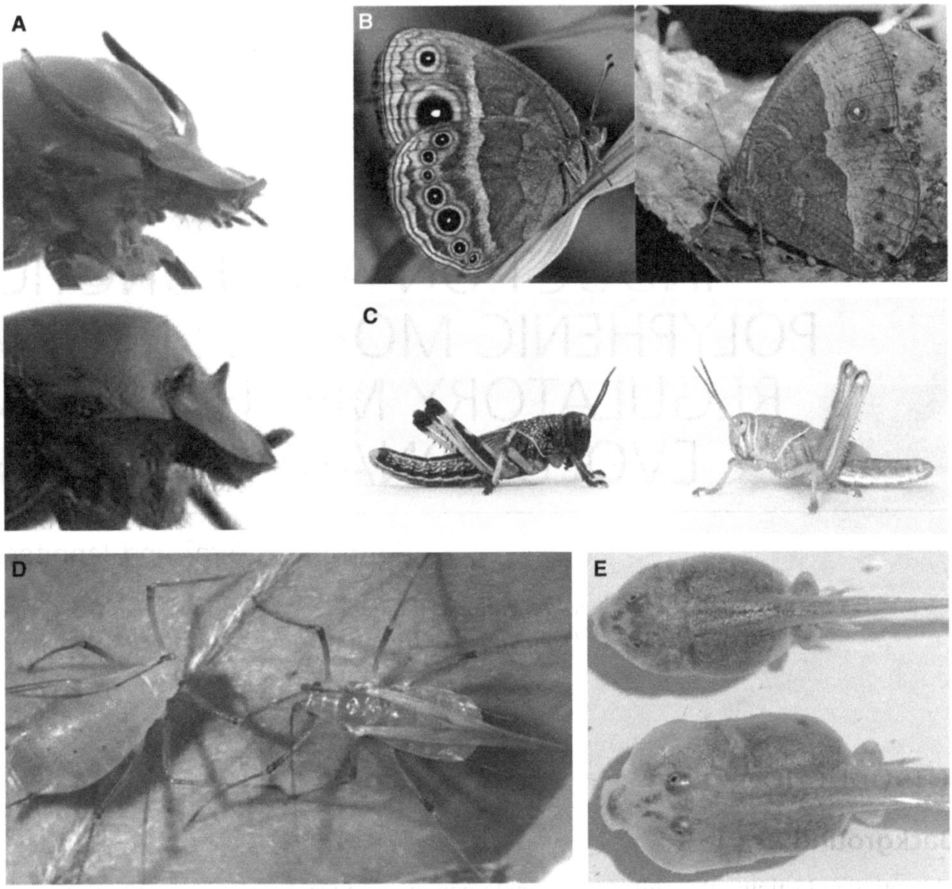

Figure 5.1. Classic examples of polyphenism. (A) Horn polyphenism in *Onthophagus* beetles (image reproduced from Valena & Moczek 2012); (B) seasonal polyphenism in *Bicyclus* butterflies (image copyright H. Piel and Antónia Monteiro); (C) phase polyphenism in *Schistocerca* locusts (image copyright Tom Fayle); (D) wing polyphenism in pea aphids; (E) feeding polyphenism in *Spea* tadpoles (image copyright David Pfennig). (See plate section for color version.)

dramatically, in an integrated suite of morphological, physiological, behavioral, and life-history traits. This results in alternate phenotypes with specific functions that are adapted to particular environments.

Polyphenism is widespread and is thought to be a major adaptive strategy for dealing with environments that are heterogeneous in time or space. Thus, polyphenisms constitute major organismal adaptations that have been molded by natural selection (Brakefield & Larsen 1984; see also Schlichting & Pigliucci 1998; West-Eberhard 2003; DeWitt & Scheiner 2004; Ghalambor *et al.* 2007: Whitman & Ananthakrishnan 2009; Piersma & van Gils 2011; Ledón-Rettig & Pfennig 2011; and Chapters 1, 2, 4, 8, & 9, this volume). Importantly, the often very large-magnitude differences in phenotype among morphs make polyphenism particularly well-suited for functional studies of plasticity. While it has become obvious that polyphenisms are adaptive and under strong selection, what is less clear is how an organism's developmental program can be environmentally modulated to produce such discrete phenotypes. To address this, and to understand how polyphenisms evolve, the underlying physiological mechanisms must be identified.

Focus of This Review

The main goal of this chapter is therefore to review recent advances in the proximate mechanisms underlying the production and functioning of alternate morphs of polyphenisms in the context of the environment in which the morphs occur. Because of space constraints, we will primarily focus on three extensively studied types of polyphenism, each of which has made important contributions to understanding specific aspects of the proximate mechanisms of polyphenism during the past 10–15 years: dispersal polyphenism in aphids, locusts and crickets; morphological polyphenism of butterfly wing patterns and beetles' horns; and metamorphic plasticity in frogs. While mainly focusing on polyphenism, we will also include a few cases of genetic polymorphism (e.g., cricket wing polymorphism), continuous plasticity (anuran plasticity in the timing of metamorphosis), and sexual dimorphism (beetle horn and mandible morphology), because these cases include mechanistic information that is not well covered in studies of traditional polyphenism and which is highly relevant to this topic. Limited space precludes discussion of many well-studied types of polyphenism, such as caste polyphenism in social insects, especially honeybees, which has been the subject of several reviews (e.g., Johnson 2010; Fischman *et al.* 2011).

We will use these examples to illustrate three different aspects of mechanistic control of polyphenic expression. Our first focus is on endocrine control of polyphenism. Prior to the last 10–15 years, physiological studies of this topic, especially insect polyphenism, focused largely on systemic endocrine regulation (i.e., variation in blood hormone levels), and much information has been obtained on this topic (Hardie & Lees 1985; Nijhout 1999; Zera *et al.* 2007). Since then, one of the most important advances in polyphenism endocrinology has been the increased focus on hormonal mechanisms that transduce specific environmental cues to internal hormonal signals that then influence morph induction. This is a key process by which adaptive phenotypes are induced to match specific environments. Of special note are studies of serotonin in locusts, corticosterone signaling in anurans, and juvenile hormone (JH) and insulin signaling in beetles with exaggerated horns or mandibles (discussed in detail below). These investigations constitute a significant expansion of endocrine studies of polyphenism to peptide hormones and biogenic amines, especially in insects, compared with earlier investigations that were primarily focused on morphogenetic hormones (JH and ecdysteroids).

Our second focus will be functional genomics approaches that investigate the molecular genetic basis of polyphenism. Genome-wide gene expression studies can now be undertaken in virtually any organism and are particularly well suited for identifying candidate genes that are involved in the initiation or maintenance of alternative phenotypes. However, detailed knowledge about what tissue and what time period to use in these experiments is the limiting step in such studies, emphasizing the importance of an intimate understanding of the polyphenic system and its functioning. We discuss these issues using examples from dispersal and morphological polyphenisms. Importantly, as discussed below, the results of transcriptome studies have augmented and extended results of hormonal studies, demonstrating how endocrine and functional genomic approaches can complement one another.

Our third focus will be on metabolic aspects of morph induction and function. For example, lipid metabolites appear to regulate phase transitions in adult locusts in unexpected ways. Multilevel studies of lipid metabolism in wing polymorphic crickets demonstrate how modulated gene expression gives rise to altered whole organism traits (triglyceride energy reserves) via modification of enzyme function and flow of metabolites through pathways of intermediary metabolism. Finally, recent studies have provided information on interactions among the insulin signaling pathway, JH regulation, and their interactions with each other and sex determining mechanisms that link external nutrient availability with plasticity in growth of sexually-selected, exaggerated traits in beetles. Each of these studies is discussed in detail below.

Why Study the Proximate Mechanisms Underlying Plasticity?

Examining the mechanistic basis of polyphenism provides the basis for understanding how these mechanisms have evolved and how they may constrain or facilitate future phenotypic modifications. As aptly stated by West-Eberhard (2003): "Confused concepts are virtually inevitable if data on real mechanisms are ignored, because one test of the adequacy of a theory is mechanistic feasibility. The confusion sired by imaginary mechanisms demonstrates the importance of proximate causation for understanding evolution." In other words, we need an accurate understanding of how polyphenisms operate at the mechanistic level not only to test frameworks about how they can evolve, but also to generate these frameworks.

For example, studies of developmental processes have determined that a developing organism can be divided into organizational subunits called modules. Particular genes operate within as well as between each module within the context of genetic networks. To affect a change in a particular phenotype, a gene must be involved in the production of that phenotype, but it also must be dissociable from its affects on other phenotypes; it must be optimally pleiotropic (Stern & Orgogozo 2009). Knowing these functional aspects of development can help allow predictions to be made about how aspects of polyphenisms can evolve. An excellent example is provided by butterfly eyespots, discussed more below. Eyespots are a developmental module, with transplanting of an eyespot focus resulting in an ectopic eyespot (Nijhout 1991, 1999). Knowing how developmental modules work allows us to predict that to gain more eyespots, a regulatory gene that initiates eyespot development must be induced in a new location. Conversely, to lose an eyespot, the opposite must happen. Evolutionary changes that result in the production of alternative phenotypes instead of a single phenotype are therefore likely to affect regulatory genes controlling the expression of a module (Moczek 2008).

In another example, Suzuki and Nijhout (2006) experimentally investigated the mechanisms by which polyphenism evolves, testing ideas put forth regarding genetic accommodation by West-Eberhard (2003). This study would not have been possible without prior detailed information on the endocrine mechanisms underlying color polymorphism in their experimental model, the moth *Manduca sexta*. Finally, studying the mechanistic basis of polyphenism provides insight into the processes that generate trait diversity, which can ultimately lead to speciation (reviewed in West-Eberhard 2003; Pfennig *et al.* 2010). This topic has been the focus of recent studies of polyphenism within, and phenotypic diversity among, species of spadefood toads (Gomez-Mestre & Buchholz 2006; Ledón-Rettig & Pfennig 2011), and is likely to be a major focus of future polyphenism studies.

BACKGROUND ON ENDOCRINOLOGY AND GENE REGULATION

Endocrinology and genome-level gene regulation approaches figure prominently in this chapter. Here we provide brief background on these topics for readers who may be less familiar with them.

Endocrinology

Hormones coordinate the expression of multiple traits and thus have long been thought to regulate the expression of traits that differ between morphs (Nijhout 1991, 1999; Zera *et al.* 2007). A hallmark of endocrine regulation is its hierarchical, integrative nature, with a single hormone controlling several subsystems thus allowing the coordinate expression of multiple traits. These subsystems can be hormone-producing glands themselves, regulated by a master hormone (e.g., CRH, corticotropin releasing factor, Figure 5.2A), or synergistic interactions among hormones in responsive cells, such as interactions between thyroxine

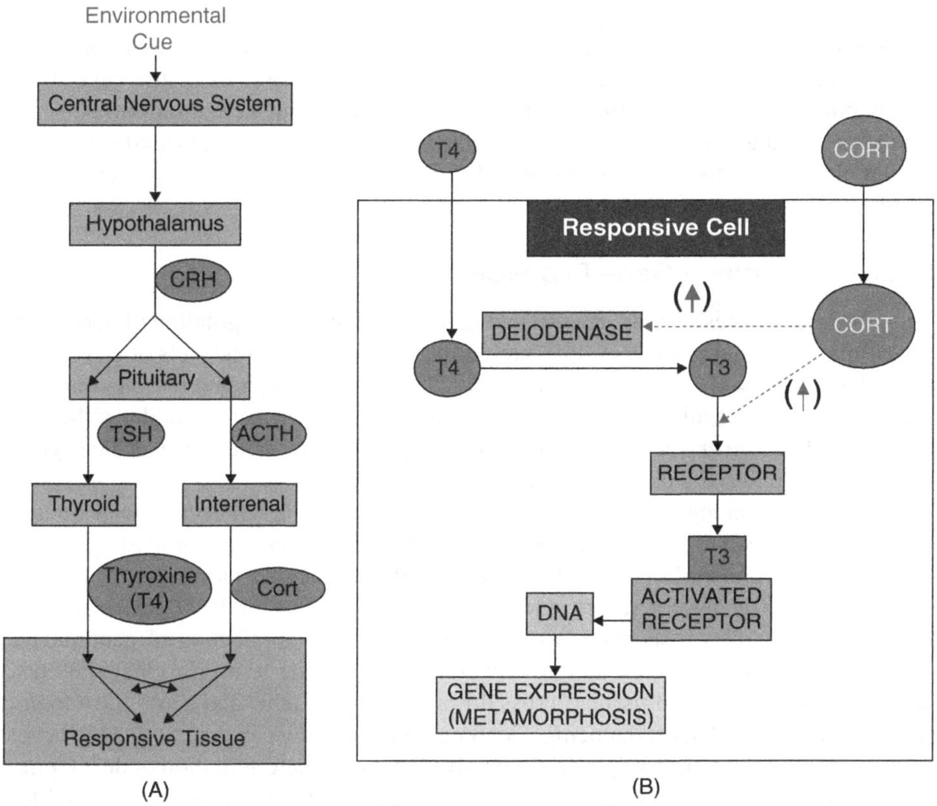

Figure 5.2. Panel (A) Simplified diagram of the endocrine regulation of the timing of metamorphosis in anurans, illustrating various aspects of endocrine regulation. CRH = corticotropin releasing hormone, TSH = thyroid-simulating hormone, ACTH = adrenocorticotropic hormone, CORT = corticosterone. As described in the text, an environmental signal is transduced through the brain into an internal endocrine signal that induces metamorphosis. Endocrine control is hierarchical with multiple control steps. Common regulators (e.g., CRH) coordinate the release of multiple hormones which interact in multiple ways to cause a tissue response (see Panel B). Panel B) Hormonal interactions in a peripheral tissue undergoing metamorphosis. T4 = thyroxine, T3 = triiodothyronine. CORT potentiates the effect of thyroxine in responsive cells by increasing the activity of the enzyme (deiodenase) that converts T4 (the prohormone, thyroxine) to the more active T3 (hormone, triidothyronine), and by upregulating the T3 nuclear receptor. There are other interactions between these hormones in peripheral tissues that are not shown. (See plate section for color version.)

derivatives and corticosterone (Figure 5.2A; see section on anuran plasticity). This gives rise to cross-talk among and integrative expression of different hormones. Alternatively, the subsystems can be multiple target tissues, thus giving rise to the coordinate expression of multiple traits.

Evolutionary change in hormone regulation can occur by modifying the signal or reception of the signal. Very different evolutionary outcomes are expected to result. For example, correlated changes in many different organs are expected when the hormone titer is varied versus more organ-specific changes when receptors are modified. Hormone titer or receptor changes can occur by changes in the rate of hormone or receptor production or degradation. Any aspect of these regulatory mechanisms can theoretically be modified to produce discontinuous phenotypes comprising polyphenisms (Zera *et al.* 2007). Finally, for hormones to regulate polyphenisms, the hormonal control of development or function must be integrated with the perception of key morph-inducing cues of the environment, such as photoperiod, temperature, nutrient levels, and habitat deterioration. Thus, neuroendocrine regulation must play a particularly important role in plasticity regulation. Most of the issues discussed above will be dealt with repeatedly in the case studies discussed below.

Molecular Studies – Gene Expression

One of the critical roles of endocrine signaling in polyphenism is regulation of gene expression. Hormones signal to an array of tissues, affecting the expression levels of genes critical for the development of the traits that differentiate alternate polyphenic morphs. Endocrine signaling and gene regulation are therefore inexorably linked, yet we address them separately here because the techniques used to assay them are different and have yielded different insights into the physiological basis of polyphenism.

Gene expression regulation of polyphenism has historically not been as well studied as nonmolecular aspects of endocrine control. Single-gene studies were limited by a lack of good candidates, and early unbiased approaches, although an important advance, yielded only a handful of genes for future study (e.g., Evans & Wheeler 1999). A revolution in molecular studies of polyphenism has come about with the application of genomic technologies in nonmodel systems. It's now possible to assay the transcriptional state of the whole organism or targeted tissues across different environments and across development using microarrays or RNA sequencing. Genome-wide profiling can then be used to develop hypotheses about the causative pathways involved in morph determination or development. For example, if genes associated with serotonin (receptors, transporters) are found to be responding to an environmental cue, then serotonin regulation is a hypothesized mechanism that can be confirmed or refuted using follow-up functional studies.

Transcriptional experiments can be designed to address two different but complementary aims. First, expression profiling can be used to capture the molecular genetic changes that occur while the organism is integrating information from the environment in order to gain insight into the molecular signals that start the developmental trajectory down one path versus the other. Second, these same approaches can be used to characterize polyphenic outcomes (i.e., alternative adult phenotypes) to determine the molecular underpinnings of the alternative phenotypes. We'll provide examples below of both of these types of experiments and the insights they've provided into the molecular genetic control of polyphenism.

CASE STUDIES

Dispersal Polyphenism in Locusts, Aphids, and Crickets

Dispersal polyphenism is a classic adaptation in which alternate morphs, adapted for dispersal versus reproduction, are induced by environmental signals such that a specific morph (e.g., flightless reproducer) matches a specific environment (e.g., stable, resource rich environment). Ecological aspects of dispersal polyphenism have been extensively studied (see references in Zera & Denno 1997 and Zera & Brisson 2012). They have also been extensively studied in the context of the mechanisms underlying morph induction and function. Early work (1960s–'80s) focused on the role of key developmental and reproductive hormones, JH and ecdysteroids (ECDs), mainly in locusts and aphids (reviewed in Hardie & Lees 1985; Zera & Denno 1997; Pener & Simpson 2009; Pener & Yerushalmi 2009). However, the small size of aphids, the complexity of morph induction in locusts, and, to some extent, over-focus on JH and ECDs to the exclusion of other potential morph-inducers, stalled progress in the 1970–1980s (Zera & Denno 1997; Pener & Simpson 2009). Application of molecular and genomic approaches substantially rejuvenated research on this topic, as did expansion of endocrine studies in locusts to various neurohormones (Pener & Simpson 2009; Pener & Yerushalmi 2009; Verlinden *et al.* 2009; Van Wielendaele *et al.* 2013). These studies have yielded significant new information on the proximate mechanisms underlying polyphenism, as have endocrine and biochemical investigations of wing polymorphism (a genetic polymorphism) conducted in *Gryllus* crickets (Zera 2009; Zera & Harshman 2011). We discuss each system below.

Locusts Migratory locusts consist of two phases, solitary and gregarious, that differ substantially in behavior, morphology, and physiology related to a solitary versus gregarious (swarming) life style. Physiological regulation of phase polyphenism is exceedingly complex, because environmental inputs at any developmental stage, and across several generations can influence phase induction, resulting in varying degrees of expression of solitary or gregarious phenotypes which are reversible (Pener & Simpson 2009). By contrast, morph induction in wing polymorphic aphids and crickets occurs during a narrow window of development and is irreversible (Sutherland 1969; Müller *et al.* 2001). Early work in locusts, beginning in the 1960s, suggested, and recent studies have confirmed, morph-differences in the JH titer (Tawfik *et al.* 2000; Pener & Simpson 2009). However, JH regulates some, but not most, differences between the phases (Pener & Yerushalmi, 2009; Pener & Simpson 2009). More recent attention has focused on the role of various peptides (corazonin, neuroparsins, adipokinetic hormone, insulin-like peptides), and various biogenic amines such as serotonin (van Wielendaele *et al.*, 2013) as regulators of various aspects of phase induction and function.

Hopper density plays a pivotal role in phase induction, and elicits its effect very quickly (hours) via direct mechano-stimulation of the hind legs, vision, and smell. Modulation of the serotonin concentration in the thoracic ganglia by mechano-stimulation appears to be a key regulator of phase induction. For example, Anstey *et al.* (2009) showed that serotonin level in the thoracic ganglia changed in concert with phase when hopper density was experimentally manipulated. Further, phase induction could be modulated by experimentally manipulating serotonin binding to receptors or by altering serotonin biosynthesis. This study represents the most thorough functional-experimental documentation of the role of biogenic amines on morph induction in any polyphenism. Many other biogenic amines differ in concentration between the phases and may also regulate phase polyphenism (Rogers *et al.* 2004), as suggested by a recent transcriptome study (see below). Importantly, prior identification of a specific environmental cue that influences phase induction (mechano-stimulation of the hind legs), was a key factor that allowed subsequent identification of the internal hormonal mechanism of morph induction.

A number of recent studies have investigated the mechanisms underlying phase polyphenism from multiple -omics perspectives. For example, a transcriptomic study corroborated the importance of biogenic amines by identifying transcript changes in the dopamine pathway that correlated with the switch between gregarious and solitarious phases (Ma *et al.* 2011). Furthermore, a chemosensory and *takeout* gene were differentially expressed in the antennae of gregarious versus solitarious phases, indicating that these genes may function in the shifts between repulsion and attraction that accompany phase transition. Finally, a recent metabolomics study of *Locusta* (Wu *et al.* 2012) identified lipid metabolites, especially carnitine, carnitine-acyl derivatives, and various glycerides as regulating phase transitions in *Locusta.* Thus, polyphenism may be regulated by mechanisms different from those involving traditional hormones or regulatory genes. Lipid metabolites could influence phase induction by altering the metabolic environment, changing the characteristics of cell membranes, or, in the case of carnitine, by functioning as neuromodulators (Wu *et al.* 2012).

Aphids Wingless aphids produce winged offspring in response to stressful environmental cues such as crowding and poor food quality (reviewed in Müller *et al.* 2001). Despite intensive investigations during the 1960s–'80s, no strong evidence supporting a role for JH in wing morph determination has been obtained (reviewed in Zera & Denno 1997; Braendle *et al.* 2006; Schwartzberg *et al.* 2008). Thus, the endocrine regulation of aphid wing morph determination remains a completely open issue, although JH possibly regulates aspects of later wing development itself in some species (Lees 1980; Ishikawa *et al.* 2013). Like the situation in locusts described above, progress in understanding the endocrine basis of aphid morph determination may come about by focusing more on non-JH hormones, such as various neuropeptides and biogenic amines.

Recent molecular studies have interrogated gene expression differences underlying the adult winged and wingless morphs (Ghanim *et al.* 2006; Brisson *et al.* 2007, 2010), identifying genes responsible for physiological differences between them. However, only one published study has profiled gene expression of wingless females as they are undergoing a crowding stress to induce winged offspring production (Ishikawa *et al.* 2012). In this study, three genes (Uba1, McrNaca, and *wingless*) were expressed at higher levels in high-density treated females relative to low-density controls, suggesting that they may play a role in transducing the wing induction signal. A RNA-Seq study identified an additional ~3500 genes that may be involved in early stages of the polyphenic response, such as genes involved in odorant binding, hormone mediated signaling pathways, and chromatin remodeling (N. Vellichirammal & J. A. Brisson, unpublished data). Thus, deciphering the molecular basis of wing-induction in aphids is still in the early stages, but promising candidate pathways have been identified.

In the future, aphid dispersal polyphenism may serve as a model for understanding how mechanisms underlying polyphenic expression constrain evolution of that polyphenism. Aphid species vary with regard to developmental stages that are susceptible to wing-inducing cues: pea aphids can only respond as embryos (Sutherland 1969) while other species (e.g., *Aphis craccivora*, Johnson 1965) can respond through the second of the four nymphal instars. No species can respond later in development. Yet, a later response may be selectively favored since it would result in more precise matching of morphology to environment. Exploring the mechanistic basis of wing development in aphids may elucidate why this apparent constraint exists.

Gryllus *Crickets* The most detailed measurements of morph differences in the JH titer (i.e., blood JH concentration) and titer regulators have been obtained for morphs of *Gryllus* crickets (Zera 2009; Zera *et al.* 2007). Importantly, the JH titer was directly measured in these studies using a well-established method, in contrast to many other unreliable investigations of insect polyphenism in which the JH titer was measured indirectly (for critique of these studies see Zera 2007). Earlier *Gryllus* studies documented differences in the blood JH titer, activity of JH titer regulators (e.g., the degradative enzyme, JH esterase) and blood ECD titer between juvenile or adult *Gryllus* morphs. In fact, Zera and Tiebel (1989) (cited in Zera *et al.* 2007) were the first to identify variation in an endocrine regulator underlying either an environmental polyphenism or a genetic polymorphism in insects. These studies have been reviewed several times and will not be discussed further here (Zera 2009; Zera & Harshman 2009; Zera *et al.* 2007).

Most recently, an unexpected morph-specific JH titer circadian cycle was identified in *G. firmus,* and other *Gryllus* species, in which the JH titer rose and fell 20–100 fold within 4–6 hours in the dispersing morph, above and below the relative constant JH titer in the flightless morph (Zhao & Zera 2004; Zera *et al.* 2007; Zera & Zhao 2009). An important message from these studies is that hormone titers in insects can be highly dynamic over very short time scales, as is known for many hormones in vertebrates (Zera *et al.* 2007). The failure to take this temporal variability into account can result in highly erroneous conclusions regarding the mechanisms of morph induction and function. For example, the correlation between JH titer and the flightless morph changes from +0.9 to –0.9 within a few hours (Zera, 2007). Comparable short-term hormone titer measurements have yet to be undertaken in other insect polyphenisms. A subsequent transcriptome study in *G. firmus* (N. Vellichirammal, J. A. Brisson & A.J. Zera, unpublished) found a morph-specific circadian change in global gene expression that is correlated with the morph-specific JH titer circadian rhythm. Morph-specific circadian change in gene expression could be a common feature of polyphenisms, but has yet to be studied outside of *Gryllus*. This topic is fertile ground for future research.

A long-studied aspect of dispersal polyphenism is morph-specific allocation of internal resources to produce the components required for morph functions (e.g., lipid flight fuel versus egg components). An important contribution of *Gryllus* studies is detailed multilevel-information on the biochemical-physiological processes involved. For example, radiotracer studies documented that the greater accumulation of triglyceride flight fuel at the expense of ovarian growth in the dispersing morph occurred by up-regulation of fatty-acid and triglyceride biosynthesis, (Zera & Harshman 2009, 2011). Subsequent enzymological, molecular, and transcriptome studies indicate that up-regulation of this pathway occurs via increased gene expression rather than by modification of catalytic properties of the enzymes (Zera & Harshman 2011; Vellichirammal *et al.* 2014). A large (10-fold) morph difference in the expression of an insulin-like-peptide gene (A. J. Zera, J. A. Brisson, unpublished data) could potentially play a role in regulating the dispersal-reproduction trade-off, as it appears to do in locusts (Van Wielendaele *et al.* 2013). These studies currently constitute the most detailed information in any polymorphism/polyphenism regarding the mechanisms by which intermediary metabolism in modified to contribute to morph adaptation.

Wing Polyphenism in *Bicyclus* Butterflies and Horn Polyphenism in Dung Beetles

Discontinuous, environmentally-induced butterfly wing patterns within lepidopteran species, especially those in which patterns match specific seasons, is a classic example of polyphenism (i.e., seasonal polyphenism) (Shapiro 1976; Nijhout 1991; Brakefield & French 1999; West-Eberhard 2003; Brakefield & Frankino 2009). As is the case with other complex polyphenisms, numerous physiological and life history traits covary with wing pattern, resulting in morphs being adaptive to different seasonal environments (Brakefield & Frankino 2009). For decades, endocrine studies of wing pattern expression contributed significantly to the development of general models of polyphenic morph induction (Nijhout 1991, 1999; Brakefield & Frankino 2009). More recent studies have focused on molecular-developmental mechanisms of wing pattern expression. Horn and mandible size polyphenisms in beetles have been even more intensively studied with respect to the molecular-developmental mechanisms underlying the expression of polyphenism.

The gene expression basis of seasonal polyphenism has been particularly well-studied in *Bicyclus* butterflies which live in regions of Africa with annual cycles of wet and dry seasons. Temperature controls the butterfly's final phenotype, with larvae raised at high (>23 °C) temperatures exhibiting the wet season morph and low (<19 °C) temperatures showing the dry season phenotype (Brakefield & Reitsma 1991). Wet season butterflies are highly active, reproductive, and have large eyespots that draw predator attacks to their wing edges. Inactive, dry season butterflies do not have this conspicuous coloration because of higher predation risk (Brakefield & Larsen 1984) and have physiological and life-history changes such as delayed reproduction and altered metabolism to deal with the stress of the season (Brakefield *et al.* 2007).

A number of studies have detailed the expression of genes responsible for production of wet versus dry season coloration. The transcription factor *Distal-less* (*Dll*) has high expression in the focus (center) of the eyespot during pupal development and is critical for eyespot formation. *Dll* disperses out from the focus, forming a gradient of expression that likely activates the expression of other genes, such as the transcription factors Engrailed/Invected (En/Inv) and *Spalt*, which can be observed in nested rings of expression that correspond to the rings of color that will be seen on the adult wing (Carroll *et al.* 1994; Brakefield *et al.* 1996; Brakefield & French 1999). Ecdysteroid levels act directly or indirectly to change the expression level of *Dll* in the pupal wing in regions that will later form the eyespots (Brakefield *et al.* 1998). By 24 hours after pupation, *Dll* expression is high in the wet season but low in the dry season morph (Brakefield *et al.* 1996). Thus, a key aspect of coloration differences between the morphs can be changed by modulating expression of this one gene. The factor or factors downstream of ecdysteroids that regulate *Dll* expression are unknown, so the gene expression switch that places an individual butterfly onto the wet season or dry season color trajectory has not been unequivocally established.

The mechanism comprising the switch between morphs is better understood in the dung beetle horn polyphenism. In dung beetles, nutrition is an environmental factor that affects sexually-selected, polyphenic horn size development. In *Onthophagus taurus*, male larvae that receive large dung balls from their mothers reach large adult body size and develop large head horns, while small dung balls result in adult males with small body sizes and no horns or rudimentary horns (Hunt & Simmons 1997; Moczek & Emlen 1999). Females do not have head horns. Large males with horns win access to females after male to male combat (Rasmussen 1994), but the tradeoff is that combat can be dangerous (Siva-Jothy 1987) and developing horns may result in a slower time to sexual maturity (Hunt & Simmons 1997).

Small males without horns are able to sneak past larger males or use alternative, unguarded tunnels to gain access to females.

Knowledge of the gene expression mechanisms underlying large versus small male horns came about with the development of genomics tools in *O. taurus*. Genome-wide transcriptional profiling of *O. taurus* horn primordia tissue using cDNA microarrays identified high transcript levels of intriguing candidate genes such as the sex determination gene *doublesex* (*dsx*) (Kijimoto *et al.* 2009). An elegant follow-up study established *dsx* as a key regulator of the sex-specific horn dimorphism (Kijimoto *et al.* 2012): *O. taurus* produces dsx alternative splice products that are sex-specific (as do other insects). Knockdown of *dsx* transcripts resulted in significantly reduced horn size in large males, little change in small-horned males, and (unexpectedly) ectopic, small horns in females. *dsx* therefore both promotes horn development in large males, demonstrating its importance in regulating the horn polyphenism, and inhibits horn development in females, showing that it is also involved in the sexual dimorphism.

The *dsx* gene therefore has dual roles, establishing sex-specific differences and horn size differences in males, which means there must be a significant interaction between the sex determination and horn development pathways (Valena & Moczek 2012). This interaction can be seen in interspecific patterns of dimorphism variation: the gain or loss of sexually dimorphic horn expression in a species is usually accompanied by the gain or loss of the male-specific horn dimorphism (Emlen *et al.* 2005). Thus, whether a species exhibits a male horn dimorphism may be constrained by whether it first evolved horn differences between the sexes. Indeed, a subsequent study of horns in sexually dimorphic (not horn polyphenic) Japanese rhinoceros beetles demonstrated a key role for *dsx* in sexually dimorphic horn formation and suggests that the recruitment of *dsx* into the horn development gene network may have been a critical event in the evolution of sexually dimorphic horns (Ito *et al.* 2013). Beetle horn dimorphisms are therefore an excellent example of how understanding the specific proximate mechanisms underlying polyphenism can provide insight into how its evolution is facilitated or constrained.

Endocrine signaling is a critical feature of butterfly color and beetle horn polyphenisms. Ecdysteroids link external temperatures to adult phenotypes through unknown mechanisms in the *Bicyclus* polyphenism (Brakefield *et al.* 1996; Koch *et al.* 1996; Oostra *et al.* 2011). Peak ecdysteroid levels exhibit a discontinuous reaction norm between 21 °C and 23 °C during pupal development, indicating that the continuous environmental signal of temperature has already been converted to a discontinuous phenotype at the level of endocrine signaling (Oostra *et al.* 2011). How this happens remains an important question for future studies. By contrast, no significant differences in the hemolymph JH titer were observed between nascent morphs (Oostra *et al.* 2011). Once the discontinuous ecdysteroid levels are established, they are directly responsible for the dichotomous resource allocation and fitness trait differences between the two morphs. For example, some immune genes are at higher expression levels in the wet season morph when mating related immune risk and microbial growth due to high temperatures are at their highest. Conversely, lipid and carbohydrate metabolism genes are most active in the dry season morph that relies more heavily on energy reserves (Oostra 2013).

In contrast to the detailed molecular-developmental information on horn development, endocrine regulation of this polyphenism is less well understood. Previous work, which indirectly inferred JH titer differences between developing horn phenotypes, and which proposed a role for JH in horn polyphenism in *O. taurus,* was criticized on methodological grounds (Zera 2007). Hemolymph JH titers are currently being directly measured in juvenile dung beetles by appropriate analytical methods (i.e., liquid chromatography-mass spectrometry; A. Moczek, personal communication), which will provide direct evidence on this issue.

By contrast, more detailed information on the endocrine regulation of exaggerated traits in male beetles (that are sexually dimorphic, or which exhibit continuous plasticity in males) has recently become available. Nutritional information may be connected to endocrine signaling through the insulin/insulin-like growth factor pathway, which was recently shown to contribute to exaggerated horn growth in the horn dimorphic rhinocerous beetle, *Trypoxylus dichotomus* (Emlen *et al.* 2012). In the stag beetle, *Cyclommatus metallifer*, which exhibits male plasticity in mandible size, modest JH titer differences occur between large-and small mandible phenotypes (Gotoh *et al.*, 2011). Interestingly, sensitivity of the developing mandibles in this species to externally-applied JH is contingent upon specific *doublesex* splice variants, which provides a mechanism whereby JH affects mandible growth plasticity in males but not in females (Gotoh *et al.* 2014).

Phenotypic Plasticity of Anuran Development

Frogs and salamanders exhibit considerable within and between species variation in juvenile growth rate, timing of metamorphosis, and morphological traits related to feeding on specific food types. Phenotypic expression of these traits is often strongly influenced by environmental variables. This leads to the adaptive production of a phenotype that matches the appropriate environment, such as the induction of metamorphosis of an aquatic tadpole to a terrestrial adult frog in response to pond drying, or the induction of carnivorous versus omnivorous morphs of spadefoot toads dependent upon the abundance of appropriate food items. Variation in anuran metamorphosis is continuous rather than polyphenic. However, studies of the endocrine mechanisms underlying plasticity in the timing of anuran metamorphosis are discussed here because they provide the most well-studied mechanistic example of how specific environmental cues are transduced to internal endocrine signals that modulate phenotypic expression. This is a central but poorly understood topic in phenotypic plasticity.

Starting with Wilber and Collins (1973), numerous evolutionary-ecological models have investigated the optimal timing of anuran metamorphosis in terms of ecological variables such as pond drying, availability of food, and predation during the juvenile and adult stages (see references in Rose 2005). However, these models largely treated proximate mechanisms regulating metamorphosis as a black box (Rose 2005). Conversely, the extensive literature on proximate mechanisms of the timing of metamorphosis (Shi 2000; Brown *et al.* 1996; Brown & Cai 2007) have largely ignored ecological factors. This situation has changed dramatically during the past decade with the advent of integrative ecological-endocrine studies of the timing of frog metamorphosis (Denver, 2009; and references therein).

In anurans, the key regulator of metamorphosis is the hormone thyroxine (actually a prohormone; see Figure 5.2). The endocrinology of thyroxine and other interacting hormones, such as corticosteroids and prolactin, is exceedingly complex, and only a very brief thumbnail sketch is given here (See Figures 5.2A and 5.2B; for detailed reviews see Shi (2000), Brown & Cai (2007), and Denver (2009).) Corticotropin releasing hormone (CRH), emanating from the brain, causes release of thyroid stimulating hormone (TSH) from the pituitary, which, in turn, causes release of thyroxine (T4) from the thyroid (Figure 5.2). T4 is taken up by target cells where it is converted to the active hormone triiodothyronine (T3), by enzymatic removal of an iodine. T3 binds to an intracellular receptor (i.e., a ligand-activated transcription factor), which binds to DNA to regulate of gene expression.

A key step in understanding how the environment specifically influences T3-T4 mediated plasticity was the discovery that the stress hormone, corticosterone, interacts with thyroxine (T4) and T3 in a number of ways to modulate the timing of metamorphosis (Denver 2009). Because stress hormones are produced in response to specific environmental signals acting on the brain, the interaction of these two hormone systems (Figures 5.2A and 5.2B) provides the mechanism by which environmental signals can be hormonally transduced to modulate the timing of developmental events that collectively constitute metamorphosis from an aquatic tadpole to a terrestrial frog. This modulation occurs in two main ways: First, in nonmammalian vertebrates, the brain neurohormone (CRH), which regulates the release of thyroxine (Figure 5.2A), simultaneously regulates the release of corticosteroids (via ACTH; Figure 5.2A). For example, simulated pond drying (or decreased food availability) increased the blood levels of CRH, thyroxine, and corticosteroids in spadefoot toads (actually species of frogs) leading to earlier metamorphosis (Boorse & Denver 2004; Denver 2009). In addition, corticosterone synergizes the effect of T3 in a variety of ways in the peripheral target tissues that undergo metamorphic change (Boorse & Denver 2004; Denver 2009; see Figure 5.2B). Another important environmental cue, an alarm pheromone that indicates predation risk, also leads to adaptive plasticity by modulating corticosterone levels (Fraker *et al.* 2009)

A related aspect of anuran developmental plasticity is termed resource polymorphism (Pfennig 1992a,b; Ledón-Rettig 2011). Many species of spadefoot toads produce two polyphenic morphs that differ dramatically in a number of traits. The omnivorous morph is much smaller, has smaller jaw muscles, smooth unserrated mouthparts, and has a longer gut, which adapts this morph to feed on detritus. The carnivorous morph, with the opposite suite of traits, feeds on shrimp as well as tadpoles. The morphs are induced by a variety of environmental cues, most notably shrimp (food) or tadpole density, as well as by the density of heterospecific competitors, genetic relatives (inhibit carnivore induction), and individual condition. Thus, similar to plastic variation in timing of metamorphosis discussed previously, the existence of resource polyphenism allows matching of a specific morph with specific ecological (food type, specific competitors) aspects of the environment. The carnivorous morph typically metamorphoses earlier than the omnivorous morph, and some aspects of morph induction are also thought to be regulated by thyroxine and corticosteroids, although only limited physiological data are available on this point (Pfennig 1992; Ledón-Rettig & Pfennig 2011).

Studies of resource polyphenism in spadefoot toads by Pfennig and colleagues are especially important with respect to illuminating the consequences of phenotypic plasticity on trait evolution (Ledón-Rettig & Pfennig 2011). For example, their studies focus on how phenotypic plasticity, arising form a variety of causes, including maternal effects and other epigenetic mechanisms, can provide the raw material for subsequent trait evolution by genetic accommodation (see West-Eberhard 2003; Gomez-Mestre & Buchholz 2006; see also Ledón-Rettig & Pfennig 2011 for a recent review).

SUMMARY AND FUTURE DIRECTIONS

Endocrine Control

Conceptual models of morph induction by ecdysteroids (ECDs) and especially JH have been a major focus of polyphenism research since the 1980s (reviewed in Nijhout 1999; Zera *et al.* 2007). Nevertheless, in many cases, the validity of the endocrine data, especially JH data, used to support these models is questionable (Zera 2007; discussed above), which makes the validity of the models based on these data uncertain (Zera *et al.* 2007). A key advance during the past decade has been the direct measurement of JH titer differences between polyphenic morphs using well-established methods (discussed above). However, this has only been done in surprisingly few cases. Thus, a major focus of future endocrine studies of polyphenism remains obtaining reliable data on the role of traditional hormones, most notably JH, in morph induction of polyphenism. Already some detailed endocrine studies (Zhao & Zera 2004) indicate a far more complex endocrine picture than suspected.

The expansion of studies of hormonal regulators of insect polyphenism beyond JH and ECDs (e.g., various peptides and biogenic amine in locusts; see above), which also was a major advance in the endocrine study of polyphenism during the past decade, will undoubtedly increase in the future. This is especially the case given the increased ability of undertake global studies of gene expression discussed above. It should not be forgotten, however, that a key feature of the endocrine control of polyphenism is the systemic (e.g., blood) hormone titer which co-ordinates the expression of traits in diverse tissues. This will necessitate follow-up physiological measures of blood titer levels, which will be challenging in many small organisms.

Finally, insect endocrine studies of polyphenism have almost exclusively focused on variation in hormone titers, with the role of variation in receptor expression largely unexplored. In this respect, insect endocrine studies lag far behind vertebrate studies of polyphenism, where the importance of receptor variation has long been appreciated (Denver 2009; Buchholz *et al.* 2011). Recent studies of the role of variation in insulin receptors in plasticity regulation (e.g., Emlen *et al.* 2012) are undoubtedly the leading edge of an explosion of future studies of receptor variability in polyphenic regulation. It will be fascinating to see how future studies untangle the relative importance of blood titer versus tissue receptor sensitivity in the control of polyphenism.

Genomic Studies of Morph Determination and Development

The role of epigenetics in polyphenism will most certainly be addressed with enthusiasm in the future. DNA methylation is the epigenetic mechanism most well-studied with respect to polyphenism. For example, changes in DNA methylation are necessary to differentiate workers from queens in honeybees (Kucharski *et al.* 2008). However, the general function(s) of DNA methylation in insects remains largely unknown. In plant and vertebrate genomes, DNA methylation works to silence transposable elements and repress gene expression via promoter methylation (Bird & Wolffe 1999), whereas in insects methylation occurs primarily in gene bodies and methylation is associated with gene expression, not gene repression (reviewed in Lyko & Maleszka 2011). Recently, honeybees provided insight into this issue, with a study demonstrating that DNA methylation at gene bodies affected alternative splicing of transcripts (Li-Byarlay *et al.* 2013). Future studies will address how epigenetic processes fit into the cascade of events that occur between an environmental signal and an alternative adult phenotype. Most likely, endocrine signaling results in the activation of altered epigenetic programs in a tissue-specific manner. Altered epigenetic states could then lead to the gene expression and alternative splicing differences that underlie alternative morph development.

Metabolism, Growth, Nutrient Acquisition

The exciting integration of endocrine regulation of polyphenic organ growth (e.g., via insulin signaling), with nutrient sensing, which has just begun during the past decade (see above) is also expected to become more prominent in future studies of plasticity.

Again, the development of methods for investigating global gene expression will likely have a profound effect by allowing this type of research to be undertaken on many interesting polyphenisms that previously were not amenable to studies of gene regulation. Locust studies (see section on 'Locusts', pp. 77 above) of fatty-acid regulators of phase indication also will likely spur on analogous future studies of the role of metabolites in morph induction. Finally, the detailed findings of phenotypic differences in intermediary metabolism between morphs of *Gryllus* and in *Bicyclus* provide the opportunity for future studies of the endocrine control of production of these metabolic differences and their integration with morph induction.

ACKNOWLEDGMENTS

A. J. Z. gratefully acknowledges support from the National Science Foundation, most recently IBN-0212486, IOS-0516973, and IOS-1122075. J. A. B. acknowledges support from NIEHS award R00ES017367.

REFERENCES

Anstey, M.L., Rogers, S.M., Ott, S.R. *et al.* (2009) Serotonin mediates behavioral gregarization underlying swarm formation in desert locusts. *Science*, **323**, 627–30.

Bird, A.P. and Wolffe, A.P. (1999) Methylation-induced repression-belts, braces, and chromatin. *Cell*, **99**, 451–4.

Boorse, G.C. and Denver, R.J. (2004) Endocrine mechanisms underlying plasticity in metamorphic timing in spadefoot toads. *Integrative and Comparative Biology*, **43**, 646–57.

Braendle, C., Davis, G.K., Brisson, J.A. and Stern, D.L. (2006) Wing dimorphism in aphids. *Heredity*, **97**, 192–9.

Brakefield, P.M. and Frankino, W.A. (2009) Polyphenisms in lepidoptera: Multidisciplinary approaches to studies of evolution and development, in *Phenotypic Plasticity of Insects. Mechanisms and Consequences* (eds D.W. Whitman and T.N. Ananthakrishnan), Science Publishers, Enfield (NH), pp. 337–68.

Brakefield, P.M. and French, V. (1999) Butterfly wings: the evolution of development of colour patterns. *Bioessays*, **21**, 391–401.

Brakefield, P.M., Gates, J., Keys, D. *et al.* (1996) Development, plasticity and evolution of butterfly eyespot patterns. *Nature*, **384**, 236–42.

Brakefield, P.M., Kesbeke, F. and Koch, P.B. (1998) The regulation of phenotypic plasticity of eyespots in the butterfly *Bicyclus anynana*. *American Naturalist*, **152**, 853–60.

Brakefield, P.M. and Larsen, T.B. (1984) The evolutionary significance of dry and wet season forms in some tropical butterflies. *Biological Journal of the Linnean Society*, **22**, 1–12.

Brakefield, P.M., Pijpe, J. and Zwaan, B.J. (2007) Developmental plasticity and acclimation both contribute to adaptive responses to alternating seasons of plenty and of stress in *Bicyclus* butterflies. *Journal of Biosciences*, **32**, 465–75.

Brakefield, P.M. and Reitsma, N. (1991) Phenotypic plasticity, seasonal climate and the population biology of *Bicyclus* butterflies (*Satyridae*) in Malawi. *Ecological Entomology*, **16**, 291–303.

Brisson, J.A., Davis, G.K. and Stern, D.L. (2007) Common genome-wide transcription patterns underlying the wing polyphenism and polymorphism in the pea aphid (*Acyrthosiphon pisum*). *Evolution and Development*, **9**, 338–46.

Brisson, J.A., Ishikawa, A. and Miura, T. (2010) Wing development genes of the pea aphid and differential gene expression between winged and unwinged morphs. *Insect Molecular Biology*, **19**, 63–73.

Brown, D.D. and Cai, L. (2007) Amphibian metamorphosis. *Developmental Biology*, **306**, 20–33.

Brown, D.D., Wang, Z., Furlow, J.D. *et al.* (1996) The thyroid hormone-induced tail resorption program during *Xenopus laevis* metamorphosis. *Proceedings of the National Academy of Sciences USA*, **93**, 1924–9.

Buchholz, D.R., Moskalik, C.L., Kulkarni, S.S. *et al.* (2011) Hormone regulation and the evolution of frog metamorphic diversity, in *Mechanisms of Life History Evolution* (eds T. Flatt and A. Heyland), Oxford University Press, Oxford, pp. 87–97.

Carroll, S.B., Gates, J., Keys, D.N. *et al.* (1994) Pattern formation and eyespot determination in butterfly wings. *Science*, **265**, 109–14.

Denver, R.J. (2009) Endocrinology of complex life cycles, in *Hormones, Brain and Behavior, 2nd edition* edn (eds D.W. Pfaff, A.P. Arnold, A.E. Etgen *et al.*), Academic Press, San Diego, pp. 707–44.

DeWitt, T.J. and Scheiner, S.M. (2004) *Phenotypic Plasticity. Functional and Conceptual Approaches*, Oxford University Press, Oxford.

Emlen, D.J., Hunt, J. and Simmons, L.W. (2005) Evolution of sexual dimorphism and male dimorphism in the expression of beetle horns: phylogenetic evidence for modularity, evolutionary lability, and constraint. *American Naturalist*, **166**, S42–S68.

Emlen, D.J., Warren, I.A., Johns, A. *et al.* (2012) A mechanism of extreme growth and reliable signaling in sexually selected ornaments and weapons. *Science*, **337**, 860–4.

Evans, J.D. and Wheeler, D.E. (1999) Differential gene expression between developing queens and workers in the honey bee, *Apis mellifera*. *Proceedings of the National Academy of Sciences of the United States of America*, **96**, 5575–80.

Fischman, B.J., Woodard, S.H. and Robinson, G.E. (2011) Molecular evolutionary analysis of insect societies. *Proceedings of the National Academy of Sciences USA*, **108**, 10847–54.

Fraker, M.E., Hu, F., Cuddapah, V. *et al.* (2009) Characterization of an alarm pheromone secreted by amphibian tadpoles that induces behavioral inhibition and suppression of the neuroendocrine stress axis. *Hormones and Behavior*, **55**, 520–9.

Ghalambor, C.K., McKay, J.K., Carroll, S.P. and Reznick, D.N. (2007) Adaptive versus non-adaptive phenotypic plasticity and the potential for contemporary adaptation in new environments. *Functional Ecology*, **21**, 394–407.

Ghalambor, C.K., Martin, L.B., & Woods, H.A. 2014. Building conceptual links between organismal biology and evolutionary ecology: Plasticity, integration, and a "theory of the individual." *Integrative Organismal Biology* (eds Martin, L.B., Ghalambor, C.K. & Woods, H.A.). John Wiley & Sons, Inc., New York. In press.

Ghanim, M., Dombrovsky, A., Raccah, B. and Sherman, A. (2006) A microarray approach identifies ANT, OS-D and takeout-like genes as differentially regulated in alate and apterous morphs of the green peach aphid *Myzus persicae* (Sulzer). *Insect Biochemistry and Molecular Biology*, **36**, 857–68.

Gomez-Mestre, I. and Buchholz, D.R. (2006) Developmental plasticity mirrors differences among taxa in spadefoot toads linking plasticity and diversity. *Proceedings of the National Academy of Sciences USA*, **103**, 19021–6.

Gotoh, H., Cornette, R., Koshikawa, S. *et al.* (2011) Juvenile hormone regulates extreme mandible growth in male stag beetles. *PLoS ONE*, **6**, e21139.

Gotoh, H., Miyakawa, H., Ishikawa, A., Ishikawa, I., Sugume, Y., Emlen, D.J., Lavine, L. C. & Miura, T. (2014) Developmental link between sex and nutrition; doublesex regulates sex-specific mandible growth via juvenile hormone signaling in stag beetles. *PLoS Genetics*. In press.

Guo, W., Wang, X., Ma, Z. *et al.* (2011) CSP and Takeout genes modulate the switch between attraction and repulsion during behavioral phase change in the migratory locust. *PLoS Genetics*, **7**, e1001291.

Hardie, J. and Lees, A.D. (1985) Endocrine control of polymorphism and polyphenism, in *Comprehensive Insect Physiology, Biochemistry and Pharmacology*, vol. **8** (eds G.A. Kerkut and L.I. Gilbert), Pergamon, New York, pp. 441–90.

Hartfelder, K. and Emlen, D.J. (2012) Endocrine control of insect polyphenism, in *Insect Endocrinology* (ed L.I. Gilbert), Elsevier, Amsterdam, pp. 464–522.

Hunt, J. and Simmons, L.W. (1997) Patterns of fluctuating asymmetry in beetle horns: an experimental examination of the honest signaling hypothesis. *Behavioral Ecology and Sociobiology*, **41**, 109–14.

Ishikawa, A., Gotoh, H., Abe, T. and Miura, T. (2013) Juvenile hormone titer and wing-morph differentiation in the vetch aphid *Megoura crassicauda*. *Journal of Insect Physiology*, **59**, 444–9.

Ishikawa, A., Ishikawa, Y., Okada, Y. *et al.* (2012) Screening of upregulated genes induced by high density in the vetch aphid *Megoura crassicauda*. *Journal of Experimental Zoology A*, **317**, 194–203.

Ito, Y., Harigai, A., Nakata, M. *et al.* (2013) The role of doublesex in the evolution of exaggerated horns in the Japanese rhinoceros beetle. *Embo Reports*, **14**, 561–7.

Johnson, B. (1965) Wing polymorphism in aphids – II. Interaction between aphids. *Entomologia Experimentalis et Applicata*, **8**, 49–64.

Johnson, B.R. (2010) Division of labor in honeybees: form, function, and proximate mechanisms. *Behavioral Ecology and Sociobiology*, **64**, 305–16.

Kang, L., Chen, X., Zhou, Y. *et al.* (2004) The analysis of large-scale gene expression correlated to the phase changes of the migratory locust. *Proceedings of the National Academy of Sciences USA*, **101**, 17611–15.

Kijimoto, T., Costello, J., Tang, Z. *et al.* (2009) EST and microarray analysis of horn development in *Onthophagus beetles*. *BMC Genomics*, **10**, 504.

Kijimoto, T., Moczek, A.P. and Andrews, J. (2012) Diversification of doublesex function underlies morph-, sex-, and species-specific development of beetle horns. *Proceedings of the National Academy of Sciences USA*, **109**, 20526–31.

Koch, P.B., Brakefield, P.M. and Kesbeke, F. (1996) Ecdysteroids control eyespot size and wing color pattern in the polyphenic butterfly *Bicyclus anynana*. *Journal of Insect Physiology*, **42**, 223–30.

Kucharski, R., Maleszka, J., Foret, S. and Maleszka, R. (2008) Nutritional control of reproductive status in honeybees via DNA methylation. *Science*, **319**, 1827–30.

Ledón-Rettig, C.C. and Pfennig, D.W. (2011) Emerging model systems in eco-evo-devo: the environmentally responsive spadefoot toad. *Evolution and Development*, **13**, 391–400.

Lees, A.D. (1980) The development of juvenile hormone sensitivity in alatae of the aphid *Megoura viciae*. *Journal of Insect Physiology*, **26**, 143–51.

Li-Byarlay, H., Li, Y., Stroud, H. *et al.* (2013) RNA interference knockdown of DNA methyl-transferase 3 affects gene alternative splicing in the honey bee. *Proceedings of the National Academy of Sciences USA*, **110**, 12750–5.

Lyko, F. and Maleszka, R. (2011) Insects as innovative models for functional studies of DNA methylation. *Trends in Genetics*, **27**, 127–31.

Ma, Z., Guo, W., Guo, X. *et al.* (2011) Modulation of behavioral phase changes of the migratory locust by the catecholamine metabolic pathway. *Proceedings of the National Academy of Sciences*, **108**, 3882–7.

Martin, L.B. & Cohen, A.A. (2014) Physiological regulatory networks: the orchestra of life? *Integrative Organismal Biology* (eds Martin, L.B., Ghalambor, C.K. & Woods, H.A.). John Wiley & Sons, Inc., New York. In press.

Mathot, K.J. & Dingemanse, N.J. (2014) Plasticity and personality. *Integrative Organismal Biology* (eds Martin, L.B., Ghalambor, C.K. & Woods, H.A.). John Wiley & Sons, Inc., New York. In press.

Moczek, A.P. (2008) On the origins of novelty in development and evolution. *Bioessays*, **30**, 432–47.

Moczek, A.P. and Emlen, D.J. (1999) Proximate determination of male horn dimorphism in the beetle *Onthophagus taurus* (Coleoptera: Scarabaeidae). *Journal of Evolutionary Biology*, **12**, 27–37.

Müller, C.B., Williams, I.S. and Hardie, J. (2001) The role of nutrition, crowding, and interspecfic interactions in the development of winged aphids. *Ecological Entomology*, **26**, 330–40.

Nijhout, H.F. (1991) *The Development and Evolution of Butterfly Wing Patterns*, Smithsonian Institution Press, Washington and London.

Nijhout, H.F. (1999) Control mechanisms of polyphenic development in insects. *BioScience*, **49**, 181–92.

Nijhout, H.F. (2003) Development and evolution of adaptive polyphenisms. *Evolution & Development*, **5**, 9–18.

Nussey, D.H. (2014) The ecological and evolutionary importance of variation in life history reaction norms. *Integrative Organismal Biology* (eds Martin, L.B., Ghalambor, C.K. & Woods, H.A.). John Wiley & Sons, Inc., New York. In press.

Oostra, V. (2013) Hormonal and transcriptional mechanisms underlying developmental plasticity of life histories in a seasonal butterfly. PhD thesis, Leiden University.

Oostra, V., de Jong, M.A., Invergo, B.M. *et al.* (2011) Translating environmental gradients into discontinuous reaction norms via hormone signalling in a polyphenic butterfly. *Proceedings of the Royal Society B.*, **278**, 278–97.

Pener, M.P. and Simpson, S.J. (2009) Locust phase polymorphism: an update. *Advances in Insect Physiology*, **36**, 1–272.

Pener, M.P. and Yerushalmi, Y. (1998) The physiology of locust phase polymorphism. *Journal of Insect Physiology*, **44**, 365–77.

Pfennig, D.W. (1992a) Polyphenism in spadefoot toad tadpoles as a locally adjusted evolutionary stable strategy. *Evolution*, **46**, 1408–20.

Pfennig, D.W. (1992b) Proximate and functional causes of polyphenism in an anuran tadepole. *Functional Ecology*, **6**, 167–74.

Pfennig, D.W., Wund, M.A., Snell-Rood, E.C. *et al.* (2010) Phenotypic plasticity's impacts on diversification and speciation. *Trends in Ecology and Evolution*, **25**, 459–67.

Piersma, T. and van Gils, J.A. (2011) *The Flexible Phenotype*, Oxford University Press, Oxford.

Rasmussen, J.L. (1994) The influence of horn and body size on the reproductive behavior of the horned rainbow scarab beetle *Panaeus difformis* (Coleoptera. Scarabaeidae). *Journal of Insect Behavior*, **7**, 67–82.

Rogers, S.M., Matheson, T., Sasaki, K. *et al.* (2004) Substantial changes in central nervous system neurotransmitters and neuromodulators accompany phase change in the locust. *The Journal of Experimental Biology*, **207**, 3603–17.

Rose, C.S. (2005) Integrating ecology and developmental biology to explain the timing of frog metamorphosis. *Trends in Ecology & Evolution*, **20**, 130–5.

Schlichting, C.D. and Pigliucci, M. (1998) *Phenotypic Evolution. A Reaction Norm Perspective*, Sinauer Associates, Inc., Sunderland, Massachusetts.

Schwartzberg, E.G., Kunert, G., Westerlund, S.A. *et al.* (2008) Juvenile hormone titres and winged offspring production do not correlate in the pea aphid, *Acyrthosiphon pisum*. *Journal of Insect Physiology*, **54**, 1332–6.

Shapiro, A.M. (1976) Seasonal Polyphenism. *Evolutionary Biology*, **9**, 259–333.

Shi, Y.-B. (2000) *Amphibian Metamorphosis. From Morphology to Molecular Biology*, Wiley-Liss, New York.

Simpson, S.J., Sword, G.A. and Lo, N. (2011) Polyphenism in insects. *Current Biology*, **21**, R738–R749.

Siva-Jothy, M.T. (1987) Mate securing tactics and the costs of fighting in the Japanese horned beetle, *Allomyrina dichotoma* L. (Scarabaeidae). *Journal of Ethology*, **5**, 165–72.

Stern, D.L. and Orgogozo, V. (2009) Is genetic evolution predictable? *Science*, **323**, 746–51.

Sutherland, O.R.W. (1969) The role of crowding in the production of winged forms by two strains of the pea aphid, *Acyrthosiphon pisum*. *Journal of Insect Physiology*, **15**, 1385–1410.

Suzuki, Y. and Nijhout, H.F. (2006) The evolution of a polyphenism by genetic accommodation. *Science*, **311**, 650–2.

Tawfik, A.I., Trieblmayr, K., Hassanali, A. and Osir, E.O. (2000) Time-course haemolymph juvenile hormone titres in solitarious and gregarious adults of *Schistocerca gregaria*, and their relation to pheromone emission, CA volumetric changes and oocyte growth. *Journal of Insect Physiology*, **46**, 1143–50.

Valena, S. and Moczek, A.P. (2012) Epigenetic mechanisms underlying developmental plasticity in horned beetles. *Genetics Research International*, **2012**, 1–14.

Van Wielendaele, P., Badisco, L. and Vanden Broeck, J. (2013) Neuropeptide regulation of reproduction in insects. *General and Comparative Endocrinology*, **188**, 23–34.

Vellichirammal, N.N., Zera, A.J., Schilder, R.J., Wehrkamp, C., Riethoven, J-J.M. & Brisson, J.A. (2014) *De novo* transcriptome assembly from fat body and flight muscles transcripts to identify morph-specific expression profiles in *Gryllus firmus*, PLoS One. In press.

Verlinden, H., Badisco, L., Marchal, E. *et al.* (2009) Endocrinology of reproduction and phase transition in locusts. *General and Comparative Endocrinology*, **162**, 79–92.

West-Eberhard, M.J. (2003) *Developmental Plasticity and Evolution*, Oxford University Press, Oxford.

Whitman, D.W. & Ananthakrishnan (eds). (2009) *Phenotypic Plasticity of Insects. Mechanisms and Consequences*. Science Publishers, Enfield (NH).

Wilber, H.M. and Collins, J.P. (1973) Ecological aspects of amphibian metamorphosis. *Science*, **182**, 1305–14.

Woods, H.A. & Wilson, J.K. (2014) An elephant in the frog: unifying concepts of physiological stasis and change. *Integrative Organismal Biology* (eds Martin, L.B., Ghalambor, C.K. & Woods, H.A.). John Wiley & Sons, Inc., New York. In press.

Wu, R., Wu, Z., Wang, X. *et al.* (2012) Metabolomic analysis reveals that carnitines are key regulatory metabolites in phase transition in locusts. *Proceedings of the National Academy of Sciences USA*, **109**, 3259–63.

Zera, A.J. (2007) Endocrine analysis in evolutionary-developmental studies of insect polymorphism: Use and misuse of hormone manipulation. *Evolution and Development*, **9**, 499–513.

Zera, A.J. (2009) Wing polymorphism in *Gryllus* (Orthoptera: Gryllidae): Proximate endocrine, energetic and biochemical mechanisms underlying morph specialization for flight vs. reproduction, in *Phenotypic Plasticity of Insects. Mechanisms and Consequences* (eds D.W. Whitman and T.N. Ananthakrishnan), Science Publishers, Enfield (NH), pp. 609–53.

Zera, A.J. and Brisson, J.A. (2012) Quantitative, physiological, and molecular genetics of dispersal and migration, in *Dispersal Ecology and Evolution* (eds J. Colbert, M. Baguette, T. Benton and J. Bullock), Oxford University Press, Oxford, pp. 63–82.

Zera, A.J. and Denno, R.F. (1997) Physiology and ecology of dispersal polymorphism in insects. *Annual Review of Entomology*, **42**, 207–31.

Zera, A.J. and Harshman, L.G. (2009) Laboratory selection studies of life-history physiology in insects, in *Experimental Evolution: Methods and Applications* (eds T. Garland Jr. and M.R. Rose), University of California Press, Berkeley, pp. 217–62.

Zera, A.J. and Harshman, L.G. (2011) Intermediary metabolism and the biochemical-molecular basis of life history variation and trade-offs in two insect models, in *Mechanisms of Life History Evolution* (eds T. Flatt and A. Heyland), Oxford University Press, Oxford, pp. 311–28.

Zera, A.J., Harshman, L.G. and Williams, T.D. (2007) Evolutionary Endocrinology: The developing synthesis between endocrinology and evolutionary genetics. *Annual Review of Ecology, Evolution and Systematics*, **38**, 793–817.

Zera, A.J. and Zhao, Z. (2009) Morph-associated JH titer diel rhythm in *Gryllus firmus*: Experimental verification of its circadian basis and cycle characterization in artificially-selected lines raised in the field. *Journal of Insect Physiology*, **55**, 450–8.

Zhao, Z. and Zera, A.J. (2004) The hemolymph JH titer exhibits a large-amplitude, morph-dependent, diurnal cycle in the wing-polymorphic cricket, *Gryllus firmus*. *Journal of Insect Physiology*, **50**, 93–102.

6

EVOLUTIONARY SYSTEMS BIOLOGY: SHIFTING FOCUS TO THE CONTEXT-DEPENDENCY OF GENETIC EFFECTS

Mihaela Pavličev[1] and Günter P. Wagner[2,3]

[1]*Cincinnati Children's Hospital, Cincinnati, OH, USA*
[2]*Department of Ecology and Evolutionary Biology, Yale University, New Haven, CT, USA*
[3]*Systems Biology Institute Yale University, West Haven, CT, USA*

> The phenotype is the name given to the results of the activities of genes. In a very simple organism, these activities may be carried out in a relatively short period, resulting in the formation, for instance, of a certain number of proteins. In more complex organisms, these proteins themselves interact with one another and with other substances so that it is a long and complex sequence of processes that the genes set in motion. (p. 9)
>
> – *C.H. Waddington, in* Towards a Theoretical Biology *v.I, 1968*

> The phenotype is the product of the harmonious interaction of all genes. There is extensive interaction not only among alleles of a locus, but also among loci. The main locale of these epistatic interactions is the developmental pathway. (p. 184)
>
> – *E. Mayr, in* Populations, Species and Evolution, *1970*

INTRODUCTION

Genes act through their many interactions during development and physiology to generate or maintain phenotypes. These interactions make gene effects mutually interdependent. In addition, genes are reused in many different contexts, generating shared genetic underpinning between the traits (pleiotropy). Thus neither genes nor traits can be seen as independent, but rather as components of more or less integrated wholes.

Integrative Organismal Biology, First Edition.
Edited by Lynn B. Martin, Cameron K. Ghalambor, and H. Arthur Woods.

This simple biological fact raises surprisingly difficult conceptual and methodological issues when describing the structure of genotype-phenotype relationships, assigning functional significance to genetic or molecular elements, and explaining evolutionary change. Indeed, many of the disagreements about the proper explanation of evolutionary change stem from this difficulty. At one extreme are gene-centered explanatory models that dominate population genetics-based theories of evolution. In these models, genes and allele substitutions are central mediators of causality. At the other extreme are a variety of developmental or systems-level explanatory models (Gould 2002). One example is the side-effect hypothesis of evolutionary novelties (Müller 1989), which claims that novel phenotypes appear as developmental side-effects of adaptive genetic changes elsewhere in the body. A related idea is the notion of developmental constraints as developed in the 1980s by Pere Alberch and others (Alberch 1983; Alberch & Gale 1983; Shubin & Alberch 1986; Alberch & Blanco 1996). In this and similar models of evolution, the loci of causal explanation are developmental interactions among embryonic traits. At both extremes, and in between, models have generated good empirical support, yet no synthetic view is available to resolve the apparent conflicts among models.

In this contribution we first argue that both approaches to causality are justified, and it is thus not a matter of choosing between right or wrong alternatives. Using simple insights from population genetics, we will show that gene-based models adequately capture situations where very similar organisms are compared, or where differences between closely related organisms are to be explained. The utility of these models, however, declines rapidly with increasing genetic distance of the organisms compared. We then summarize recent work from our labs that suggests an alternative way of capturing biological causality. Specifically, we propose that causal agency should be assigned to *specific interactions* among molecular entities rather than to the molecular entities themselves (proteins, RNA molecules, etc.). From this view flows a set of consequences for how we understand features of the genotype-phenotype map, such as modularity, epistasis, and pleiotropy. Finally, we introduce an alternative to the 1996 Wagner-Altenberg model of variational modularity (Wagner & Altenberg 1996).

TWO ENDS OF THE SPECTRUM IN EVOLUTIONARY BIOLOGY

What are the genetic causes of evolutionary change? In the context of the neo-Darwinian synthesis, which we do not question here, evolutionary change is synonymous with genetic change leading to phenotypic differences. Ultimately, evolutionary phenotypic differences are rooted in genetic differences among individuals and species. The goal of evolutionary genetics is the same as that of any other branch of genetics, only at a larger scale of genetic differences, in particular when it comes to comparisons between species or more distantly related taxa. So one may ask, what constitutes a genetic explanation of a heritable phenotypic difference between individuals, populations and species?

Consider a simple case: a difference in eye color between two closely related individuals associated with a difference in alleles influencing eye color. In the natural sciences, a "cause" is any material fact that is a "difference maker" (Love 2011), and the allele clearly plays this role. Hence it is appropriate to identify the allelic difference as the cause of the observed difference in phenotype. The gene, or better the allele substitution, is the cause of the phenotype difference. An extension of the "genes as cause of phenotype" is the notion of additive effects in population genetics. The above also points out that population genetics focuses only on loci at which alleles segregate in populations.

A large portion of population genetics focuses on additive effects because the amount of additive genetic variance determines the responses of populations to natural and artificial selection. The predictive value of these models, however, is limited to one or a few generations – because the effects of allele substitutions change with genetic and environmental contexts (Hansen 2013). Dependence of the effect of allele substitution on environment is called gene-environment interaction, phenotypic plasticity, or reaction norm (see Chapter 5, this volume, for the review of mechanisms). Dependence of allele substitutions (henceforth *gene effects*) on genetic changes preceding them at other loci in the genome is called epistasis. Other contexts are known to influence gene effects, such as which other allele is present at the locus itself (dominance), the sex of an individual in which allele substitution occurs (sex-specific effects), the parent from which the allele was inherited (maternal/paternal effects), the organelles within the cell (Rand *et al.* 2004), and even combinations of these; many of these contexts also include a genetic component. We focus here only on the effects of genes at other loci, but other contexts can be thought of in similar ways.

Conditional effects become more apparent when considering distantly related taxa. For example, alleles known to be lethal in some species are fixed in others (Kondrashov *et al.* 2002). A particularly impressive example is recent work on the evolution of ligand specificity of nuclear receptors (Bridgham *et al.* 2009). In brief, amino acid substitutions that cause the derived ligand specificity are functional only in the context of certain other residues that "enable" the functional effect of focal amino acid changes. The enabling amino acid residues were present in the ancestral protein, but once the derived function is established, the enabling amino acid residues are free to mutate and are absent in the derived receptor proteins. In these cases it becomes more difficult to assign causal agency to particular mutations. Due to their context dependency, gene effects cannot be simply added up to account for the phenotypic difference.

At the population level, however, these effects are often small. The statistical route around this complication is to average gene effects across contexts and then to treat the resulting effects as additive (e.g., Falconer & Mackay 1996). This is a legitimate approach since it turns out that these average (so-called additive) effects in fact explain the heritability of quantitative trait differences. If we can assume invariable context, then the additive approximation is exact. In other words, all else equal, the single allele substitution really does explain the phenotypic difference. This approach is thus sufficient when the change in the background is small, e.g., at microevolutionary scales.

So far we have considered cases where phenotypic differences among closely related individuals can be explained by contributions of individual alleles. At the other end of the continuum is evolutionary developmental biology, in which a central issue is to understand how particular evolutionary changes depend on previous changes in the lineage. Evolutionary developmental biology addresses relatively large genetic/phenotypic differences, often using the comparative method, rather than addressing within-population variation. Whereas population geneticists downplay context-dependency, evolutionary developmental biologists focus squarely on it (Gould & Lewontin 1979).

The two perspectives on causal explanation of phenotypic differences are complementary. Across short mutational distances, causal agents clearly are the individual allele substitutions; at slightly greater genetic distances the picture gets blurred because of epistatic interactions among multiple allele substitutions; and at the meso- and macro-evolutionary scales, the effects of individual allele substitutions fade, and differences among larger scale entities like gene regulatory networks are the locus of explanations. Although there is clear mechanistic continuity – in the sense that any macro-evolutionary difference started out as a variant within a population – there is also a conceptual discontinuity in the various explanatory models.

Is there a way to unify the achievements of both, the gene-centered and the network centered/developmental approaches? And is there a way to avoid the conceptual discontinuity in explaining both small and large-scale differences? In the next sections, we synthesize ideas emerging from recent work in quantitative genetics and functional molecular evolution. This synthesis then leads us to first steps toward a unified model of evolutionary change and genetic causation.

CONTEXT-DEPENDENCY IN POPULATION GENETICS: EPISTASIS

First, a brief note on terminology, as genetic interaction or epistasis has many different meanings that are often misunderstood. Genetic context-dependency, or epistasis, means that the phenotypic effect of an allele substitution at one locus depends on the genotypes at other loci (Figure 6.1). We distinguish two aspects of epistasis. One is the mere dependency of gene-effects on genotypes at other loci, as measured when comparing mean phenotypes of individual genotype classes. This kind of epistasis is called *physiological* or *functional* epistasis (Cheverud & Routman 1995). The second is a statistical measure of epistasis, which incorporates both the size of the functional epistatic effect and the influence of allele frequencies in a given population. This so-called *statistical* epistasis is expressed as epistatic variance. The population genetic approach is to decompose phenotypic variance into environmental, additive genetic, and residual interaction sources (Fisher 1930). The interplay between allele frequencies and estimates of interaction and additive variance can be best seen when the frequencies of alleles in a genetic background go to fixation. In this case, the effect of substitution at a locus, which was variable across backgrounds, becomes invariable in the limit of small allele frequencies. Such is the case in population bottlenecks, in which variation at some (background) loci is lost due to genetic drift, causing effects that were previously epistatic to become additive (Goodnight 1987, 1988).

Cheverud and Routman (1995; see also Stadler *et al.* 2000) have shown that by treating interaction variance as residual, some of the physiological epistasis contributes toward statistical *additive* variance. This means not only that the residual interaction variance is smaller than expected from the nonlinearities in development and physiology, but that these nonlinearities affect the portion of genetic variance that is heritable (additive genetic variance). To what extent epistasis affects additive variance depends on further aspects of genetic architecture that cannot be sufficiently addressed here (see Hansen & Wagner 2001; Carter *et al.* 2005; Hansen *et al.* 2006; Pavličev *et al.* 2010; Hansen 2013).

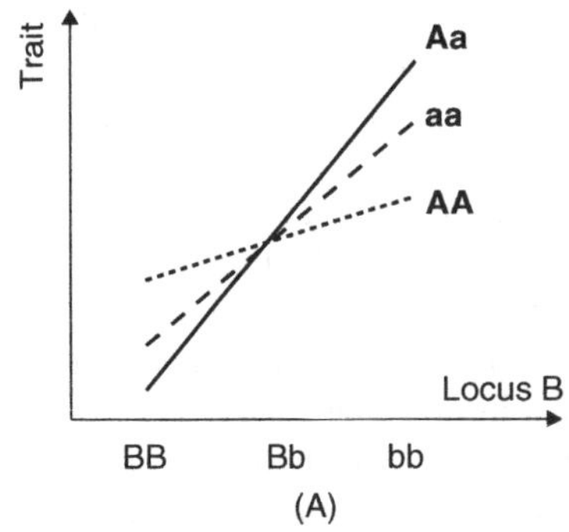

Figure 6.1. Epistasis: genetic context-dependency of phenotypic effect of the allele substitution. Depending on a genotype at the locus B, the substitution of an allele **A** with an allele **a** at the A locus has different effect on the phenotype.

This misalignment of functional and statistical epistasis often confuses discussions of how important epistasis is in evolution (Hill *et al.* 2008; but see Gjuvsland *et al.* 2007).

Note also that gene interaction is a potentially misleading term. Whether or not we account for allele frequencies, the "interaction" in transmission and population genetic vocabulary means that the total effect of substitutions at all loci is not equal to the summed effects measured at single loci (all else equal). Thus the term *interaction* refers to how the phenotypic effects combine, not to the actual underlying processes. How interactions are defined depends on the scale on which the traits values are measured (Wagner 2010). Such an effect may arise due to a variety of mechanisms (Gibson 1996; Lehner 2011), direct physical interaction between the genes or their products being only one possibility.

GENETIC EFFECTS STRUCTURED BY DEVELOPMENT: PLEIOTROPY

Evolutionary developmental biology has been traditionally focused on context-dependency, especially in the notion of evolutionary constraint. The intuitive evidence for constraint is that variation is not present uniformly in all directions of phenotypic space. Rather, even if the mutation at the genetic level is random with respect to environmental requirements, developmental processes map genetic variation nonuniformly onto the phenotype. For example, some genes affect some portions of the phenotype but not others, a phenomenon called *variational modularity* (Wagner *et al.* 2007). Moreover, some traits are frequently affected together, a phenomenon called morphological integration. These patterns can certainly be caused also by external selection, and numerous studies address the relative importance of selection and constraint (Rooney & Zhang 1999; Chenoweth *et al.* 2010; Marroig & Cheverud 2010). Here, we are interested in constraint arising from the structure of developmental interactions, often referred to as genotype-phenotype map (GP map). Pleiotropy of mutation, i.e., the number of traits affected by a mutation is an aspect of the GP map structure which integrates affected traits, while others stay unaffected by mutations (Wagner & Zhang 2011). Pleiotropy of genes involved in producing a phenotype therefore reflects the structure of development and influences the distribution of mutational effects on the phenotype.

The *mechanism* that limits the effect of allele substitutions is the developmental process. One can describe it as the pleiotropic structure of the GP map; the instructions that determine which mutations affect which suites of traits. This structure is thus itself genetically determined. If the structure is invariant in a population, the pleiotropic effects will be invariant as well.

However, this is a static view. The GP maps themselves also evolve as taxa diverge. In other words, constraint due to pleiotropy can also evolve as the alleles at background loci change.

WHAT ARE THE CONSEQUENCES OF CONTEXT-DEPENDENCY?

If genetic effects differ across backgrounds, they can evolve. Likewise, if pleiotropic effects vary across genetic backgrounds, pleiotropy can evolve, and with it the constraints determining how mutational variation (including covariation) maps into phenotypic space. Here we consider an example, and subsequently examine the consequences in terms of a model.

An example of context dependency of pleiotropy has been documented by Monteiro *et al.* (2007), who studied eyespots on butterfly wings (*Bicyclus anynana*). The authors studied two mutations, *missing* and *spotty*, that both have large effects. When introduced into the wild type genetic background, *missing* reduces or removes two eyespots on the hind wing. *Spotty* introduces the homologous eyespots on the forewing that are absent in wild type. However, when the mutations are introduced jointly, *missing* reduces not only the eyespots in the hind wing, but also the size of homologous eyespots on the forewing produced by *spotty*. No effect of *spotty* is seen on the hind wing. Thus the mutation *missing* is not pleiotropic when it occurs on the wild type background, but it is when it happens with *spotty*. Here pleiotropy refers to the wings affected (one or both). This example also shows that the pleiotropy at the genetic level can differ from pleiotropy at the phenotypic level, which is modified during development.

How Prevalent Is Variation in Pleiotropy?

The evidence for context-dependency of pleiotropy is accumulating as more studies address multiple characters, and as they search explicitly for context-effects (e.g., Lenski 1988; Ehrich *et al.* 2003; Cheverud *et al.* 2004; Pepin *et al.*, 2006; Kim *et al.* 2009). A detailed review of this topic will be provided elsewhere (Pavličev & Cheverud, in prep.). Briefly, a QTL mapping method (so-called relationship-QTL mapping; Figure 6.2) was developed to map genomic loci at which pleiotropy varies across genotypes in a population (Cheverud *et al.* 2004). This method was further extended to identify single epistatic interactions involved to affect covariance between the traits (Pavličev *et al.* 2008; Leamy *et al.* 2009; Lawson *et al.* 2011).

Recent theoretical work suggests further that variation in genetic background, which affects pleiotropy of the loci, can indeed respond to selection (Pavličev *et al.* 2011). Given existing variation in pleiotropic effects, evolution can increase or decrease covariance between traits. The mutational variance in the phenotype is produced by a different combination of traits and, because the distribution of heritable phenotypic variance is altered, selection can take the population in a different direction (Schluter 1996; Pavličev *et al.* 2011; Guillaume & Otto 2012; Pavličev *et al.* 2013; Watson *et al.* 2014).

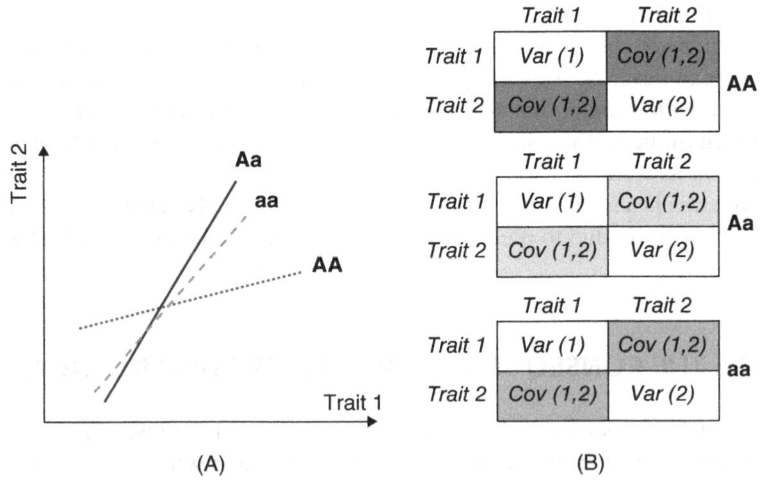

Figure 6.2. A scheme of relationship QTL. (A) the regression slope of the values of one trait on the values of a second trait differs among the genotypes at this locus (here genotypes AA, Aa and aa). (B) the phenotypic covariance also differs between the subpopulations sharing a genotype at this locus. (See plate section for color version.)

What Mechanism Underlies Genetic Context-Dependency of Pleiotropy?

We recently proposed that background-dependent variation in pleiotropy provides a window into how pleiotropic effects are modified in trait-specific ways (Pavličev & Wagner 2012). The loci uncovered by the population mapping studies thus modify pleiotropy of other loci. If pleiotropy varies across backgrounds, it can evolve. The minimal toy model is that of a common gene product involved in pathways leading to two or more traits, and a gene that modifies an effect on a single trait (Figure 6.3). The pleiotropic effects of a new allele at the shared gene can be modified in trait-specific ways by gene products involved in trait-specific pathways. Such modifications increase the independence of variation, or the *autonomy* of the two traits. Thus the primary *genic* pleiotropy may differ from the pleiotropy we detect in G-P associations, the level at which the autonomy of variation is measured. Like any model, this one simplifies a complex reality. However several lines of evidence, presented below, support it.

Apart from revealing the genetic architecture of variational trait autonomy that occurs in spite of gene sharing, this model also allows inferences about the evolution of the genetic architecture itself. Modifying interactions may well underlie the evolution of constraint arising from correlation between traits. But under what circumstances do modifying interactions occur? The literature suggests that genomes frequently evolve compensatory changes, which minimize the effects of otherwise deleterious mutations (Schrag *et al.* 1997; Kulathinal *et al.* 2004; Maisnier-Patin & Andersson 2004; Poon & Chao 2005; Howe & Denver 2008; Ichinose *et al.* 2013). Deleterious mutations may become fixed by chance, if selection efficiency is low (Muller's ratchet; Muller 1932). But mutations that improve overall fitness may also have deleterious pleiotropic effects on some traits, and compensatory mutations may evolve that alleviate those problems specifically. We recently developed this idea into the *Selection-Pleiotropy-Compensation* model (*SPC*, Pavličev & Wagner 2012).

The SPC model addresses a fundamental principle of evolution that may apply to many levels of biological organization, from cells to populations, and ultimately to species-level traits. The most interesting cases are associated with decoupling of developmentally linked traits, enabling evolutionary differentiation of traits and thus an increase in organismal complexity. However this mode of differentiation also applies at other levels, such as the evolution of cell types arising from a common lineage, or of species arising from common ancestors. Exploration of all these levels goes beyond the scope of this chapter. Here we illustrate the conceptual utility of the model by two examples.

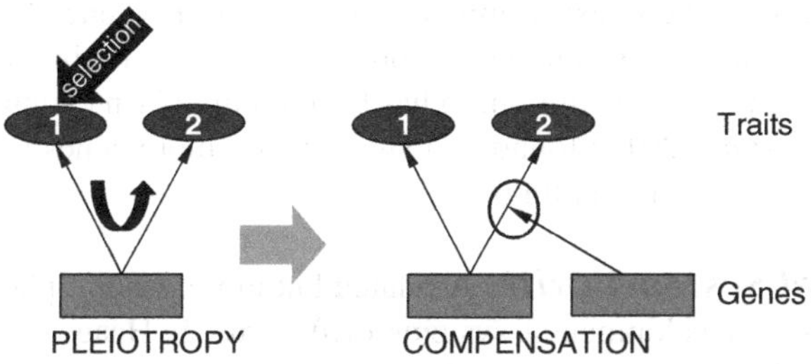

Figure 6.3. Schematic representation of compensation for pleiotropic effect: as a consequence of selection on trait 1, an allele at the common gene affects traits 1 and 2 (the propagation of the phenotypic effect is represented by the U-arrow), upon which the effect of this gene on a trait 2 is modified by a genetic interaction on the right-hand side of the figure (encircled).

Modification of Serially Homologous and Symmetrical Body Parts Many body plans are metameric – they consist of repeated, serially homologous elements. This is obvious in annelid segments, but vertebrates too have serially homologous limbs and vertebrae. Furthermore, in bilaterally symmetrical organisms, left and right sides can be considered repeated elements, in the sense that they implement largely common developmental programs. Divergence of repeated structures may implement the same mechanism, namely local modification of the shared genes.

Sexual dimorphism is another instance, at the population level, where this model may be useful. Dimorphism is common, despite males and females sharing quite similar genomes. Studies commonly find sex-specific gene effects (Vaughn *et al.* 1999; Norgard *et al.* 2008; Thach *et al.* 2009; Delph *et al.* 2010). Homologous traits in different sexes can be formalized as pleiotropic (Lande 1980), and thus dimorphism is easily thought of as sex-specific modification of pleiotropic developmental genes. The variable context is the sex of the individual, and its genetic basis, at least in species with genetic sex determination, are the sex-determining genes and pathways (Williams & Carroll 2009). A nice example – in which a dimorphic character is generated by interactions among sex-determining genes and other genes involved more directly in producing a trait – is the sex comb in *Drosophila* (Kopp 2011). Sex combs are melanized structures that develop in males from bristles on prothoracic leg. Kopp (2011) demonstrated that the crucial part of the genetic regulatory network necessary to generate sex combs and restrict them to the males is the interaction between male splice variant of *Doublesex (dsx)* and the hox gene *Scr*. Certainly upstream genes are involved in positioning the sex comb, but the evolution of this particular interaction established the sex-specific comb, and it is also the modification of this interaction that causes diversity among the *Drosophila* species with comb.

Implications of SPC for Developmental Complexity and Population Variation

The SPC model has several implications. Some correspond to known phenomena that have defied explanation. The following examples illustrate these ideas.

Genetic Complexity of Traits In the SPC model, evolution proceeds by secondary modification in traits under stabilizing selection, of deleterious side effects that arise due to traits under directional selection. A prediction of this model, which contradicts those from other models, is that genetic modifications will accumulate particularly in phenotypically invariant traits – leading to increased genetic complexity in these traits (for a parallel reasoning derived in different context, compare Woods 2009). Evidence for such an effect has indeed been documented in a recent gene expression study of murine fore- and hind limbs (Taher *et al.* 2011). In mice, the hind limb has been modified (in evolution) more than has the forelimb. Taher *et al.* (2011) found that significantly more genes are expressed during development in fore- than in hind limb.

Developmental Systems Drift A related but more general phenomenon in developmental genetics is developmental systems drift (True & Haag 2001). Developmental systems drift describes instances where the genetic developmental basis of a trait evolves even though the phenotype is conserved. A good example of developmental systems drift is the nematode vulva, which in different species develops from the same cells yet the cells are under different genetic control (Kiontke *et al.* 2007; Sommer 2012). One explanation is that there are many ways for development to produce the same feature, so there is some "developmental drift." An alternative proposed by Johnson and Porter (2001), and that can be considered an example of the SPC model, is that changes in developmental genes (of traits under stabilizing selection) are triggered by selection on other traits affected by the same genes, and their compensation. Thus development of traits evolves to accommodate changes in other traits that share a developmental genetic basis.

Genomic Red Queen The proposed model implies that the evolution of the GP map is a consequence of selection against deleterious pleiotropic effects. Successful evolutionary changes thus involve a combination of adaptive, fitness increasing mutations, deleterious side effects and compensatory changes ameliorating these deleterious side effects.

A further evolutionary consequence is genetic change stemming from coevolution between interacting loci. Even under stabilizing selection, a trait's genetic basis evolves to counteract degradation by changes in genes shared with evolving traits. This dynamical aspect is analogous to a model of evolutionary dynamics in ecological communities dubbed the "Red Queen" by Van Valen (Van Valen 1973; Liow *et al.* 2011). The Red Queen captures the notion that species are parts of each other's environments, such that evolutionary improvement by one species may negatively affect other species, such that the average level of adaptation in the community stays constant. The proposed *Genomic Red Queen* parallels this idea in that a genetic basis of a character evolves due to entanglement with other characters, while maintaining its phenotype.

Large Mutational Target Variational modularity can be based not only on secondary modification of pleiotropy during development, but also by restricting the number of primary pleiotropic effects. As pointed out by Hansen (2003), a potential drawback of such restricted pleiotropy is that by reducing the number of characters affected by a gene, the number of genes affecting a character will also be reduced. A character with a genetic basis composed of a lower number of genes can also be affected by a lower number of mutations (i.e., has smaller mutational target), thus reducing the amount of variation that can be generated and the potential for advantageous mutations (Pavličev & Hansen 2011). The contrary is the case in the SPC model – modularity due to local modification increases the number of genes involved and the potential to generate variation.

Evolution of Evolvability The structure of the G-P map determines the directions in phenotypic space in which variation is produced (canalized). Evolutionary potential is proportional to the availability of heritable variation in the direction of selection, less the entanglement with the variation in other directions. If this distribution differs across genetic backgrounds, canalization and constraint also vary across backgrounds. In other words, evolvability can itself evolve (epistasis). Such effects have been modeled (Pavličev *et al.* 2011; Watson *et al.* 2014).

Atavisms, De-differentiation Secondary modification of pleiotropy may also allow its reversal. Such a phenomenon has been observed in yucca moths (Pellmyr & Krenn 2002). Several species of *Tegeticula* are mutualists with yuccas (Agavaceae). The plant provides the moth with food for larvae, whereas the moth help fertilize the plant's ovules. To collect pollen, the moths use tentacles present only in females. Pellmyr and Krenn (2002) found that in species having secondarily abandoned the mutualism, tentacles are vestigial and their remnants frequently occur in both sexes. Thus as selection on the functionality of tentacles relaxed, the sex-specificity was also lost. We interpret this as evidence that tentacle-generating pathways are present in both sexes, but their sex-specific modification has broken down in those species where the tentacles lost their function. According to our model, the sex-specificity of a tentacle is maintained by a co-adaptive fit between the character genes and their modifiers. Thus as the character genes deteriorate, the sex-specific modification is also lost.

GENERALIZING SPC MODEL: INTERACTION OF GENE EFFECTS, OR EFFECTS OF INTERACTIONS?

In theoretical biology, the systems approach to complexity has a long tradition. One idea that strongly influenced evolutionary biology is the near-decomposability of systems (Simon 1962; Bechtel & Richardson 1993). In near-decomposable systems, sets of highly connected units can be treated as nearly independent from other sets, and the relationships between these sets studied without special attention to their components. Such an organization may be particularly malleable (evolvable), conferring local stability of subsystems. Based on the observation of near-decomposability of phenotypes (Olson & Miller 1958; Berg 1960), it has been suggested that gene effects are also decomposable into corresponding sets (Wagner & Altenberg 1996). This means that sets of genes are shared among sets of (therefore) covarying traits but do not contribute to other sets of traits (Figure 6.4A). Variational modularity of characters is in this model achieved by a restricted pleiotropy, resulting in a quasi independent genetic underpinning of modules.

Recent work on genetic networks, however, shows that while variation in phenotypes is modular, the sets of genes involved are more overlapping (e.g., Bossi & Lehner 2009; see more supporting evidence below). Most genes play many functional roles at global and at very specific scales, contradicting the Wagner-Altenberg (1996) model of variational modularity. The model we presented here addresses this contradiction by proposing that variational modularity arises by the modularity of gene interaction, rather than by the modularity of gene expression (Figure 6.4B).

In this model, the variational phenotypic modules do not correspond to sets of genes dedicated to each module as in the model by Wagner and Altenberg (1996). In the Wagner-Altenberg model one may think of genes with restricted expression domains, so that each variational module has its own set of "private" genes. This model runs into problems when we consider the degree of pleiotropy and the limited number of genes in the genome (Stern & Orgogozo 2008; Rockman 2012).

The alternative model proposed here assumes that each variational module is produced and influenced by a unique set of interactions between the physical products of genes. Modules arise from module specific gene interactions rather than module-specific sets of genes (Figure 6.4B; see "*epistatic pleiotropy*" Wolf *et al.* 2005, 2006). This model reconciles the observations that genes tend to have a larger number of pleiotropic roles while showing fewer pleiotropic effects of mutations.

The model implies a conceptual shift away from the "contributions" of individual genes. Rather, the interactions among genes are the primary functional unit underlying character phenotype and character autonomy. Thus, it seems more meaningful to assign biological roles to specific interactions rather than to assign specific biological roles to genes and their physical products directly (see Chapter 9, this volume, for a parallel conceptualization for physiological systems).

Several lines of evidence support this suggestion. First, the same genes are reused in surprisingly many contexts, and even global patterning genes are involved in very specific,

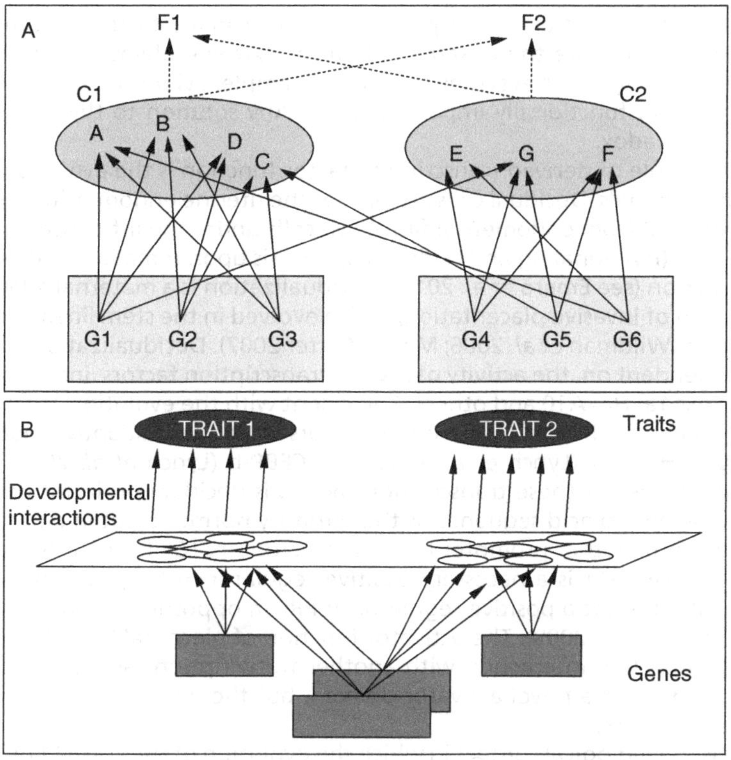

Figure 6.4. Genotype-phenotype maps underlying variational modularity. (A) GP map after Wagner and Altenberg (1996), in which characters have relatively separate genetic bases. (B) alternative GP map proposed here, in which the genes may be shared but interactions (the intermediate plane in which interactions are shown as circles) are character-specific, such that arising variation is nevertheless modular. *Source*: Wagner, G.P. & Altenberg, L. (1996) Complex adaptations and evolution of evolvability. *Evolution*, **50**, 967–76. Copyright © John Wiley & Sons, Inc., 1996.

local interactions (Bossi & Lehner 2009). Also, work on the evolution of transcription factors (Box 6.1) shows that derived functions of transcription factors can be realized through derived cell-specific interactions with other transcription factors.

Box 6.1. Mechanistic Interlude: How Transcription Factors Escape the Pleiotropy Constraint

One line of research influencing our views on gene interactions is work on the evolution of transcription factors. Transcription factors are proteins that bind to specific DNA binding sites, recruit other transcription factors and co-factors, and either enhance or repress the transcription of target genes. It has long been recognized that transcription factors have conserved domains dedicated to core activities, like DNA binding and nuclear localization (Gehring 1987). At the same time, however, their regulatory activities, and thus their functions in gene expression, are known to evolve. On first principles, evolution of transcription factors is puzzling. Most are used in both developing and adult tissues and, therefore, any

(*continued*)

changes to a transcription factor potentially affect many tissues. In other words, transcription factors are the poster-children for extreme pleiotropy, which should limit their evolvability. Nevertheless, there is ample evidence that transcription factors evolve in functionally important ways. Any solution to this problem must resolve this paradox.

One example of derived transcription factor function is the evolution of decidual cells in mammals. Decidual cells are part of the uterine endometrium. In humans they are derived from endometrial fibroblast cells under the influence of increased progesterone (P4) signaling and undergo apoptosis upon removal of P4, as it occurs in menstruation (see Emera *et al.* 2012). Decidualization is a maternal adaptation to the evolution of invasive placentation, which evolved in the stem lineage of placental mammals (Wildman *et al.* 2006; Mess & Carter 2007). Decidualization is regulated by, and dependent on, the activity of several transcription factors, including CEBP-b, HoxA11, Foxo1a, HoxA10 and others. Coincident with the evolution of invasive placentation, some of the key transcription factors underwent adaptive modification, in particular HoxA11 (Lynch *et al.* 2008) and CEBP-b (Lynch *et al.* 2011). The best understood target of these transcription factors is decidual prolactin (PRL), which has the same amino acid sequence as the pituitary hormone but in decidual cells is regulated from another promoter than in the pituitary (Gellersen & Brosens 2003). For instance, HoxA11 is a necessary positive regulator of PRL, but only in placental mammals; it is *not* a positive regulator of PRL in opposum, chicken, or ancestral therians (Lynch *et al.* 2008). The activator function of placental HoxA11 is the result of a protein-protein interaction with another transcription factor, Foxo1a. Hence, what evolved is not a novel activator domain, but the ability to be an activator in the presence of Foxo1a.

Here is the biological context in which the evolution of the HoxA11 protein took place. In endometrial fibroblast cells the Foxo1a protein is largely localized in the cytoplasm. Decidualization leads to the activation of the c-AMP signal transduction pathway which leads to a modification of Foxo1a which in turn gets translocated into the nucleus. In the nucleus, Foxo1a binds to enhancers at which HoxA11 already is present and the HoxA11:Foxo1a complex is an activator of gene expression (transcription). Ancestral HoxA11 also forms a complex with Foxo1a but this complex is repressive as is HoxA11 in the absence of Foxo1a. The evolutionary change in HoxA11 thus led to an activating switch in the presence of Foxo1a, or more precisely, a switch from repression to transcription in the presence of a signal that makes Foxo1a enter the nucleus.

The point is that the evolutionary change in HoxA11 apparently only affects cells in which Foxo1a is localized to the nucleus. Such a context-sensitive change may limit its effects to that subset of cells in which appropriate interaction partners are present, thereby limiting pleiotropic effects. It is also notable, that the evolutionary change in HoxA11 created a novel functional interaction (not a physical one, which evolved earlier, Brayer *et al.* 2011), which led to a derived function.

At the population genetic level, gene-mapping studies often detect only a handful of gene effects, which explain only a modest portion of heritable variation. In human height, this lack of explained heritability is suggested to be due to a great number of SNPs with very small effects (Yang *et al.* 2010). Furthermore, gene-mapping studies of the same characters in different populations give different results (Scoville *et al.* 2011; Rockman 2012). The latter is no big surprise given that different experimental populations harbor polymorphisms in different genomic regions (and frequencies) and only polymorphic regions can be detected. Together, however, these observations indicate that many genes contribute to the genetic basis of a character. Owing to limited number of genes in the genome, sets of genes underlying single characters must overlap and single genes must therefore be involved in generating many characters (Hansen 2003; Paaby & Rockman 2013). Nevertheless, single mutations (and even single-gene knockouts) appear to affect only a modest number of characters (Wagner *et al.* 2008; Wang *et al.* 2010; Wagner & Zhang 2011). How is this possible?

Interactions of genetic effectors may explain these seemingly contradictory facts. The variation appearing at the phenotypic level may be detected as a direct effect of substitution at single genes if only one interacting partner is variable, but it will reveal epistatic interactions if both genes are variable. At the same time, interactions may shield some mutations from appearing in the phenotype.

Perhaps the most accessible conceptualization is in terms of coadaptation between genes. Interacting genes may not physically interact, but jointly contribute to a functional unit in a nonadditive way that makes their changes dependent on each other. Although physical interaction among gene products is not logically necessary, it is nevertheless likely and may be also more robust than indirect systemic interactions. At least the work on HoxA11 and CEBP evolution (see Box 1) shows that direct protein–protein interaction is a way to realize an interaction-based character specificity.

Note that with respect to the population genetics, the proposed shift in perspective may require a stronger focus on the role of epistasis, but not necessarily an entirely new framework.

To summarize, the proposed genetic architecture based on gene interaction explains modularity of variation and robustness of the character against mutational change. Variational modularity is due to interactions modifying the effects of genes with broad pleiotropic effects, whereas robustness of a character is due to the potential of interacting genes to buffer genetic changes. Moreover, given that these characteristics are associated with variational individuality of the character, interactions are possibly also involved in the character origin. Indeed, evolutionary novelty has been associated with establishment of a genetic interaction between *Scr* and *Doublesex* in the case of sex comb origin (Kopp 2011).

CONCLUSION

Genes don't *make* phenotypes, yet little evolution happens without genetic change playing a role. *Mutational effects* are obviously involved, yet they are not inherent properties of the genes mutated but rather of the genetic and developmental contexts in which those genes act. Here, we related context-dependency in population genetics (epistasis) to evolutionary developmental biology, trying to address the conceptual discontinuity between these disciplines. In population genetics, gene effects vary depending on the genetic and environmental background. In evolutionary developmental biology, context-dependency of gene effects refers to developmental constraints.

Although context dependency is evident in both micro- and macro-evolution, its importance for explaining evolutionary dynamics depends on the scale of genetic differences involved. Established developmental processes result in the recognizable anatomical structures. The importance of these structures becomes clear when they are already individualized, but the question is how they become individualized (Wagner 2014). This question is central neither to population genetics nor to evolutionary developmental studies, but instead resides in, and potentially mediates, precisely the gap between them.

We thus propose that modular variational structures can be established by character-specific interactions (i.e., by modification of shared genes), rather than by restricted, character-specific gene expression. Variational modularity is thus generated by the modular interaction between genes. Conceptualizing evolutionary processes in terms of evolving gene interactions rather than contributions of individual genes connects micro-evolutionary to macro-evolutionary processes. Certainly, two-way interactions may not fully bridge the two approaches. Interactions may themselves be context-dependent (Hansen & Wagner 2001). How far we get by accounting for pairwise dependencies is an empirical question.

More generally, this concept promotes an alternative to the gene-within-a-context view, namely to think of an interaction between genes rather than of an interaction between a gene and a context, without *a priori* distinction between the context and the actor.

As the quotes of two influential representatives of evolutionary population and evolutionary developmental genetics at the beginning of this chapter imply, the two fields set out from a similar standpoint, albeit they diverged later. Many of the ideas here have their beginnings in Waddington's and Mayr's time. The present conceptualization proposes how to integrate them into a research program that accommodates approaches from both population genetics and evolutionary developmental biology.

REFERENCES

Alberch, P. (1983) Development and evolution: embryos, genes, and evolution. *Science*, **221**, 257–8.

Alberch, P. and Gale, E.A. (1983) Size dependence during the development of the amphibian foot. Colchicine-induced digital loss and reduction. *Journal of Embryology and Experimental Morphology*, **76**, 177–97.

Alberch, P. and Blanco, M.J. (1996) Evolutionary patterns in ontogenetic transformation: from laws to regularities. *International Journal of Developmental Biology*, **40**, 845–58.

Bechtel, W. and Richardson, R.C. (1993) *Discovering complexity. Decomposition and Localization as Strategies in Scientific Research*, Princeton University Press, New Jersey.

Berg, R. (1960) The ecological significance of correlation pleiades. *Evolution*, **17**, 171–80.

Bossi, A. and Lehner, B. (2009) Tissue specificity and the human protein interaction network. *Molecular Systems Biology*, **5**, 260.

Brayer, K.J., Lynch, V.J. and Wagner, G.P. (2011) Evolution of a derived protein–protein interaction between HoxA11 and Foxo1a in mammals caused by changes in intramolecular regulation. *Proceedings of the National Academy of Science U S A*, **108**, E414–20.

Bridgham, J.T., Ortlund, E.A. and Thornton, J.W. (2009) An epistatic ratchet constrains the direction of glucocorticoid receptor evolution. *Nature*, **461**, 515–19.

Carter, A.J., Hermisson, J. and Hansen, T.F. (2005) The role of epistatic gene interactions in the response to selection and the evolution of evolvability. *Theoretical Population Biology*, **68**, 179–96.

Chenowcth, S.F., Rundle, H.D. and Blows, M.W. (2010) The contribution of selection and genetic constraints to phenotypic divergence. *American Naturalist*, **175**, 186–96.

Cheverud, J.M. and Routman, E.J. (1995) Epistasis and its contribution to genetic variance components. *Genetics*, **139**, 1455–61.

Cheverud, J.M., Ehrich, T.H., Vaughn, T.T. *et al.* (2004) Pleiotropic effects on mandibular morphology II: differential epistasis and genetic variation in morphological integration. *Journal of Experimental Zoology B: Molecular and Developmental Evolution*, **302**, 424–35.

Delph, L.F., Arntz, A.M., Scotti-Saintagne, C. and Scotti, I. (2010) The genomic architecture of sexual dimorphism in the dioecious plant Silene latifolia. *Evolution*, **64**, 2873–86.

Ehrich, T.H., Vaughn, T.T., Koreishi, S.F. *et al.* (2003) Pleiotropic effects on mandibular morphology I. Developmental morphological integration and differential dominance. *Journal of Experimental Zoology B: Molecular and Developmental Evolution*, **296**, 58–79.

Emera, D., Romero, R. and Wagner, G. (2012) The evolution of menstruation: a new model for genetic assimilation: explaining molecular origins of maternal responses to fetal invasiveness. *Bioessays*, **34**, 26–35.

Falconer, D.S. and Mackay, T.F.C. (1996) *Introduction to Quantitative Genetics*, 4th edn, Pearson, Prentice Hall, Essex, UK.

Fisher, R.A. (1930) *The Genetical Theory of Natural Selection*, Dover, New York.

Gehring, W.J. (1987) Homeo boxes in the study of development. *Science*, **236**, 1245–52.

Gellersen, B. and Brosens, J. (2003) Cyclic AMP and progesterone receptor cross-talk in human endometrium: a decidualizing affair. *Journal of Endocrinology*, **178**, 357–72.

Gibson, G. (1996) Epistasis and pleiotropy as natural properties of transcriptional regulation. *Theoretical Population Biology*, **49**, 58–89.

Gjuvsland, A.B., Hayes, B.J., Omholt, S.W. and Carlborg, O. (2007) Statistical epistasis is a generic feature of gene regulatory networks. *Genetics*, **175**, 411–20.

Goodnight, C.J. (1987) On the effect of founder events on the epistatic genetic variance. *Evolution*, **41**, 80–91.

Goodnight, C.J. (1988) Epistasis and the effect of founder events on the additive genetic variance. *Evolution*, **42**, 441–54.

Gould, S.J. (2002) *The Structure of Evolutionary Theory*, Belknap Press of Harvard University Press, Cambridge, Massachusetts.

Gould, S.J. and Lewontin, R.C. (1979) The spandrels of San Marco and the Panglossian paradigm: a critique of the adaptationist programme. *Proceedings of the Royal Society of London B Biological Science*, **205**, 581–98.

Guillaume, F. and Otto, S.P. (2012) Gene functional trade-offs and the evolution of pleiotropy. *Genetics*, **192**, 1389–1409.

Hansen, T.F. (2003) Is modularity necessary for evolvability? Remarks on the relationship between pleiotropy and evolvability. *Biosystems*, **69**, 83–94.

Hansen, T.F. (2013) Why epistasis is important for selection and adaptation? *Evolution*, early online.

Hansen, T.F., Alvarez-Castro, J.-M., Carter, A. *et al.* (2006) Evolution of genetic architecture under directional selection. *Evolution*, **60**, 1523–36.

Hansen, T.F. and Wagner, G.P. (2001) Modeling genetic architecture: a multilinear theory of gene interaction. *Theoretical Population Biology*, **59**, 61–86.

Hill, W.G., Goddard, M.E. and Visscher, P.M. (2008) Data and theory point to mainly additive genetic variance for complex traits. *PLoS Genetics*, **4**, e1000008.

Howe, D.K. and Denver, D.R. (2008) Muller's Ratchet and compensatory mutation in Caenorhabditis briggsae mitochondrial genome evolution. *BMC Evolutionary Biology*, **8**, 62.

Ichinose, M., Iizuka, M., Kusumi, J. and Takefu, M. (2013) Models of compensatory molecular evolution: Effects of back mutation. *Journal of Theoretical Biology*, **323C**, 1–10.

Johnson, N.A. and Porter, A.H. (2001) Toward a new synthesis: population genetics and evolutionary developmental biology. *Genetica*, **112–13**, 45–58.

Kim, H.S., Huh, J. and Fay, J.C. (2009) Dissecting the pleiotropic consequences of a quantitative trait nucleotide. *FEMS Yeast Research*, **9**, 713–22.

Kiontke, K., Barriere, A., Kolotuev, I. *et al.* (2007) Trends, stasis, and drift in the evolution of nematode vulva development. *Current Biology*, **17**, 1925–37.

Kondrashov, A.S., Sunyaev, S. and Kondrashov, F.A. (2002) Dobzhansky-Muller incompatibilities in protein evolution. *Proceedings of the National Academy of Science U S A*, **99**, 14878–83.

Kopp, A. (2011) Drosophila sex combs as a model of evolutionary innovations. *Evolution and Development*, **13**, 504–22.

Kulathinal, R.J., Bettencourt, B.R. and Hartl, D.L. (2004) Compensated deleterious mutations in insect genomes. *Science*, **306**, 1553–4.

Lande, R. (1980) Sexual dimorphism, sexual selection and adaptation in polygenic characters. *Evolution*, **38**, 292–305.

Lawson, H.A., Cady, J.E., Partridge, C. *et al.* (2011) Genetic effects at pleiotropic loci are context-dependent with consequences for the maintenance of genetic variation in populations. *PLoS Genetics*, **7**, e1002256.

Leamy, L.J., Pomp, D. and Lightfoot, J.T. (2009) Genetic variation in the pleiotropic association between physical activity and body weight in mice. *Genetics, Selection, Evolution*, **41**, 41.

Lehner, B. (2011) Molecular mechanisms of epistasis within and between genes. *Trends in Genetics*, **27**, 323–31.

Lenski, R.E. (1988) Experimental studies of pleiotropy and epistasis in *Escherichia coli*. II. Compensation for maladaptive effects associated with resistance to virus T4. *Evolution*, **42**, 433–40.

Liow, L.H., Van Valen, L. and Stenseth, N.C. (2011) Red Queen: from populations to taxa and communities. *Trends in Ecology and Evolution*, **26**, 349–58.

Love, A.C. (2011) Hierarchy, causation and explanation: ubiquity, locality and pluralism. *Interface Focus*, **2**, 115–25.

Lynch, V.J., May, G. and Wagner, G.P. (2011) Regulatory evolution through divergence of a phosphoswitch in the transcription factor CEBPB. *Nature*, **480**, 383–6.

Lynch, V.J., Tanzer, A., Wang, Y. *et al.* (2008) Adaptive changes in the transcription factor HoxA-11 are essential for the evolution of pregnancy in mammals. *Proceedings of the National Academy of Science U S A*, **105**, 14928–33.

Maisnier-Patin, S. and Andersson, D.I. (2004) Adaptation to the deleterious effects of antimicrobial drug resistance mutations by compensatory evolution. *Research in Microbiology*, **155**, 360–9.

Marroig, G. and Cheverud, J.M. (2010) Size as a line of least resistance II: Direct selection on size or correlated response due to constraints? *Evolution*, **64**, 1470–88.

Mayr, E. (1970) *Populations, Species, and Evolution; An Abridgment of Animal Species and Evolution*, Belknap Press of Harvard University Press, Cambridge, Massachusetts.

Mess, A. and Carter, A.M. (2007) Evolution of the placenta during the early radiation of placental mammals. Comparative biochemistry and physiology. *Part A, Molecular & Integrative Physiology*, **148**, 769–79.

Monteiro, A., Chen, B., Scott, L.C. *et al.* (2007) The combined effect of two mutations that alter serially homologous color pattern elements on the fore and hindwings of a butterfly. *BMC Genetics*, **8**, 22.

Müller, G. (1989) Ancestral patterns in bird limb development: A new look at Hampé's experiment. *Journal of Evolutionary Biology*, **2**, 31–47.

Muller, H.J. (1932) Some genetic aspects of sex. *American Naturalist*, **66**, 118–38.

Norgard, E.A., Roseman, C.C., Fawcett, G.L. *et al.* (2008) Identification of quantitative trait loci affecting murine long bone length in a two-generation intercross of LG/J and SM/J Mice. *Journal of Bone Mineral Research*, **23**, 887–95.

Olson, E.C. and Miller, R.L. (1958) *Morphological Integration*, University of Chicago Press, Chicago.

Paaby, A.B. and Rockman, M.V. (2013) The many faces of pleiotropy. *Trends in Genetics*, **29**, 66–73.

Pavličev, M. and Hansen, T.F. (2011) Genotype-phenotype maps maximizing evolvability: modularity revisited. *Evolutionary Biology*, **38**, 371–89.

Pavličev, M. and Wagner, G.P. (2012) A model of developmental evolution: selection, pleiotropy and compensation. *Trends in Ecology and Evolution*, **27**, 316–22.

Pavličev, M., Cheverud, J.M. and Wagner, G.P. (2011) Evolution of adaptive phenotypic variation patterns by direct selection for evolvability. *Proceedings of the Royal Society of London B Biological Science*, **278**, 1903–12.

Pavličev, M., Le Rouzic, A., Cheverud, J.M. *et al.* (2010) Directionality of epistasis in a murine intercross population. *Genetics*, **185**, 1489–1505.

Pavličev, M., Wagner, G.P., Noonan, J.P. *et al.* (2013) Genomic correlates of the loci associated with fore and hind limb differentiation in mice. *Genome Biology and Evolution*, **10**, 1926–36.

Pavličev, M., Kenney-Hunt, J.P., Norgard, E.A. *et al.* (2008) Genetic variation in pleiotropy: differential epistasis as a source of variation in the allometric relationship between long bone lengths and body weight. *Evolution*, **62**, 199–213.

Pellmyr, O. and Krenn, H.W. (2002) Origin of a complex key innovation in an obligate insect-plant mutualism. *Proceedings of the National Academy of Science U S A*, **99**, 5498–5502.

Pepin, K.M., Samuel, M.A. and Wichman, H.A. (2006) Variable pleiotropic effects from mutations at the same locus hamper prediction of fitness from a fitness component. *Genetics*, **172**, 2047–56.

Poon, A. and Chao, L. (2005) The rate of compensatory mutation in the DNA bacteriophage phiX174. *Genetics*, **170**, 989–99.

Rand, D.M., Haney, R.A. and Fry, A.J. (2004) Cytonuclear coevolution: the genomics of cooperation. *Trends in Ecology and Evolution*, **19**, 645–53.

Rockman, M.V. (2012) The QTN program and the alleles that matter for evolution: all that's gold does not glitter. *Evolution*, **66**, 1–17.

Rooney, A.P. and Zhang, J. (1999) Rapid evolution of a primate sperm protein: relaxation of functional constraint or positive Darwinian selection? *Molecular Biololgy and Evolution*, **16**, 706–10.

Schluter, D. (1996) Adaptive radiation along genetic lines of least resistance. *Evolution*, **50**, 1766–74.

Schrag, S.J., Perrot, V. and Levin, B.R. (1997) Adaptation to the fitness costs of antibiotic resistance in Escherichia coli. *Proceedings of the Royal Society of London B Biological Science*, **264**, 1287–91.

Scoville, A.G., Lee, Y.W., Willis, J.H. and Kelly, J.K. (2011) Explaining the heritability of an ecologically significant trait in terms of individual quantitative trait loci. *Biological Letters*, **7**, 896–8.

Shubin, N.H. and Alberch, P. (1986) *A Morphogenetic Approach to the Origin and Basic Organization of the Tetrapod Limb*, Plenum Press, New York, NY, USA.

Simon, H. (1962) The architecture of complexity. *Proceedings of the American Philosophical Society*, **106**, 467–82.

Sommer, R.J. (2012) Evolution of regulatory networks: nematode vulva induction as an example of developmental systems drift. *Advances in Experimental Medicine and Biology*, **751**, 79–91.

Stadler, P.F., Seitz, R. and Wagner, G.P. (2000) Evolvability of complex characters: population dependent Fourier decomposition of fitness landscapes over recombination spaces. *Bulletin of Mathematical Biology*, **62**, 399–428.

Stern, D.L. and Orgogozo, V. (2008) The loci of evolution: how predictable is genetic evolution? *Evolution*, **62**, 2155–77.

Taher, L., Collette, N.M., Murugesh, D. *et al.* (2011) Global gene expression analysis of murine limb development. *PLoS One*, **6**, e28358.

Thach, B.T., Kenney-Hunt, J.P., Simon, T.C. *et al.* (2009) Sex-specific quantitative trait loci linked to autoresuscitation failure in SWR/J mice. *Heredity*, **103**, 469–75.

True, J.R. and Haag, E.S. (2001) Developmental system drift and flexibility in evolutionary trajectories. *Evolution and Development*, **3**, 109–19.

Van Valen, L.M. (1973) A new evolutionary law. *Evolutionary Theory*, **1**, 1–30.

Vaughn, T.T., Pletscher, L.S., Peripato, A. *et al.* (1999) Mapping quantitative trait loci for murine growth: a closer look at genetic architecture. *Genetics Research*, **74**, 313–22.

Waddington, C.H., and International Union of Biological Sciences (1968) *Towards a Theoretical Biology*, An IUBS symposium. Aldine Pub. Co., Chicago.

Wagner, G.P. (2010) The measurement theory of fitness. *Evolution*, **64**, 1358–76.

Wagner, G.P. (2014) *Homology, Genes and Evolutionary Innovation*. Princeton University Press.

Wagner, G.P. and Altenberg, L. (1996) Complex adaptations and evolution of evolvability. *Evolution*, **50**, 967–76.

Wagner, G.P., Pavličev, M.P. and Cheverud, J.M. (2007) The road to modularity. *Nature Reviews Genetics*, **8**, 921–31.

Wagner, G.P. and Zhang, J. (2011) The pleiotropic structure of the genotype-phenotype map: the evolvability of complex organisms. *Nature Reviews Genetics*, **12**, 204–13.

Wagner, G.P., Kenney-Hunt, J.P., Pavličev, M. *et al.* (2008) Pleiotropic scaling of gene effects and the "cost of complexity.". *Nature*, **452**, 470–2.

Wang, Z., Liao, B.Y. and Zhang, J. (2010) Genomic patterns of pleiotropy and the evolution of complexity. *Proceedings of the National Academy of Science U S A*, **107**, 18034–9.

Watson, R.A., Wagner, G.P., Pavličev, M. *et al.* (2014) The evolution of phenotypic correlations and "developmental memory.". *Evolution*, **68**, 1124–38.

Wildman, D.E., Chen, C., Erez, O. *et al.* (2006) Evolution of the mammalian placenta revealed by phylogenetic analysis. *Proceedings of the National Academy of Science U S A*, **103**, 3203–8.

Williams, T.M. and Carroll, S.B. (2009) Genetic and molecular insights into the development and evolution of sexual dimorphism. *Nature Reviews Genetics*, **10**, 797–804.

Wolf, J.B., Leamy, L.J., Routman, E.J. and Cheverud, J.M. (2005) Epistatic pleiotropy and the genetic architecture of covariation within early and late-developing skull trait complexes in mice. *Genetics*, **171**, 683–94.

Wolf, J.B., Pomp, D., Eisen, E.J. *et al.* (2006) The contribution of epistatic pleiotropy to the genetic architecture of covariation among polygenic traits in mice. *Evolution and Development*, **8**, 468–76.

Woods, H.A. (2009) Evolution of homeostatic physiological systems. Chapter in *Phenotypic Plasticity of Insects: Mechanisms and Consequences*. Eds. DW Whitman & TN Ananthakrishnan, Science Publishers, Enfield, NH.

Yang, J., Benyamin, B., McEvoy, B.P. *et al.* (2010) Common SNPs explain a large proportion of the heritability for human height. *Nature Genetics*, **42**, 565–9.

7

THE ROLE OF ECOLOGICAL EPIGENETICS IN INTEGRATIVE BIOLOGY

Aaron W. Schrey[1], Joshua Banta[2], Holly J. Kilvitis[3], and Christina L. Richards[3]

[1]*Department of Biology, Armstrong State University, Savannah, GA, USA*
[2]*Department of Biology, University of Texas at Tyler, Tyler, TX, USA*
[3]*Department of Integrative Biology, University of South Florida, Tampa, FL, USA*

INTRODUCTION

Molecular biology has revolutionized how we study organisms. Inspired by promises from the Modern Synthesis that "genes" provide the "blueprint" to understanding organismal complexity (Pigliucci 2010), researchers have collected increasing amounts of sequence data on an increasing number of organisms. However, despite the big data now available, we still don't understand how organisms construct phenotypes from genotypes (Pigliucci 2010; Martin *et al.* 2011).

Epigenetics is one major new area in an Extended Synthesis, which expands the scope of the Modern Synthesis, in an effort to better explain organismal complexity (Pigliucci & Müller 2010). The term epigenetics was coined by Waddington to mean the total developmental system that generates an adult organism (Waddington 1942a). This definition has narrowed over time, as we have learned about the molecular basis of epigenetic variation (Jablonka & Lamb 2002), to mean specifically changes in gene expression, including mitotic inheritance of patterns of gene expression (Holliday 2006), and "the study of nuclear inheritance which is not based on changes in DNA sequence (Holliday 2006)."

More recently, epigenetics has been defined as the study of heritable changes in gene expression (Richards 2006) or stably heritable phenotypes (Berger *et al.* 2009) that occur without alterations in DNA sequence. Notably, epigenetic variation has the potential to be independent of genetic variation. Richards (2006) defined three categories of

Integrative Organismal Biology, First Edition.
Edited by Lynn B. Martin, Cameron K. Ghalambor, and H. Arthur Woods.

epigenetic variation: (1) obligate, where epigenetic variation depends on genetic variation, (2) facilitated, where the genotype determines the effect of the epigenotype in a probabilistic manner, and (3) pure, which is largely independent of genetic variation. Because epigenetic mechanisms within and across generations can have different implications for ecology and evolution, Jablonka and Raz (2009) distinguish between epigenetics and epigenetic inheritance, which "occurs when phenotypic variations that do not stem from variations in DNA base sequences are transmitted to subsequent generations of cells or organisms." The debate over the definition of epigenetics remains unresolved, but understanding epigenetic variation among individuals and populations will have a profound impact on the study of ecology.

Epigenetics, Phenotype, and Inheritance

Epigenetic modifications contribute to phenotypic variation (Jablonka 2013), which may have wide ecological implications (Table 7.1), including phenotypic plasticity in response to different environments (He & Amasino 2005; Nicotra *et al.* 2010). There are several kinds of epigenetic modifications, although the most studied so far is DNA methylation (Jaenisch & Bird 2003). Methylation generates phenotypic variation in several taxa. In a mutant form of the plant *Linaria vulgaris*, flower symmetry is changed from bilateral to radial by methylation of the *Lcyc* gene (Cubas *et al.* 1999). The epigenetic modification is heritable, and demethylation of the *Lcyc* gene causes a reversion to wildtype. In mice, epigenetic modifications of the *agouti* locus alter coat color (Morgan *et al.* 1999) and modifications of the *axin* locus generate a kinked-tail phenotype (Rakyan *et al.* 2003). A spontaneous mutation in methylation to the promoter of the *colorless nonripening* locus in the tomato (*Solanum lycopersicum*) prevents fruit ripening (Manning *et al.* 2006). Because epigenetic modifications are more dynamic and reversible than DNA mutations (Slatkin 2009), more rapid flexibility can be generated than is possible with genetically-based adaptation alone.

An excellent example of epigenetic mechanisms underling phenotypic plasticity is found in vernalization (the acceleration of flowering after prolonged cold temperature) of *Arabidopsis thaliana*. Vernalization causes changes in histone methylation that alter chromatin structure and ultimately allows for plasticity in flowering (Bastow *et al.* 2004). Another example is found in the leaf characteristics of holly trees in response to herbivory (Herrera & Bazaga 2013). Here, mammalian browsing is correlated to a prickly leaf phenotype, and DNA methylation differs between prickly and nonprickly leaves. Further, experiments on nearly isogenic epigenetic recombinant inbred lines of *A. thaliana* demonstrate that variation in DNA methylation may contribute substantially to heritability in plant traits and the plasticity of these traits (Zhang *et al.* 2013).

TABLE 7.1. A brief Review of the Major Areas where Epigenetic Variation has been shown to have Ecological Importance

Epigenetic Variation Can	References
alter gene expression	Nätt *et al.* 2012
generate phenotypic variation	Rakyan *et al.* 2003; Weaver *et al.* 2004; Kucharski *et al.* 2008
underlie phenotypic plasticity	Bastow *et al.* 2004; He & Amasino 2005; Herrera *et al.* 2012; Zhang *et al.* 2013
respond to environmental stress	Lira-Medeiros *et al.* 2010; Paun *et al.* 2010; Verhoeven *et al.* 2010; Herrera & Bazaga 2010, 2011; Herrera *et al.* 2012
facilitate invasive species	Chwedorzewska & Bednarek 2012; Richards *et al.* 2012, Schrey *et al.* 2012

As gene expression clearly contributes to phenotype, it is critical to learn how epigenetic variation affects gene expression. Recently, Nätt *et al.* (2012) examined the relationship between DNA methylation and differential gene expression in two closely related lineages, the domesticated chicken and its wild ancestor, the red junglefowl (*Gallus gallus*). They identified more than 3000 genes were differently expressed between the two subspecies, and they screened promoter sequences for variation in methylation over generations. While 84% of the gene expression that was significantly different in the first generation was also different in the second, there was no consistent relationship between methylation pattern and gene regulation pattern. Sometimes methylation was associated with up-regulation, sometimes with down-regulation, and often with no change (Nätt *et al.* 2012). Thus, untangling the complicated relationship between epigenetic variation and changes in gene expression is a significant problem to overcome.

Some epigenetic characters can be stably transmitted across generations (Jablonka & Raz 2009; Feng *et al.* 2010; Jablonka 2013). Further, epigenetic inheritance can occur by multiple mechanisms, which could be inherited in different ways (i.e., direct induction of the germ line, parallel induction of the somatic and germ line, and somatic induction; Jablonka & Raz 2009). Recently, Jablonka (2013) reviewed empirical evidence of epigenetic inheritance and found strong support for the transmission of epigenetic information across generations. Further, the information transmitted and number of generations of stable transmission may be affected by external and internal conditions. Thus, environmentally induced phenotypic variation mediated in some part by epigenetics can be a mechanism that generates adaptation (Jablonka 2013).

For example, a study of genetically identical dandelion plants (*Taraxacum officinale*) showed that variation in methylation was induced by different stresses, and that the majority of induced changes were transmitted to the next generation (Verhoeven *et al.* 2010). A study of the epigenetically regulated *agouti* locus went even further to show the potential evolutionary importance: artificial selection for a purely epigenetically controlled coat color phenotype increased the prevalence of that phenotype in an experimental population (Cropley *et al.* 2012). Thus, this study finds that selection on epigenetically-controlled traits can lead to shifts in phenotypes in populations, a kind of epigenetic evolution. While multiple points of evidence exist to support epigenetic inheritance, the mechanisms and implications of these mechanisms are not fully known.

Models of Epigenetic Variation

Epigenetic mechanisms could be profoundly important to evolutionary processes; however, to date, modeling efforts have been inhibited by the scarcity of data. In the adaptive landscape metaphor, epigenetic mechanisms may allow for fitness peaks to be reached more quickly, facilitate peak shifts, or facilitate the transition between fit genotypic states (Pál 1998; Pál & Miklos 1999). Epigenetic modifications could facilitate canalization or genetic assimilation by maintaining an advantageous phenotype for multiple generations, thereby allowing time for genetic mutations to stabilize the phenotype (Waddington 1942b, 1953; West-Eberhard 2005).

Several models demonstrate that epigenetic mechanisms add potential outcomes that are not possible with genetic variation alone. For example, Day and Bonduriansky (2011) provide a theoretical framework and mathematical model that combines genetic and nongenetic mechanisms of inheritance. Nongenetic mechanisms are broadly defined and include epigenetic mechanisms. The model clearly indicates that genetic and nongenetic inheritance mechanisms can interact, and if there is an association between the two, selection can directly and indirectly act on both. Also, the expected change in the genetic component of inheritance from parent to offspring may be affected by nongenetic mechanisms (i.e., methylation). Conversely, some DNA sequences may be more likely to acquire differences in methylation; thus, the expected change in the nongenetic inheritance component can be directly affected by the genetic component. This model allows for some unexpected outcomes: the inheritance of acquired traits, transmission of spontaneously arising factors, and an increase in phenotypic plasticity. Ultimately, the model demonstrates that including nongenetic inheritance generates a broad range of evolutionary possibilities.

ECOLOGICAL EPIGENETICS

Ecological epigenetics is the study of epigenetics in natural populations. Characterizing the distribution of epigenetic variation among individuals and populations is a fundamental task in ecological epigenetics. Two important characteristics of that variation have already been identified: (1) there is a lot of it, and (2) it does not correspond precisely to patterns of genetic variation. While ecological epigenetics has not often been studied in animal populations (Ledón-Rettig *et al.* 2013), more progress has been made in plants (Schrey *et al.* 2013); and already multiple studies have identified the importance of epigenetic variation in an array of ecological processes (Table 7.1). We focus on three areas where epigenetic variation has and will likely continue to provide important insights for ecology.

Epigenetics and Invasive Species Biology

Epigenetic variation may facilitate the spread of invasive species, in part by enabling organisms to adjust rapidly to changing environments (Rando & Verstrepen 2007). If some species (or populations) are better able to regulate expression of genes via epigenetic mechanisms, which then causes the expression of ecologically important phenotypes, they may be better able to colonize new areas or expand their native ranges (Bossdorf *et al.* 2008; Richards *et al.* 2006; Nicotra *et al.* 2010). Epigenetic mechanisms also may compensate for decreases in genetic variation associated with colonizations.

Recent studies on two systems, Japanese knotweeds (*Fallopia japonica* and *F.* × *bohemica*) and house sparrows (*Passer domesticus*), show the importance of epigenetic variation in invasive species biology. Japanese knotweeds have been introduced into diverse habitats in Europe and North America. Introduced populations have almost no genetic diversity (Richards *et al.* 2008, 2012); yet several phenotypic traits show significant differences within and among populations. There was nearly five times as much epigenetic variation as genetic variation among Japanese knotweeds in the northeastern United States, and differentiation at some epigenetic loci could be attributed to different habitats (Richards *et al.* 2012). Interestingly, another very successful introduced species, the house sparrow, also shows a great deal of epigenetic variation (Schrey *et al.* 2012: Liebl *et al.* 2013). Recently introduced birds had a higher proportion of methylation than did birds in older populations (Schrey *et al.* 2012). Also, a study of epigenetic variation across a recent introduction in Kenya detected a potential compensatory relationship between epigenetic and genetic diversity, and between epigenetic diversity and inbreeding (Liebl *et al.* 2013).

Epigenetics and Response to the Environment

Epigenetics may contribute to an organism's ability to respond to changing environments. If environmentally-induced epigenetic states are inherited, they may be adaptive in changing environments (Jablonka *et al.* 1995; Lachmann & Jablonka 1996). This is highlighted by research on asexual and clonal organisms, which have little genetic variation, yet are able to persist in, and adapt to, changing environments (Castonguay & Angers 2012; Herrera *et al.* 2012). One of the best cases described thus far is found in the nectar-living yeast, *Metschnikowia reukaufii* (Herrera *et al.* 2012), which occurs in clonal populations. This yeast had environmentally-induced changes in DNA methylation in response to different nectar environments. There was greater epigenetic than genetic variation, and experimental strains demonstrated that nectar characteristics generated changes in DNA methylation. Further, these epigenetic changes allowed *M. reukaufii* to exploit different resources. Similarly, epigenetic variation separated the asexual hybrid fish *Chrosomus eos-neogaeus* into groups that corresponded to pH differences among lakes (Massicotte *et al.* 2011). These fish populations had greater epigenetic than genetic variation among individuals, and genetic and epigenetic markers had different spatial patterns of differentiation (Massicotte & Angers 2012). Thus, epigenetically controlled traits, which may have higher phenotypic variance, could enhance the adaptive possibilities of asexual or low-diversity taxa (Jablonka & Lamb 1989; Geoghegan & Spencer 2012).

Epigenetics and Development

Confirming the role of epigenetic effects on phenotypic variation in natural populations would enhance our understanding of how developmental mechanisms evolve. Epigenetic processes play an important role in the development of most organisms (Bird 2002; Tang & Ho 2007; Angers *et al.* 2010). One well-known example of the role of epigenetics in development is in the differentiation between queen and worker honey bees (*Apis mellifera*) in response to diet (Kucharski *et al.* 2008). Experimentally silencing a DNA methyltransferase in honey bees altered larval development, such that they were more likely to develop into queens (Kucharski *et al.* 2008). Thus, differential methylation may underlie many of the differences between queens and workers.

In mammals, the genome is epigenetically reprogrammed throughout ontogeny, including nearly global erasure of methylation patterns during gametogenesis followed by their re-establishment later in development (Tang & Ho 2007; Dolinoy *et al.* 2007; Feng *et al.* 2010). Epigenetic processes are typically regulated by intrinsic developmental signals; yet, studies indicate that environments experienced during critical periods of development can alter methylation (Angers *et al.* 2010; Dolinoy *et al.* 2007). In rats, maternal variation in licking and grooming are associated with differences in anxiety-like behaviors, stress responsiveness, and changes in methylation of the glucocorticoid receptor in the hippocampus in their offspring, such that offspring that were licked and groomed more frequently were less fearful and had dampened stress responses as adults (Weaver *et al.* 2004). Also, maternal nutrition or exposure to toxins can alter patterns of DNA methylation and influence the disease susceptibility of offspring (Tost 2010). Expression of the *avy* epiallele of the murine *agouti* gene is mediated by methylation (Morgan *et al.* 1999; Jirtle & Skinner 2007; Skinner *et al.* 2010) and can be altered by maternal diet (Waterland & Jirtle 2003, 2004; Dolinoy *et al.* 2006, Dolinoy *et al.* 2007). Altered methylation in rats exposed to environmental levels of bisphenol A (BPA), a residue in many plastics, has been associated with increased susceptibility to cancer (Ho *et al.* 2006; Skinner *et al.* 2010). In addition, even transient embryonic exposure to fungicides and pesticides led to reduced spermatogenesis and male infertility, which stemmed from altered methylation and was transmitted transgenerationally (Anway *et al.* 2005; Skinner *et al.* 2010).

These studies provide insight into how epigenetics can be altered by environmental factors. However, the ecological significance of these relationships has not been considered

(Ledón-Rettig *et al.* 2013). But by doing so, we could begin to understand how phenotypic variation in natural populations is associated with developmental differences in epigenetic programming.

CASE STUDY: *ARABIDOPSIS THALIANA*

The mustard *Arabidopsis thaliana* provides a unique opportunity to demonstrate how inherited epigenetic polymorphisms contribute to complex phenotypes, and how they may underlie the evolution of development in natural populations. Recent studies have shown natural epigenetic variation (Vaughn *et al.* 2007) and geographically structured variation for a wide variety of traits (e.g., Atwell *et al.* 2010; Banta *et al.* 2012). We used published data on methylation polymorphisms in 16 loci across 76 natural accessions of *A. thaliana* (Vaughn *et al.* 2007; Figure 7.1), along with their published flowering times (Atwell *et al.* 2010), to perform epigenetic mapping of flowering time. We controlled for genetic similarity in our analysis to rule out that these epi-polymorphic sites were merely markers for genetic polymorphisms (Atwell *et al.* 2010; Brachi *et al.* 2010). Thus only epi-polymorphic sites that are decoupled from genetic similarity should be significant, although residual population structure can still remain (Atwell *et al.* 2010; Brachi *et al.* 2010) and the genetic loci in close linkage disequilibrium with the epi-polymorphic sites cannot be ruled out as the causative drivers at this preliminary stage. This analysis revealed that the methylation state at one locus was associated with variation in flowering time. Accessions that were

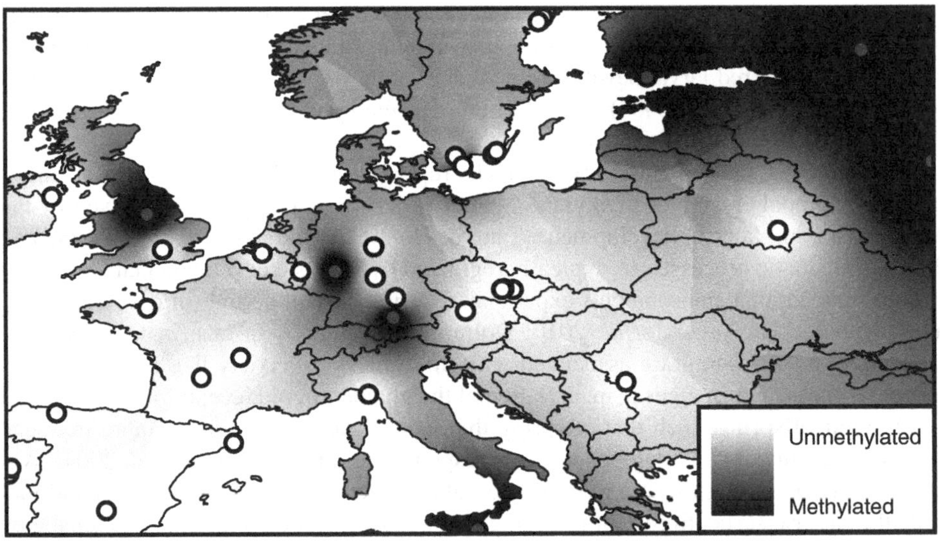

Figure 7.1. The chromosomal fragment around the gene *MEE57* in *Arabidopsis thaliana* has two different, presumably stable epi-alleles (methylated versus unmethylated) that segregate in the wild. The spatially interpolated frequencies of the two contrasting epi-alleles across the landscape are represented here, based on the locations of collection of the 76 accessions in our study and their epi-genotypes at the *MEE57* fragment. Circles shaded grey represent accessions with the methylated epi-allele, and circles shaded white represent accessions with the unmethylated epi-allele. Black areas indicate a high frequency of the methylated epi-allele, and white areas indicate a high frequency of the unmethylated epi-allele. Many points are obscured by other points. (See plate section for color version.)

methylated at this locus flowered much earlier than those that were not. The only gene within the interval of this locus (AT4G13610.1: *MATERNAL EFFECT EMBRYO ARREST 57*) functions in DNA (cytosine-5-)-methyltransferase activity (Pagnussat *et al.* 2005). This is consistent with a scenario where the gene is differentially methylated in nature, and the epi-polymorphism affects flowering time among different individuals in the wild, perhaps by controlling methylation at other loci in the genome. We believe this intriguing candidate gene (and candidate epi-gene) are worthy of follow-up investigations to test the hypothesis that cascading epigenetic effects influence variation in flowering time in the wild.

CONCLUSION

Integrative biology recognizes that understanding biological diversity is a complex problem that requires studies from the molecular level through the level of ecosystems, and at all levels in between. Over the last few decades, molecular studies of epigenetic phenomena have revolutionized our understanding of genome function and developmental biology, but these processes also challenge our understanding of heritable phenotypic variation at the level of whole organisms and even the process of evolution by natural selection (Jablonka & Lamb 1989; Danchin *et al.* 2011). During its short history, ecological epigenetics has proven to be very promising even though key questions remain unanswered. Empirical studies in natural populations will help answer and refine major questions about the roles that epigenetics play in organismal biology. While a great deal of work remains, epigenetics provides a promising lens through which to view organismal biology at a variety of supraorganismal levels.

REFERENCES

Angers, B., Castonguay, E. and Massicotte, R. (2010) Environmentally induced phenotypes and DNA methylation: how to deal with unpredictable conditions until the next generation and after. *Molecular Ecology*, **19**, 1283–95.

Anway, M.D., Cupp, A.S., Uzumcu, M. and Skinner, M.K. (2005) Epigenetic transgenerational actions of endocrine disruptors and male fertility. *Science*, **308**, 1466–9.

Atwell, S., Huang, Y.S., Vilhjálmsson, B.J. *et al.* (2010) Genome-wide association study of 107 phenotypes in *Arabidopsis thaliana* inbred lines. *Nature*, **465**, 627–31.

Banta, J.A., Ehrenreich, I.M., Gerard, S. *et al.* (2012) Climate envelope modeling reveals intraspecific relationships among flowering phenology, niche breadth, and potential range size in *Arabidopsis thaliana*. *Ecology Letters*, **15**, 769–77.

Bastow, R., Mylne, J.S., Lister, C. *et al.* (2004) Vernalization requires epigenetic silencing of FLC by histone methylation. *Nature*, **427**, 164–7.

Berger, S.L., Kouzarides, T., Shiekhattar, R. and Shilatifard, A. (2009) An operational definition of epigenetics. *Genes & Development*, **23**, 781–3.

Bird, A. (2002) DNA methylation patterns and epigenetic memory. *Genes & Development*, **16**, 6–21.

Bossdorf, O., Richards, C.L. and Pigliucci, M. (2008) Epigenetics for ecologists. *Ecology Letters*, **11**, 106–15.

Brachi, B., Faure, N., Horton, M. *et al.* (2010) Linkage and association mapping of *Arabidopsis thaliana* flowering time in nature. *PLoS Genetics*, **6** (5), e1000940.

Castonguay, E. & Angers, B. (2012) The key role of epigenetics in the persistence of asexual lineages. *Genetics Research International*, **2012**, Article ID 534289, 9 pages. doi:10.1155/2012/534289

Chwedorzewska, K.J. and Bednarek, P.T. (2012) Genetic and epigenetic variation in a cosmopolitan grass *Poa annua* from Antarctic and Polish populations. *Polish Polar Research*, **33**, 63–80.

Cropley, J.E., Dang, T.H.Y., Martin, D.I.K. and Suter, C.M. (2012) The penetrance of an epigenetic trait in mice is progressively yet reversibly increased by selection and environment. *Proceedings of the Royal Society B*, **279**, 2347–53.

Cubas, P., Vincent, C. and Coen, E. (1999) An epigenetic mutation responsible for natural variation in floral symmetry. *Nature*, **401**, 157–61.

Danchin, E., Charmantier, A., Champagne, F.A. *et al.* (2011) Beyond DNA: integrating inclusive inheritance into an extended theory of evolution. *Nature Reviews Genetics*, **12**, 475–86.

Day, T. and Bonduriansky, R. (2011) A unified approach to the evolutionary consequences of genetic and nongenetic inheritance. *American Naturalist*, **178**, E18–E36.

Dolinoy, D.C., Weidman, J.R. and Jirtle, R.L. (2007) Epigenetic gene regulation: linking early developmental environment to adult disease. *Reproductive Toxicology*, **23**, 297–307.

Dolinoy, D.C., Weidman, J.R., Waterland, R.A. and Jirtle, R.L. (2006) Maternal genistein alters coat color and protects avy mouse offspring from obesity by modifying the fetal epigenome. *Environmental Health Perspectives*, **114**, 567–72.

Feng, S., Cokus, S.J., Zhang, X. *et al.* (2010) Conservation and divergence of methylation patterning in plants and animals. *Proceedings of the National Academy of Science U.S.A*, **107**, 8689–94.

Geoghegen, J.L. and Spencer, H.G. (2012) Population-epigenetic models of selection. *Theoretical Population Biology*, **81**, 232–42.

He, Y.H. and Amasino, R.M. (2005) Role of chromatin modification in flowering-time control. *Trends in Plant Science*, **10**, 30–5.

Herrera, C.M. and Bazaga, P. (2010) Epigenetic differentiation and relationship to adaptive genetic divergence in discrete populations of the violet *Viola cazorlensis*. *New Phytologist*, **187**, 867–76.

Herrera, C.M. and Bazaga, P. (2011) Untangling individual variation in natural populations: ecological, genetic and epigenetic correlates of long-term inequality in herbivory. *Molecular Ecology*, **20**, 1675–88.

Herrera, C.M. and Bazaga, P. (2013) Epigenetic correlates of plant phenotypic plasticity: DNA methylation differs between prickly and nonprickly leaves in *heterophyllous Ilex aquifolium* (Aquifoliaceae) trees. *Botanical Journal of the Linnean Society*, **171**, 441–52.

Herrera, C.M., Pozo, M.I. and Bazaga, P. (2012) Jack of all nectars, master of most: DNA methylation and the epigenetic basis of niche width in a flower living yeast. *Molecular Ecology*, **21**, 2602–16.

Ho, S.M., Tang, W.Y., Belmonte de Frausto, J. and Prins, G.S. (2006) Developmental exposure to estradiol and bisphenol A increases susceptibility to prostate carcinogenesis and epigenetically regulates phosphodiesterase type 4 variant 4. *Cancer Research*, **66**, 5624–32.

Holliday, R. (2006) Epigenetics: A historical overview. *Epigenetics*, **1**, 76–80.

Jablonka, E. (2013) Epigenetic inheritance and plasticity: the responsive germline. *Progress in Biophysics and Molecular Biology*, **111**, 99–107.

Jablonka, E. and Lamb, M.J. (1989) The inheritance of acquired epigenetic variations. *Journal Theoretical Biology*, **139**, 69–83.

Jablonka, E. and Lamb, M.J. (2002) The changing concept of epigenetics. *Annuals of the New York Academy of Science*, **981**, 82–96.

Jablonka, E., Oborny, B., Molnar, I. *et al.* (1995) The adaptive advantage of phenotypic memory in changing environments. *Philosophical Transactions of the Royal Society London Series B*, **350**, 133–41.

Jablonka, E. and Raz, G. (2009) Transgenerational epigenetic inheritance: prevalence, mechanisms, and implications for the study of heredity and evolution. *Quarterly Review of Biology*, **8**, 131–76.

Jaenisch, R. and Bird, A. (2003) Epigenetic regulation of gene expression: how the genome integrates intrinsic and environmental signals. *Nature Genetics*, **33**, S245–S254.

Jirtle, R.L. and Skinner, M.K. (2007) Environmental epigenomics and disease susceptibility. *Nature Reviews Genetics*, **8**, 253–62.

Johannes, F., Porcher, E., Teixeira, F.K. *et al.* (2009) Assessing the impact of transgenerational epigenetic variation on complex traits. *PLoS Genetics*, **5**, e1000530.

Kucharski, R., Maleszka, J., Foret, S. and Maleszka, R. (2008) Nutritional control of reproductive status in honeybees via DNA methylation. *Science*, **319**, 1827–30.

Lachmann, M. and Jablonka, E. (1996) The inheritance of phenotypes: an adaptation to fluctuating environments. *Journal of Theoretical Biology*, **181**, 1–9.

Ledón-Rettig, C., Richards, C.L. and Martin, L.B. (2013) Epigenetics for behavioral ecologists. *Behavioral Ecology*, **24**, 311–24.

Liebl, A.L., Schrey, A.W., Richards, C.L. and Martin, L.B. (2013) Patterns of DNA methylation throughout a range expansion of an introduced songbird. *Integrative and Comparative Biology*, **53**, 351–8.

Lira-Medeiros, C.F., Parisod, C., Fernandes, R.A. *et al.* (2010) Epigenetic variation in mangrove plants occurring in contrasting natural environment. *PLoS One*, **5**, e10326.

Manning, K., Tor, M., Poole, M. *et al.* (2006) A naturally occurring epigenetic mutation in a gene encoding an SBP-box transcription factor inhibits tomato fruit ripening. *Nature Genetics*, **38**, 948–52.

Martin, L.B., Liebl, A.L., Trotter, J.H. *et al.* (2011) Integrators: physiological determinants of phenotypic plasticity. *Integrative and Comparative Biology*, **51**, 514–27.

Massicotte, R., Whitelaw, E. and Angers, B. (2011) DNA methylation: a source of random variation in natural populations. *Epigenetics*, **6**, 421–7.

Massicotte, R. & Angers, B. (2012) General-purpose genotype or how epigenetics extend the flexibility of a genotype. *Genetics Research International*, **2012**, Article ID 317175, 7 pages doi:10.1155/2012/317175.

Morgan, H.D., Sutherland, H.G.E., Martin, D.I.K. and Whitelaw, E. (1999) Epigenetic inheritance at the agouti locus in the mouse. *Nature Genetics*, **23**, 314–18.

Nätt, D., Rubin, C.J., Wright, D. *et al.* (2012) Heritable genome-wide variation of gene expression and promoter methylation between wild and domesticated chickens. *BMC Genomics*, **13**, 59.

Nicotra, A.B., Atkin, O.K., Bonser, S.P. *et al.* (2010) Plant phenotypic plasticity in a changing climate. *Trends in Plant Science*, **15**, 684–92.

Pagnussat, G.C., Hee-Ju, Y., Ngo, Q. *et al.* (2005) Genetic and molecular identification of genes required for female gametophyte development and function in Arabidopsis. *Development*, **132**, 603–14.

Pál, C. (1998) Plasticity, memory and the adaptive landscape of the genotype. *Proceedings of the Royal Society B*, **265**, 1319–23.

Pál, C. and Miklós, I. (1999) Epigenetic inheritance, genetic assimilation and speciation. *Journal of Theoretical Biology*, **200**, 19–37.

Paun, O., Bateman, R.M., Fay, M.F. *et al.* (2010) Stable epigenetic effects impact adaptation in allopolyploid orchids (Dactylorhiza: Orchidaceae). *Molecular Biology & Evolution*, **27**, 2465–73.

Pigliucci, M. (2010) Genotype-phenotype mapping and the end of the "genes as blueprint" metaphor. *Philosophical Transactions of the Royal Society B*, **365**, 557–66.

Pigliucci, M. and Müller, G.B. (2010) *Evolution – The Extended Synthesis*, MIT Press.

Rakyan, V.K., Chong, S., Champ, M.E. *et al.* (2003) Trangenerational inheritance of epigenetic states at the murine Axin(Fu) allele occurs after maternal and paternal transmission. *Proceedings of the Natlional Academy of Science U.S.A*, **100**, 2538–43.

Rando, O. and Verstrepen, K. (2007) Timescales of genetic and epigenetic inheritance. *Cell*, **128**, 655–68.

Richards, C.L., Walls, R., Bailey, J.P. *et al.* (2008) Plasticity in salt tolerance traits allows for invasion of salt marshes by Japanese knotweed s.l. (*Fallopia japonica* and *F.* × *bohemica*, Polygonaceae). *American Journal of Botany*, **95**, 931–42.

Richards, C.L., Schrey, A. and Pigliucci, M. (2012) Invasion of diverse habitats by few Japanese knotweed genotypes is correlated with epigenetic differentiation. *Ecology Letters*, **15**, 1016–25.

Richards, E.J. (2006) Inherited epigenetic variation – revisiting soft inheritance. *Nature Reviews Genetics*, **7**, 395–401.

Schrey, A., Coon, C., Grispo, M., Awad, M., McCoy, E., Mushinsky, H., Richards, C.L., & Martin, L.B. (2012) Epigenetic variation may compensate for decreased genetic variation with introductions: a case study using house sparrows (*Passer domesticus*) on two continents. *Genetics Research International*, **2012**, Article ID 979751, 7 pages. doi:10.1155/2012/979751.

Schrey, A.W., Alvarez, M., Foust, C.M. *et al.* (2013) Ecological epigenetics: Beyond MS-AFLP. *Integrative and Comparative Biology*, **53**, 340–50.

Skinner, M.K., Manikkam, M. and Guerrero-Bosagna, C. (2010) Epigenetic transgenerational actions of environmental factors in disease etiology. *Trends in Endocrinology & Metabolism*, **21**, 214–22.

Slatkin, M. (2009) Epigenetic inheritance and the missing heritability problem. *Genetics*, **182**, 845–50.

Sung, S. and Amasino, R.M. (2004) Vernalization in Arabidopsis thaliana is mediated by the PHD finger protein VIN3. *Nature*, **427**, 159–64.

Tang, W.Y. and Ho, S.M. (2007) Epigenetic reprogramming and imprinting in origins of disease. *Reviews in Endocrine & Metabolic Disorders*, **8**, 173–82.

Tost, J. (2010) DNA methylation: an introduction to the biology and the disease-associated changes of a promising biomarker. *Molecular Biotechnology*, **44**, 71–81.

Vaughn, M.W., Tanurdzic, M., Lippman, Z. *et al.* (2007) Epigenetic natural variation in *Arabidopsis thaliana*. *PLoS Biology*, **5**, 1617–29.

Verhoeven, K.J.F., Jansen, J.J., van Dijk, P.J. and Biere, A. (2010) Stress-induced DNA methylation changes and their heritability in asexual dandelions. *New Phytologist*, **185**, 1108–18.

Waddington, C.H. (1942a) The epigenotype. *Endeavour*, **1942**, 18–20.

Waddington, C.H. (1942b) Canalization of development and the inheritance of acquired characters. *Nature*, **150**, 563–5.

Waddington, C.H. (1953) Genetic assimilation of an acquired character. *Evolution*, **7**, 118–26.

Wang, J.L., Tian, L., Madlung, A. *et al.* (2004) Stochastic and epigenetic changes of gene expression in Arabidopsis polyploids. *Genetics*, **167**, 1961–73.

Waterland, R.A. and Jirtle, R.L. (2003) Transposable elements: targets for early nutritional effects on epigenetic gene regulation. *Molecular & Cellular Biology*, **23**, 5293–5300.

Waterland, R.A. and Jirtle, R.L. (2004) Early nutrition, epigenetic changes at transposons and imprinted genes, and enhanced susceptibility to adult chronic diseases. *Nutrition*, **20**, 63–8.

Weaver, I.C.G., Cervoni, N., Champagne, F.A. *et al.* (2004) Epigenetic programming by maternal behavior. *Nature Neuroscience*, **7**, 847–54.

West-Eberhard, M.J. (2005) Developmental plasticity and the origin of species differences. *Proceedings of the National Academy of Science U.S.A*, **102**, 6543–9.

Zhang, Y.-Y., Fischer, M., Colot, V. and Bossdorf, O. (2013) Epigenetic variation creates potential for evolution of plant phenotypic plasticity. *New Phytologist*, **197**, 314–22.

8

AN ELEPHANT IN THE FOG: UNIFYING CONCEPTS OF PHYSIOLOGICAL STASIS AND CHANGE

H. Arthur Woods and J. Keaton Wilson

Division of Biological Sciences, University of Montana, Missoula, MT, USA

INTRODUCTION

Organisms face (at least) two inescapable realities. First, for any organism, no single set of phenotypes can possibly perform well in all environments it encounters. Environments are complex in too many ways, and they vary across too many temporal and spatial scales. The problem is one of extreme multivariate optimization, in which the organism must try to generate good, or at least adequate, complex phenotypes in most environments most of the time. Second, organisms operate as integrated wholes (see the Introduction to this volume). They are closely coordinated modules of interacting parts, unified by shared resources. As a consequence, they must control their physiological processes across multiple levels of organization, and there must be tradeoffs among modules for shared resources. When the organism is performing well, and it has adequate access to energy, materials, and information, these systems and tradeoffs can create finely tuned responses to environmental variation. By contrast, when organisms lack adequate resources, or when they encounter novel or statistically unusual patterns of environmental variation, those tradeoffs, interacting with complex phenotypic and environmental cross-correlations, can lead to costs paid, apparently poor regulation, and the suites of pathological syndromes that can result in sickness or poor performance.

The physiological systems used by organisms for responding to changing environments are, like the environments themselves, complex and multivariate. Nevertheless, at some basic level, all physiological systems can be described as containing complementary elements of stasis and change. Individuals, for example, can maintain stable gradients

Integrative Organismal Biology, First Edition.
Edited by Lynn B. Martin, Cameron K. Ghalambor, and H. Arthur Woods.

(stasis) of ions that are far from any biochemical equilibrium – which is possible only by expending energy and by closely coordinating fluxes of materials and information (change). Likewise, at higher levels of organization, organisms may be able to hold a particular phenotype constant (stasis) only by exhibiting plasticity in other traits (change). This tension between stasis and change is pervasive. Moreover, the balance changes frequently, both because environments impinge on organisms with new opportunities and challenges and because life histories unfold.

This chapter compares a set of concepts, some of them proposed just in the last few years, about physiological stasis and change (see also Schulte 2014). All of the concepts are derived from the basic idea of homeostasis. Our comparison leads to two main conclusions. The first is that physiological stability absolutely requires physiological change and, conversely (and counterintuitively), change requires stability. The second conclusion is that all of the concepts related to homeostasis are similar in that they identify aspects of a single underlying phenomenon, and we argue that it is time to step back and articulate its key features. Furthermore, that phenomenon can and should be understood in relation to the more general umbrella of ideas known as phenotypic plasticity and canalization (Pigliucci 2001; Debat & David 2001; West-Eberhard 2003; Piersma & Drent 2003; Ghalambor *et al.* 2007; Piersma & van Gils 2010; see Chapter 2, this volume). Such synthesis can lead to productive cross-pollination across fields – concepts well known in one field can fertilize rapid change in others; we identify several examples where such progress may be possible.

ELABORATIONS OF STASIS AND CHANGE

Below we briefly describe eight different stasis concepts, starting with homeostasis. There are other concepts, certainly, but for brevity we do not cover them here. For fuller explanations, the interested reader should consult the main references we cite. Subsets of these concepts have been compared by other authors, especially by those interested in relating physiological disturbances to the biological concept of stress (Koolhaas *et al.* 2011; Schulte, 2014).

1. Homeostasis

Homeostasis was the first fully realized, coherent concept of physiological regulation and stasis. It traces to the seminal work of Claude Bernard in the late 19th century (Bernard 1865, 1878; Cooper 2008) and was crystallized and invigorated by Walter Cannon in the late 1920s and early 1930s (Cannon 1929, 1932). Cannon provided the conceptual unification of different types of regulation, and he elucidated many important homeostatic mechanisms. Cannon viewed homeostasis as a process in which a negative feedback loop (Wiener 1948) regulates the level of some physiological factor (pH, blood osmolality, body temperature, etc.). The system achieves stability (homeostasis) by measuring the deviation of the physiological factor from some setpoint and then directing other physiological parts, the organs, tissues, and cells, to counteract the deviation. The looping structure (Figure 8.1A) can be broken conceptually into the parts that integrate and process information (Woods 2009; Martin *et al.* 2011) and those that carry out the actual physiological work of moving molecules or ions, pumping fluids, modifying membrane potentials, etc. A related idea is "homeokinesis," which refers to steady fluxes rather than to steady levels (Mangum & Towle 1977).

Homeostasis continues to play a central role in physiology, as is apparent from any modern textbook on physiology (Guyton & Hall 2010; Hill *et al.* 2013), and searching Web

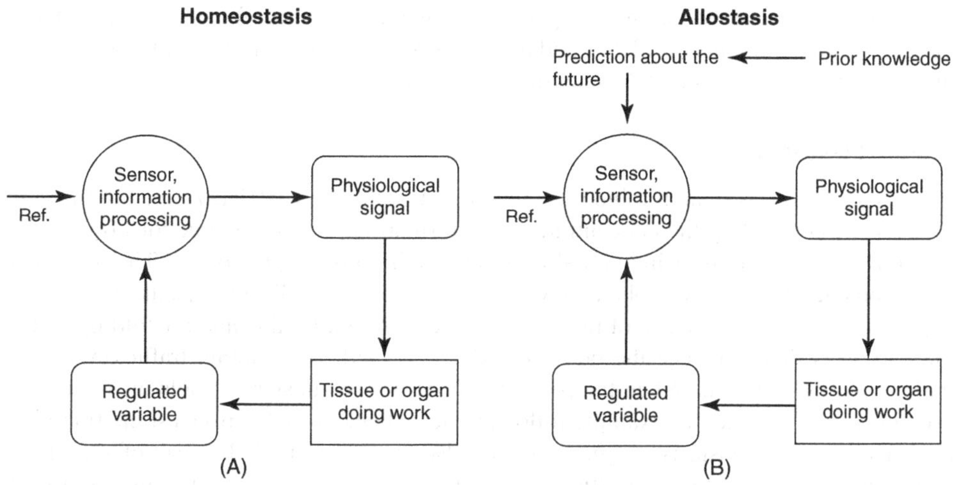

Figure 8.1. Homeostasis and allostasis. (A) Schematic of homeostatic feedback loop, from Woods and Wilson (2013) but similar layouts can be found in many sources. The sensor or information processor compares the level of the regulated variable to some reference or setpoint, and produces a signal whose level is proportional to the perceived deviation and that directs other entities (cells, tissues, organs) to do counteracting physiological work. *Source*: Woods, H.A. & Wilson, J.K. (2013) An information hypothesis for the evolution of homeostasis. *Trends in Ecology & Evolution*, **28**, 283–9. Copyright © Elsevier, 2013. (B) In rheostasis and allostasis, the emphasis shifts to modification of setpoints in response to current or anticipated conditions. This panel redrawn from Sterling (2004).

of Science with the topic term "homeostasis" brings up over 95 000 records since 1929 when Cannon proposed the term (Cannon 1929). Nevertheless, in the past 20 years, the basic concept has been modified in various ways to account for its perceived inadequacies. These new ideas are aimed at giving a clearer view of the functions and evolution of physiological systems as integrated wholes, rather than just for single, suborganismal systems. In this sense, we think of homeostatic systems as fundamental units of physiology in the same way that genes are fundamental entities of genetics or that populations are of ecology: they are real and necessary – but insufficient by themselves to explain all the phenomena we observe.

2. Enantiostasis

Enantiostasis, proposed and coined by Mangum and Towle (Mangum & Towle 1977), addresses the observation that not all internal factors remain constant during environmental change. Rather, they often change in coordinated ways such that the effects of change in one factor are offset by changes in another. Indeed, the word itself is composed of Greek parts meaning "a counterbalanced state of opposing conditions" (Mangum & Towle 1977). The consequence is that internal physiological conditions are unstable with respect to levels of individual factors (nonhomeostasis), but those changes give stability in some other factor. For example, the denaturing effects of rising levels of urea in shark blood can be offset by rising levels of TMAO (Hochachka & Somero 2001); the concentrations of the two compounds show enantiostasis whereas relative protein stability shows homeostasis. Another example: decreasing body temperature, in ectotherms, often depresses the activity of individual enzymes, but the effects on metabolic fluxes can be offset by a variety of mechanisms, including isozyme substitution, changes in the enzyme's affinity for its substrate, and by increases in the amounts of mRNA and enzyme (Powers & Schulte 1998). In this case, the fluxes show homeostasis and the enzymes enantiostasis.

3. Proteostasis

Proteostasis, proposed and coined by Balch and colleagues (Balch *et al.* 2008), is the process of "controlling the concentration, conformation, binding interactions (quaternary structure), and location of individual proteins making up the proteome." The idea arises from considering protein pools as having some degree of collective health. Proteostasis depends both on the influence of intracellular chemistry on folding and misfolding, and on the pathways that influence the pool, including biosynthesis, folding, trafficking, aggregation, and degradation. When the proteome becomes too unstable, pools of intracellular proteins can aggregate, for example following heat shock (which unfolds some fraction of proteins). Poor proteostasis in aging humans also appears to underlie a set of serious and common diseases, including cystic fibrosis, Huntington's disease, Alzheimer's disease, and amyotrophic lateral sclerosis (Balch *et al.* 2008).

4. Rheostasis

Rheostasis, summarized in a 1990 book (Mrosovsky 1990), builds on homeostasis by emphasizing that regulation of physiological factors occurs around shifting setpoints. The prefix rheo is used to denote something flowing rather than something more stable like homeo (though Cannon's use of homeo also conveys much the same thing). The maintenance of a particular level of a factor can depend on the time of day or year, or the developmental stage of an organism. Mrosovsky breaks these time-dependent changes in setpoints into two broad categories: reactive and programmed. He also outlines a number of problems associated with more standard views of homeostasis, namely lack of emphasis on feed-forward and positive feedback systems (regulatory mechanisms that have been shown to be important in physiological systems (Rutter 1999; Abdel-Sater 2011)), competition between different feedback systems, and the frequent lack of an evolutionary perspective.

5. Allostasis

Allostasis – a concept (and associated term) first proposed by Sterling and Eyer (Sterling & Eyer 1988) and expanded by a number of authors subsequently (Schulkin *et al.* 1994; McEwen 1998; McEwen & Wingfield 2003; Schulkin 2011; Sterling 2012) – builds from the simpler idea of homeostasis. First, like homeostasis, allostasis refers to systems that regulate particular physiological factors (e.g., body temperature, pH, blood osmolality). Second, like rheostasis, it emerged from the observation that setpoints for almost all physiological systems vary over time, and that the variation is coordinated in response to context and need (Figure 8.1B). Third, allostasis acknowledges that regulatory systems incur costs, which collectively are called the allostatic load. In general, allostasis is coordinated at substantially higher organizational levels than is homeostasis, which acts more locally and tends to ignore broader organismal contexts.

The idea of allostatic load refers to the costs that organisms pay for activating systems, and it couches the costs in terms of their effects on interactions among multiple physiological systems. McEwen and Wingfield (McEwen & Wingfield 2003) describe two types of allostatic load. Type 1 load occurs when organisms have insufficient energy available for maintaining homeostasis, even if they otherwise have appropriate sensors and systems.

In this situation, organisms may be forced into energetic tradeoffs, such that allocation of energy to one system occurs makes it unavailable to others (see section below on 'Costs and Constraints', pp. 130–131). The second kind of allostatic load (Type 2) occurs when the long-term activation of an allostatic system causes cascading pathological problems in other systems. In particular, chronic activity (or inactivity) of allostatic systems can ripple across systems via shared signaling and metabolic pathways. The decrement in performance arises, in such situations, not from lack of energy (or other resources), but from complex systems of regulatory linkages being presented with unusual or rare kinds of environments or stresses. For example, humans experiencing bouts of stress, which activate the hypothalamic-pituitary-adrenal axis and secrete the stress hormone corticosterone into circulation, can derive short-term benefits from the response in some circumstances (e.g., when our bodies need to be physiologically prepared to flee or fight). But if the stress is repeated too often, or is chronic, human physiological systems can pay a high price in terms of stress-related diseases (Sterling & Eyer, 1988). In both types of allostatic load, the "load" is a measure of current state of challenge, and high loads can make it more difficult for organisms to respond to subsequent stressors.

6. The Reactive Scope Model

The reactive scope model, proposed by Romero *et al.* (Romero *et al.* 2009), responds to perceived problems in allostasis. Those problems include that allostasis is cast too narrowly in terms of energy budgets (Walsberg 2003) and glucocorticoids, that it fails to account adequately for the long-lasting effects of early developmental events (Caldji *et al.* 2001), and that it doesn't represent enough of an advance over the basic idea of homeostasis to warrant a new name and associated set of constructs (Dallman 2003). (These problems are disputed by key authors working on allostasis (McEwen & Wingfield 2010).) This leads Romero and colleagues to propose a separate, graphical model, which incorporates behavioral and physiological responses to stimuli and describes long-term pathological effects of overstimulation of allostatic systems. The model addresses the levels of mediators (i.e., physiological mediators like prostaglandin, glucocorticoids or neurotransmitter concentrations as well as behavioral mediators like locomotion, foraging or migration); mediators are either "summary molecules" that integrate physiological information from multiple sources into the circulating levels of particular molecules, or they are the means by which organisms relieve their allostatic loads (Romero *et al.* 2009). The reactive scope model envisions three zones of stress on allostatic and homeostatic systems – normal reactive scope, homeostatic overload, and homeostatic failure. In the normal reactive scope, a mediator's level may vary over time with few consequences, as in rheostasis, but damage may occur if it passes into either of the other zones. The model also explicitly includes elements of reactive and predictive (or anticipatory) homeostasis.

7. Physiological Regulatory Networks (PRNs)

Physiological regulatory networks, proposed recently (Martin *et al.* 2011; Cohen *et al.* 2012), provide a framework for examining multiple systems simultaneously, with particular emphasis on the evolution of physiological systems (see also Chapter 9, this volume). The premise is that regulatory systems, in individual organisms, often are interconnected by shared integrator molecules and control networks (Cohen *et al.* 2012). Changes in one set of regulatory molecules, or the states of control networks, can have far-reaching effects on other systems, representing a kind of pleiotropy that can constrain or expand the potential for evolution (Ketterson & Nolan Jr 1999; Gerhart & Kirschner 2007;

McGlothlin & Ketterson 2008; Ketterson *et al.* 2009). Additionally, because of cross-system integration, phenotypes at the organismal scale may not be predictable from examining subnetworks alone. Martin and Cohen (Chapter 9, this volume) advocate basing whole-organism studies explicitly in network theory, which provides formal methods for characterizing complexity. Though the authors acknowledge the difficulty in studying PRNs, they also point out that ecologists have made significant progress on similarly daunting food-web networks over the last 100 years.

8. Homeostatic noise

Homeostatic noise, also proposed recently (Woods & Wilson 2013), focuses on physiological noise stemming from extrinsically-driven variation in internal environments, and on the effects that noise has on communication systems in organisms. Noise in biological systems can be both a blessing and a curse. Over the past 15 years, increasingly diverse studies have shown that noise can play constructive, or adaptive, roles for organisms (Viney & Reece 2013), by increasing intra-organismal physiological diversity (H.A. Woods 2014) or by driving adaptive transitions between bi- or multistable states in complex networks. However, noise in biological systems can also be destructive, especially if it distorts signals. The ideas proposed by Woods & Wilson merge older ideas about homeostasis with the field of information theory, developed in the late 1940s to describe information transfer and noise in human-built systems (Shannon 1948). In their view, the primary function of homeostasis is to control sources of physiological noise arising as distortion from extrinsic sources. For example, variation in blood pH, which in most vertebrates is under tight homeostatic control, can have profound consequences on signaling in other systems, because the protein and RNA components of those systems usually are sensitive to pH. Thus, uncontrolled variation in blood pH can result cascading detrimental effects on other internal signaling systems. Similar arguments can be constructed for the other major physiological factors that are under tight homeostatic control in most organisms (fluid levels of calcium, glucose, osmolality, and temperature). Woods and Wilson propose that the factors that are tightly regulated via homeostasis are also factors for which other methods of noise-control (e.g., filtering, redundancy) may not be effective. In addition, although the other stasis concepts seek primarily to explain the mechanisms by which organisms regulate internal factors (the try to understand *how*), homeostatic noise is an alternative explanation for the evolution of homeostasis (it proposes a new *why*).

AXES OF (DIS)SIMILARITY

The eight concepts above are for the most part distinct, even though some of them share substantial overlap. Nevertheless, their differences can be categorized along just a few important axes. Below, we identify those axes and array the concepts along them; not all concepts occur on every axis.

Fixed versus Changing Setpoints

Homeostasis is often thought to imply, or even require, fixed setpoints. By contrast, rheostasis, allostasis, and reactive scope explicitly focus on changing setpoints. These models posit many adaptive explanations for changing setpoints, but the general reason is that no single set of setpoints is best across all potential internal or external conditions. Enantiostasis and proteostasis occupy a middle ground in that they are concerned with how changes in two or more underlying factors (i.e., changes in the setpoints of those factors) offset one another to give stability (homeostasis) for some other factor. Both PRNs and homeostatic noise are also concerned with changing setpoints, but more in the context of control and coordination by complex networks.

This axis is haunted somewhat by the straw man of fixed setpoints. Cannon himself raised the possibility that homeostatic setpoints could change, which is one reason he chose the word "homeostasis" over "homostasis" (Cannon 1929; Mrosovsky 1990). An even-handed assessment would acknowledge that homeostasis can incorporate changing setpoints, and that what differs about many of the newer models is their more explicit focus on them.

Reactive versus Anticipatory

Homeostatic systems are reactive in that they measure deviations from setpoints and initiate counteracting physiological work. Similarly, enantiostasis and proteostasis are largely reactive, though not as explicitly. There are, however, many circumstances in which setpoints are modified in anticipation of future events or conditions – for example, premigratory fattening in birds (King & Farner 1963), cycles of weight gain and loss in hibernating mammals (Barnes & Mrosovsky 1974), and various animal anorexias (Mrosovsky & Sherry 1980). Of the stasis concepts besides homeostasis, rheostasis is most clearly focused on anticipation, though the reactive scope, allostasis, and PRN concepts incorporate it as well. Homeostatic noise is not explicitly concerned with anticipation.

Simple versus Complex (or Low-level versus High-level Biological Organization)

Homeostasis provides simple regulatory models. The typical textbook system is stripped to its essentials, containing one or a few sensors and integrators, some control signals, and the membranes, tissues, or organs doing the physiological work. Such idealized feedback loops are fundamental units of physiology, and recognizing and describing them was a major triumph for 19th- and 20th-century physiology (Bernard 1865, 1878; Cannon 1932; Cooper 2008).

The other stasis concepts contain additional complexity aimed at extending the functionality of homeostasis. For example, enantiostasis requires simultaneous control of two or more offsetting factors (e.g., urea and TMAO), which themselves interact with each other (to give stable enzyme kinetics). Rheostasis, allostasis, and reactive scope incorporate the idea that organisms can change setpoints of factors, which implies more complicated, higher-level structures for controlling the lower-level systems. The physiological regulatory networks model aims explicitly at these high-level structures, describing the control networks that communicate with, and coordinate the actions of, multiple lower-level networks. The networked linkages in PRNs have consequences both for the functions of integrated physiological systems in organisms and for the evolution of physiological systems – especially as facilitated or constrained by pleiotropy among subsystems (see also Ketterson & Nolan Jr 1999).

Homeostatic noise invokes a different kind of modest complexity – linkages among systems mediated by physiological noise (Woods & Wilson 2013). In particular, Woods and Wilson argue that homeostatic systems are important precisely because they control factors whose variation has universal effects on other aspects of organismal biochemistry, i.e., that homeostatic systems are linked by a kind of diffuse background noise that they collectively create. For example, pH has universal effects on functional molecules in organisms, especially enzymes, receptors, and RNAs. Poor pH homeostasis thus can have cascading effects on the homeostatic abilities of other physiological systems, because those other systems also are constructed from molecules whose functions depend, inescapably, on pH.

Short versus Long Time Scales

Many of the lower-level concepts (homeostasis, enantiostasis, proteostasis, homeostatic noise) can refer to processes occurring on short time scales – here and now, with immediate consequences for the performance of the organism – although they can also focus on long-term effects of dysregulation (e.g., degenerative diseases stemming from poor proteostasis (Balch *et al.* 2008)). Other concepts mostly play out over longer times, especially those that focus on mechanisms used to anticipate future conditions (rheostasis, allostasis, reactive scope, and PRNs). Because of its importance, we return to the issue of time scale below.

CONSTRUCTING A CONCEPTUAL ELEPHANT

Our central argument: each of the eight stasis concepts is a context-specific version of a single, broad phenomenon. The individual concepts differ in whether they focus more on stability or change, whether they look forward or backward in time, whether they are simple or complex, whether they work over short time scales or long, and whether they invoke lower or higher levels of biological organization. Nevertheless, all of the concepts describe how organisms simultaneously stabilize and modify their physiological phenotypes in response to changing internal states and external environments. Although this conclusion may seem simple – even self-evident in retrospect – the various "stasis communities" do not cross-pollinate frequently and have failed so far to construct any unified vision of physiological stasis and change.

We suggest that reciprocal coupling between stability and change is a universal aspect of organismal physiology. In one direction, such coupling means that stability depends on change. This relationship has already been claimed for allostasis by McEwan and Wingfield (McEwen & Wingfield 2003), which they characterize as "stability through change," and for the process of developmental canalization (Flatt 2005). We propose that this phrase in fact applies to all stasis concepts. The environment around any organism changes constantly. In the absence of active control by the organism, physiological stability of particular factors would be extremely unlikely. Thus, there must be active control. But active control in turn requires flexibility and responsiveness to external (and other internal) conditions, and well-timed modifications of fluxes and concentrations in internal spaces. In other words, the only way for an organism to stabilize a physiological trait is by altering other, related traits.

The converse – "change through stability" – emerges from the concept of homeostatic noise (Woods & Wilson 2013). The authors proposed that one function of homeostatic systems is to minimize variation in a set of key factors that, if poorly controlled, would constitute a kind of global physiological noise that disrupts both communication through other physiological channels and homeostatic regulation by other feedback loops. Organisms that successfully stabilize those factors provide quiet physiological backgrounds against which environmental information is processed more efficiently and interpreted better, and in which the multitude of interacting physiological systems, such as those envisioned in PRNs (Cohen *et al.* 2012), can communicate effectively with one another. This idea can be thought of as the physiological equivalent of developmental stability: much as core parts of developmental systems need to be highly stereotyped against environmental or genetic variation in order for subtler forms of plasticity to arise in appropriate ways, physiological systems work best in the absence of distorting forms of physiological noise. In this sense, organisms can make beneficial changes to their setpoints (change) only when they exhibit stability in the basic physiological factors interacting with all other intra-organismal information processing.

What to call the single, broad concept? We're reluctant to coin a new term, or set of terms, in a field already suffering from a glut (see Dallman 2003). A more productive way forward is to recognize that the stasis concepts are physiology-specific versions of the broader phenomena of plasticity and canalization (see Chapters 1–5), and also are related to the concept of phenotypic flexibility.

Phenotypic plasticity refers to single genotypes producing multiple phenotypes across sets of environments (Bradshaw 1965; Pigliucci 2001; West-Eberhard 2003; see earlier chapters in this book). Studies of plasticity often focus on irreversible developmental plasticity. Plasticity can also refer to reversible, rapid physiological change. This kind of plasticity has been called "phenotypic flexibility" by Piersma and colleagues (Piersma & Lindström 1997; Piersma & Drent 2003; Piersma & van Gils 2010), which they define as "reversible within-individual variation, [as] a function of environmental conditions varying predictably (e.g., with season), or of more stochastic fluctuations in the environment" (Piersma & Drent 2003). In the stasis concepts described above, most change in physiological setpoints, levels, signals, and fluxes can be reasonably described as phenotypic flexibility. Other kinds of change are best described as physiological instances of developmental plasticity, such as the long-term effects of allostatic load analyzed by allostasis and the reactive scope model.

More broadly, physiological stasis is a kind of short-term, physiological canalization. The idea of canalization traces to Waddington's work on developmental stability (Waddington 1942, 1953), and the word and concept have been contentious because they have come to mean a number of different things (Debat & David 2001; Nijhout & Davidowitz 2003; Dworkin 2005). In addition, canalization can be conceptually partitioned into different types, for example environmental versus genetic (Wagner *et al.* 1997). Here, we use Flatt's relatively broad definition (Flatt 2005: 290): "Canalization is the reduced sensitivity of a phenotype to changes or perturbations in the underlying genetic and nongenetic factors that determine its expression." In this sense, physiological stasis is to canalization as phenotypic flexibility is to phenotypic plasticity – i.e., homeostasis and phenotypic flexibility both focus on shorter-term, more reversible, physiological processes. Framing stasis this way makes available, for physiological studies, the substantial body of theoretical work already done on the consequences of canalization for the evolution of phenotypes (Waddington 1953; West-Eberhard 2003), and of robustness (de Visser *et al.* 2003).

UNIFICATION DIFFUSES CONCEPTS AND TOOLS

Placing the stasis concepts firmly under the umbrella of plasticity, flexibility, and canalization provides novel perspectives on the functions and evolution of physiological systems. We discuss four examples below.

Relating Physiological Stasis to Fitness

The casual reader, dipping into the (large) literature on stasis concepts, would conclude that the core purpose of physiological systems is to provide some kind of stability.

The problem is to identify the stabilized thing. Stability sometimes refers to measurable physiological parameters, such as concentrations or fluxes of metabolites (as in homeostasis and homeokinesis, respectively). But stability is ill-defined in other concepts. For example, allostasis can be characterized as "stability through change." "Change" has a straightforward meaning – alteration to subsystems by the higher-level structures responsible for integration and control. But what does "stability" mean? We gather that it is something like "organismal integrity." If so, a better approach would be to invoke fitness directly (Sterling 2004, 2012). Fitness, not stability, is the most important currency for understanding the processes shaping physiological systems over evolutionary time.

The broader point here is that the different models focus on different metrics of good and bad regulation. The older models tend to be concerned with physiological mechanisms and low-level integration of simple systems; for these models, actual stability in one or more physiological factors may indeed be an important criterion. But in the more complex models, which attempt to explain integration of multiple systems and, especially, the evolution of those systems, the currency of stability rapidly becomes untenable. In these cases, the models should focus instead on fitness. This is a broadening of the argument made already for allostasis, by Sterling (Sterling 2004, 2012). There is, of course, a middle ground, which is to understand how stability and change in individual factors is related to performance and fitness. Analyzing this middle ground will require both the mechanistic expertise to measure and manipulate components and the ecological and evolutionary expertise to evaluate the effects on fitness.

Extensible, Overlapping Time Scales

Above, we discussed briefly how the eight stasis concepts can be arrayed by time scale. In fact, this point deserves more attention, and we revisit it now that we have linked the stasis concepts more formally to plasticity and canalization. The eight stasis concepts can be arrayed by the time scale to which they are most pertinent; the plasticity concepts fall onto this same scale (Figure 8.2).

Homeostasis covers processes primarily in the range of one second to one day, and generally describes the fastest stasis events in organisms. The next three (enantiostasis, proteostasis, and rheostasis) cover slightly longer time frames, because they require coordination of rapid-acting but separate components and thus cannot operate at the fastest scales. Although most of the mechanisms underlying proteostasis occur at short time scales, the idea is explicitly concerned with relationships between the poor health of protein pools and end-of-life diseases. Additionally, rheostasis covers longer time scales because it includes physiological regulation that anticipates the future (particularly cycles of programmed physiological change over the course of an organism's lifetime). Allostasis, reactive scope, and physiological regulatory networks occupy another middle range that is shifted to still longer time scales, reflecting both that they invoke more complex networks of interacting parts and that they predict later-life dysfunction when early regulation is disturbed beyond some bounds. Phenotypic flexibility is concerned with longer-scale but reversible physiological remodeling, e.g., changes in organ size. Thus, phenotypic flexibility generally requires a few days to more than a month.

Finally, there are the longer-term processes of phenotypic developmental plasticity and predictive adaptive responses. Developmental plasticity is the archetypal form, in which genotypes interact with environments during development to produce phenotypes (Pigliucci 2001). Because the phenotype arises during development, developmental plasticity operates over significantly longer time scales. Predictive adaptive response

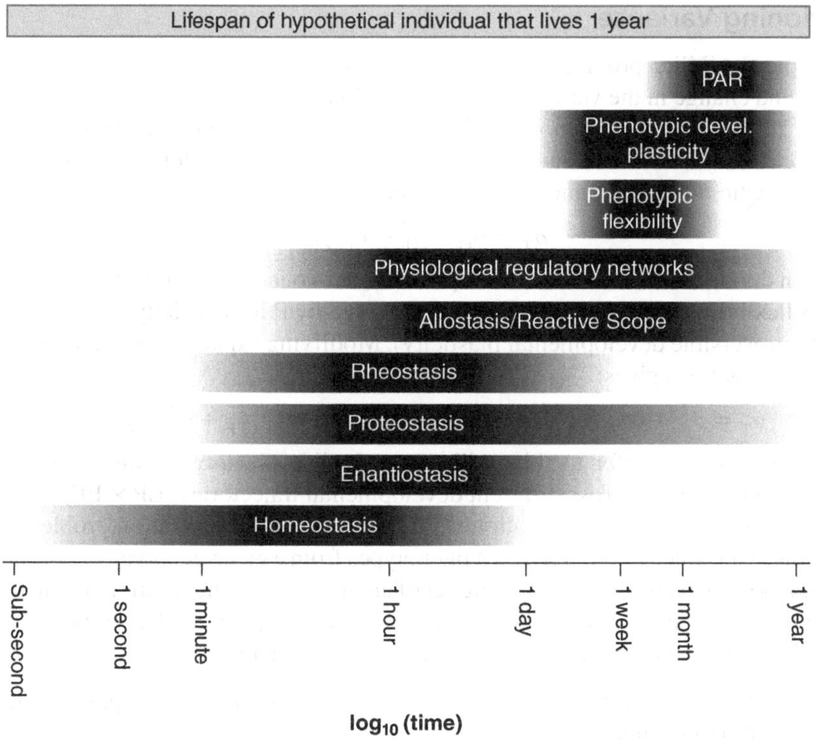

Figure 8.2. Time scales of phenotypic modification. To illustrate time scales over which the different stasis and plasticity concepts operate, we map the concepts onto the lifespan of an organism that lives for one year (and is evolved to have a one-year life span; i.e., it didn't die unexpectedly early). It is reasonable to include both stasis and plasticity concepts on the same graph because, as we argue in the text, in all cases stasis and plasticity are reciprocally dependent – they require each other. In the plot, the x-axis is on a log scale, which expands short times and compresses long ones. Note that the x-axis functions in two ways. First, it maps out the lifespan of the organism. Second, it provides the time scale over which the stasis and plasticity concepts are relevant. In other words, the graph is not meant to suggest that the organism engages in homeostasis only for the first hour of its life – rather, that homeostatic processes operate, throughout the organism's life, over time scales of several hours or less. The ranges for each concept are approximate, and the reader probably can think of exceptions for each one; nevertheless, the time scales of primary concern for each concept are in the right range.

(PARs) (Gluckman *et al.* 2005) are a form of developmental plasticity that emphasizes (i) the adaptive value of phenotypic changes, and (ii) the idea that although costs may be paid early, the payoff (adaptive benefit) may not appear until late in life. In principle, we could add even longer-term processes to Figure 8.2, such as cross-generational effects arising from maternal investment or epigenetic legacies (see Chapter 7, this volume). To keep the scope reasonable, we refrain.

What does this exercise in time scales provide? It reinforces the idea that the unified concept proposed above is real. In effect, all of the stasis and plasticity concepts reflect a single phenomenon: the process of genotypes, residing in individual organisms, altering the phenotypes of their carriers in response to, or in anticipation of, changes in their environments.

Partitioning Variance

Phenotypic flexibility provides conceptual tools that could help to analyze problems of stability and change in the various stasis models. In particular, Piersma and Drent (Piersma & Drent 2003) show how phenotypic flexibility can be analyzed in a quantitative genetics framework. They start with a standard equation for decomposing total phenotypic variance (T) into genetic (G), environmental (E), and interactions (G × E), plus error (e)

$$\sigma_{\mathrm{T}}^2 = \sigma_{\mathrm{G}}^2 + \sigma_{\mathrm{E}}^2 + \sigma_{\mathrm{G\times E}}^2 + e. \tag{8.1}$$

They then divide the environmental component into two factors – a reversible part (denoted by F for flexibility) and an irreversible part (denoted by P for plasticity, which in their case refers to irreversible developmental plasticity). Modifying Equation (8.1) and including all possible interactions gives:

$$\sigma_{\mathrm{T}}^2 = \sigma_{\mathrm{G}}^2 + \sigma_{\mathrm{EP}}^2 + \sigma_{\mathrm{EF}}^2 + \sigma_{\mathrm{G\times EP}}^2 + \sigma_{\mathrm{G\times EF}}^2 + \sigma_{\mathrm{EP\times EF}}^2 + \sigma_{\mathrm{G\times EP\times EF}}^2 + e \tag{8.2}$$

Here the interactions refer to things like how flexible phenotypes vary across different genotypes (G × EF) or across different developmental trajectories (EP × EF).

This framework provides a path toward analyzing the optimization problem faced by organisms: to produce a complex set of phenotypes from a given genotype that is interacting with a complex and constantly changing set of environments. The solution, for an organism, is to modify its phenotype. Other than G and e (which have standard definitions), the kinds of contributions to its current phenotype state can be envisioned as:

EP = phenotypes reflecting fixed effects from prior environments; the part not influenced by variation in genotype;

EF = phenotypes reflecting reversible physiological state in response to current environment;

G × EP = standard developmental plasticity, or the combined effects of the organism's genotype interacting with its prior environments,

G × EF = genotype-dependent reversible physiological plasticity in response to current environments;

EP × EF = the influence of developmental plasticity on short-term reversible plasticity; although this could also mean the converse (effects of short-term reversible plasticity on developmental plasticity), in practice we think the first influence will be stronger in the first direction;

G × EP × EF = genotype-dependent effects of developmental plasticity on short-term reversible plasticity.

This framework is general enough to handle the difference between reacting to prior or current environments versus anticipating future environments. Moreover, it provides a compact, quantitative way to analyze the long-term effects of different kinds of stresses on short-term physiological changes. For example, the reactive scope model is concerned in part with long-term (developmental) effects of early stress on later reactivity of the glucocorticoid-mediated stress response. This would be variation that is included in EP × EF or, if one is interested in different genotypes, in G × EP × EF.

COSTS AND CONSTRAINTS

The literature on plasticity also provides key ideas about costs and constraints, which are largely missing from current discussions of stasis. This is true of simple homeostatic models, which are often analyzed without reference to costs (true too of most discussions of enantiostasis, proteostasis, and rheostasis). Clearly, however, costs could be significant, and could influence the evolution of homeostasis. For example, large amounts, or high rates, of physiological work done by tissues or organs may incur significant energetic costs.

Potential costs are more explicitly addressed by allostasis and the reactive scope model. First, it takes energy to activate and control homeostatic systems, and energy might be in short supply. For example, when food availability was reduced, white-crowned sparrows (Zonotrichia leucophrys pugetensis) lost body mass and suppressed reproduction (Wingfield *et al.* 1983). This result indicates that maintaining energy homeostasis, which is undoubtedly important to surviving food shortage, exacts costs from other aspects of fitness. More generally, short-term activation of regulatory systems can stress other systems in organisms, which may select for rapid deactivation once the disturbance has passed (Romero *et al.* 2009). Finally, there may be long-term detrimental effects associated with upregulating systems either early in development or for too long or too frequently later in the life cycle. For example, elevated stress hormones during early development can depress condition later in life (Nowicki *et al.* 2002; Spencer *et al.* 2003).

At a higher organizational level, costs and constraints arise from the interconnectedness of regulatory systems (Ketterson *et al.* 2009). Considering individual organisms, both allostasis and the reactive scope model focus on the effects of activating regulatory systems either for much longer or with much greater frequency than is typical – and they conclude that such activation can cause serious pathology in other regulatory systems. In effect, the pathologies (costs) arise when interconnected sets of regulatory systems encounter conditions that are statistically rare (see Gluckman *et al.* 2005).

Finally, costs and constraints affect the evolutionary trajectories of lineages, a possibility that is most explicitly analyzed by physiological regulatory networks (Martin *et al.* 2011; Cohen *et al.* 2012). The idea is that the organization of physiological systems, into subnetworks controlled by integrator molecules (which can be thought of as master-control knobs that simultaneously affect the functions of multiple parts), constrains both the kinds of phenotypic variation generated and the evolutionary trajectories of evolving populations (Ketterson & Nolan 1999; Gerhart & Kirschner 2007; Cohen *et al.* 2012).

Thus, although it is clear that many of the stasis concepts invoke costs, there has been no general treatment of the nature and magnitude of those costs. By contrast, costs and constraints have played a key role for many years in analyses of plasticity (Dewitt *et al.* 1998; Auld *et al.* 2010). We suggest that any study discussing the integration or evolution of stasis systems will need to consider the costs and constraints carefully and would profit from using the frameworks already defined by work on developmental plasticity.

CONCLUSION

In science, whatever the field, there is often tension between "lumpers" and "splitters" (Dobzhansky 1964). The lumpers are the practitioners who search for simplicity, who focus on large-scale patterns, who sometimes ignore exceptions, at least when they're not too common or extreme. The splitters are those who revel in diversity, who create new categories to describe novel variants, and who deal in complexity. In biology, several subfields have lumper-splitter schisms (e.g., phylogenetics, allometry). Often the differences in approach reflect intentional lumping and splitting, in which practitioners of both approaches are at least aware of the other point of view. There is, however, a second kind of lumping-vs-splitting that arises from incomplete knowledge. A well-known allegory for this situation: the group of blind people not deducing the fact of the elephant from their individual, restricted observations of it. Each person – our biologists – comes away with his or her own theory about what it is, and the theories turn out to be profoundly different, based on the different data each one obtained from contact with the elephant. Many areas of biology surely are still in this state.

Apparently we (the authors) are lumpers. Moreover, we've seen an elephant in the fog – an elephant that turns out to be a set of physiological instances of canalization and plasticity. The diversification of stasis concepts is understandable, and perhaps even desirable in the sense that the concepts each identify a part of some conceptual whole. Nevertheless, there is a real risk of making the field more complicated than it needs to be. We prefer parsimony and a focus on the ideas that unify threads.

A strength of our view is that it simplifies and clarifies terminology, which is no small thing. Two additional strengths are perhaps more important. First, by linking subfields, conceptual unification facilitates the transfer of ideas from one part to another, which should benefit all fields by diffusing good ideas widely. In the case of phenotype-environment matching, unification provides clear links between the various stasis concepts and better-developed theoretical bases underlying the concepts of plasticity and canalization, and their evolution. Second, unification will put the stasis concepts on a more stable evolutionary footing. Collectively, these shifts will bring the various physiological models firmly into the mainstream.

ACKNOWLEDGMENTS

We thank Creagh Breuner, Cameron Ghalambor, Ellen Ketterson, Marty Martin, Trish Schulte, Raul Suarez, and Bertanne Visser for discussion and comments on previous drafts. This work was supported by NSF grant IOS-0844916 to HAW. HAW also thanks the faculty and staff of l'Institut de Recherche sur la Biologie de l'Insecte, Université François Rabelais, Tours, France, for supporting a sabbatical stay, during which this chapter was written.

REFERENCES

Abdel-Sater, K. (2011) Physiological positive feedback mechanisms. *Am J Biomed Sci*, **3**, 145–55.

Auld, J.R., Agrawal, A.A. and Relyea, R.A. (2010) Re-evaluating the costs and limits of adaptive phenotypic plasticity. *Proceedings of the Royal Society B: Biological Sciences*, **277**, 503–11.

Balch, W.E., Morimoto, R.I., Dillin, A. and Kelly, J.W. (2008) Adapting proteostasis for disease intervention. *Science*, **319**, 916–19.

Barnes, D.S. and Mrosovsky, N. (1974) Body weight regulation in ground squirrels and hypothalamically lesioned rats: slow and sudden set point changes. *Physiology & Behavior*, **12**, 251–8.

Bernard, C. (1865) *Introduction à l'Etude de La Médicine Expérimentale*, Ballière, Paris.

Bernard, C. (1878) *Leçons Sur Les Phénomènes de La Vie Communs Aux Animaux et Aux Végétaux*, Baillière, Paris.

Bradshaw, A. (1965) Evolutionary significance of phenotypic plasticity in plants. *Advances in Genetics*, **13**, 115–55.

Caldji, C., Liu, D., Sharma, S. *et al.* (2001) Development of individual differences in behavioral and endocrine responses to stress: role of the postnatal environment, in *Handbook of Physiology* (eds B. McEwen and H. Goodman), Oxford University Press, New York, pp. 271–92.

Cannon, W. (1929) Organization for physiological homeostasis. *Physiological Reviews*, **9**, 399–431.

Cannon, W. (1932) *The Wisdom of the Body*, WW Norton, New York.

Cohen, A.A., Martin, L.B., Wingfield, J.C. *et al.* (2012) Physiological regulatory networks: ecological roles and evolutionary constraints. *Trends in Ecology & Evolution*, **27**, 428–35.

Cooper, S.J. (2008) From Claude Bernard to Walter Cannon. Emergence of the concept of homeostasis. *Appetite*, **51**, 419–27.

Dallman, M.F. (2003) Stress by any other name ... ? *Hormones and Behavior*, **43**, 18–20.

Debat, V. and David, P. (2001) Mapping phenotypes: canalization, plasticity and developmental stability. *Trends in Ecology & Evolution*, **16**, 555–61.

Dewitt, T.J., Sih, A. and Wilson, D.S. (1998) Costs and limits of phenotypic plasticity. *Trends in Ecology & Evolution*, **13**, 77–81.

Dobzhansky, T. (1964) Biology, molecular and organismic. *American Zoologist*, 443–52.

Dworkin, I. (2005) A study of canalization and developmental stability in the sternopleural bristle system of Drosophila melanogaster. *Evolution*, **59**, 1500–9.

Flatt, T. (2005) The evolutionary genetics of canalization. *The Quarterly Review of Biology*, **80**, 287–316.

Gerhart, J. and Kirschner, M. (2007) The theory of facilitated variation. *Proceedings of the National Academy of Sciences of the United States of America*, **104** (Suppl 1), 8582–9.

Ghalambor, C.K., McKay, J.K., Carroll, S.P. and Reznick, D.N. (2007) Adaptive versus non-adaptive phenotypic plasticity and the potential for contemporary adaptation in new environments. *Functional Ecology*, **21**, 394–407.

Gluckman, P.D., Hanson, M.A. and Spencer, H.G. (2005) Predictive adaptive responses and human evolution. *Trends in Ecology & Evolution*, **20**, 527–33.

Guyton, A. and Hall, J. (2010) *Textbook of Medical Physiology*, 12th edn, Saunders, Philadelphia.

Hill, R., Wyse, G. and Anderson, M. (2013) *Animal Physiology*, 3rd edn, Sinauer, Sunderland, MA.

Hochachka, P. and Somero, G. (2001) *Biochemical Adaptation: Mechanism and Process in Physiological Evolution*, Oxford University Press, Oxford.

Ketterson, E.D., Atwell, J.W. and McGlothlin, J.W. (2009) Phenotypic integration and independence: Hormones, performance, and response to environmental change. *Integrative and Comparative Biology*, **49**, 365–79.

Ketterson, E.D. and Nolan, V. Jr. (1999) Adaptation, exaptation, and constraint: a hormonal perspective. *American Naturalist*, **154**, S4–S25.

King, J. and Farner, D. (1963) The relationship of fat deposition to zugunruhe and migration. *The Condor*, **65**, 200–23.

Koolhaas, J.M., Bartolomucci, A., Buwalda, B. *et al.* (2011) Stress revisited: a critical evaluation of the stress concept. *Neuroscience and Biobehavioral Reviews*, **35**, 1291–1301.

Mangum, C. and Towle, D. (1977) Physiological adaptation to unstable environments: inconstancy of the internal milieu in an animal may be a regulatory mechanism. *American Scientist*, **65**, 67–75.

Martin, L.B., Liebl, A.L., Trotter, J.H. *et al.* (2011) Integrator networks: illuminating the black box linking genotype and phenotype. *Integrative and Comparative Biology*, **51**, 514–27.

McEwen, B.S. (1998) Stress, adaptation, and disease: allostasis and allostatic load. *Annals of the New York Academy of Sciences*, **840**, 33–44.

McEwen, B.S. and Wingfield, J.C. (2003) The concept of allostasis in biology and biomedicine. *Hormones and Behavior*, **43**, 2–15.

McEwen, B.S. and Wingfield, J.C. (2010) What's in a name? Integrating homeostasis, allostasis and stress. *Hormones and Behavior*, **57**, 1–16.

McGlothlin, J.W. & Ketterson, E.D. (2008) Hormone-mediated suites as adaptations and evolutionary constraints. *Philos Trans R Soc Lond B Biol Sci.*, **363**, 1611–20.

Mrosovsky, N. (1990) *Rheostasis: The Physiology of Change*, Oxford University Press, Oxford.

Mrosovsky, N. and Sherry, D. (1980) Animal anorexias. *Science*, **207**, 837–42.

Nijhout, H.F. and Davidowitz, G. (2003) Developmental perspectives on phenotypic variation, canalization, and fluctuating asymmetry, in *Developmental Instability: Causes and Consequences* (ed M. Polak), Oxford University Press, Oxford, pp. 3–13.

Nowicki, S., Searcy, W.A. and Peters, S. (2002) Brain development, song learning and mate choice in birds: a review and experimental test of the "nutritional stress hypothesis.". *Journal of Comparative Physiology. A, Neuroethology, Sensory, Neural, and Behavioral Physiology*, **188**, 1003–14.

Piersma, T. and Drent, J. (2003) Phenotypic flexibility and the evolution of organismal design. *Trends in Ecology & Evolution*, **18**, 228–33.

Piersma, T. and van Gils, J. (2010) *The Flexible Phenotype: a Body-centered Integration of Ecology, Physiology, and Behaviour*, Oxford University Press, Oxford.

Piersma, T. and Lindström, A. (1997) Rapid reversible changes in organ size as a component of adaptive behaviour. *Trends in Ecology and Evolution*, **12**, 134–8.

Pigliucci, M. (2001) *Phenotypic Plasticity: Beyond Nature and Nurture*, Johns Hopkins University Press.

Powers, D. and Schulte, P. (1998) Evolutionary adaptations of gene structure and expression in natural populations in relation to a changing environment: a multidisciplinary approach to address the million-year saga of a small fish. *Journal of Experimental Zoology*, **282**, 71–94.

Romero, L.M., Dickens, M.J. and Cyr, N.E. (2009) The Reactive Scope Model – a new model integrating homeostasis, allostasis, and stress. *Hormones and Behavior*, **55**, 375–89.

Rutter, G. (1999) Insulin secretion: feed-forward control of insulin biosynthesis? *Current Biology*, **9**, R443–R445.

Schulkin, J. (2011) Social allostasis: anticipatory regulation of the internal milieu. *Frontiers in Evolutionary Neuroscience*, **2**, 111.

Schulkin, J., McEwen, B.S. and Gold, P.W. (1994) Allostasis, amygdala, and anticipatory angst. *Neuroscience and Biobehavioral Reviews*, **18**, 385–96.

Schulte, P.M. (2014) What is environmental stress? Insights from fish living in a variable environment. *The Journal of Experimental Biology*, **217**, 23–34.

Shannon, C. (1948) A mathematical theory of communication. *The Bell System Technical Journal*, **27**, 379–423, 623–56.

Spencer, K.A., Buchanan, K.L., Goldsmith, A.R. and Catchpole, C.K. (2003) Song as an honest signal of developmental stress in the zebra finch (*Taeniopygia guttata*). *Hormones and Behavior*, **44**, 132–9.

Sterling, P. (2004) Principles of allostasis: optimal design, predictive regulation, pathophysiology, and rational therapeutics, in *Allostasis, Homeostasis, and the Costs of Adaptation* (ed J. Schulkin), Cambridge University Press, Cambridge, pp. 1–37.

Sterling, P. (2012) Allostasis: a model of predictive regulation. *Physiology & Behavior*, **106**, 5–15.

Sterling, P. and Eyer, J. (1988) Allostasis: A new paradigm to explain arousal pathology, in *Handbook of Life Stress, Cognition and Health* (eds S. Fisher and J. Reason), John Wiley & Sons, pp. 629–49.

Viney, M. and Reece, S.E. (2013) Adaptive noise. *Proceedings of the Royal Society B: Biological Sciences*, **280**, 20131104.

De Visser, J.A.G.M., Hermisson, J., Wagner, G.P. *et al.* (2003) Perspective: Evolution and detection of genetic robustness. *Evolution*, **57**, 1959–72.

Waddington, C. (1942) Canalization of development and the inheritance of acquired characters. *Nature*, **150**, 563–5.

Waddington, C. (1953) The genetic assimilation of an acquired character. *Evolution*, **7**, 118–26.

Wagner, G., Booth, G. and Bagheri-Chaichian, H. (1997) A population genetic theory of canalization. *Evolution*, **51**, 329–47.

Walsberg, G.E. (2003) How useful is energy balance as a overall index of stress in animals? *Hormones and Behavior*, **43**, 16–17.

West-Eberhard, M. (2003) *Developmental Plasticity and Evolution*, Oxford University Press, Oxford.

Wiener, N. (1948) *Cybernetics*, MIT Press, Cambridge, MA.

Wingfield, J., Moore, M. and Farner, D. (1983) Endocrine responses to inclement weather in naturally breeding populations of white-crowned sparrows (*Zonotrichia leucophrys pugetensis*). *The Auk*, **100**, 56–62.

Woods, H.A. (2009) Evolution of homeostatic physiological systems, in *Phenotypic Plasticity of Insects: Mechanisms and Consequences* (eds D. Whitman and T. Ananthakrishnan), Science Publishers, Enfield, NH.

Woods, H.A. and Wilson, J.K. (2013) An information hypothesis for the evolution of homeostasis. *Trends in Ecology & Evolution*, **28**, 283–9.

Woods, H.A. (2014) Mosaic physiology from developmental noise: within-organism physiological diversity as an alternative to phenotypic plasticity and phenotypic flexibility. *The Journal of Experimental Biology*, **217**, 35–45.

9

PHYSIOLOGICAL REGULATORY NETWORKS: THE ORCHESTRA OF LIFE?

Lynn B. Martin[1] and Alan A. Cohen[2]

[1]*Department of Integrative Biology, University of South Florida, Tampa FL, USA*
[2]*Groupe de recherche PRIMUS, Department of Family Medicine, University of Sherbrooke, CHUS-Fleurimont, Sherbrooke, QC, Canada*

"Tangled intricacy is the mother of nature's invention."

– *Denis Noble*

INTRODUCTION

Linear thinking and reductionism have been and will continue to be valuable in physiology, but there is a growing appreciation that much in biology depends on interactions (Chapter 6, 11, & 18, this volume). To date, the study of interactions in physiology has been restricted to single functions or systems; immunologists, for example, focus on how various cells and proteins control infections and heal wounds, and reproductive physiologists study how the brain and gonads are coordinated to achieve fertility via a few hormones. In general, cross-talk between these and other systems has been little considered. In this chapter, we advocate an explicit focus on physiological interactions at the scale of the whole individual. Our point of view emerges from advances in other disciplines focusing on interactions among physiological systems including psychoneuroimmunology (Ader & Cohen 1975; Glaser 2005; Segerstrom 2007), behavioral endocrinology (Ricklefs & Wikelski 2002; Lessells 2008; McGlothlin & Ketterson 2008), and ecological immunology (Folstad & Karter 1992; Martin *et al.* 2008; Martin 2009). These disciplines have

Integrative Organismal Biology, First Edition.
Edited by Lynn B. Martin, Cameron K. Ghalambor, and H. Arthur Woods.

demonstrated that physiological elements conventionally attributed to one process (e.g., growth, immunity, metabolism) actually work pleiotropically on many endpoints.

Whereas the discovery of linkages among physiological systems has both biomedical and evolutionary implications, much of physiology lacks grounding in network theory, the set of mathematical methods developed to describe and understand complex interactions. Network theory underpins most other biological research devoted to interactions such as systems biology, developmental genetics, and community ecology. In these disciplines, network thinking has led to insights not provided by reductionism or even by studies of simple interactions. Key examples include gene epistasis and its effects on phenotypic variation (Wagner 2007; Natarajan *et al.* 2013), and species interactions and community productivity (Tilman *et al.* 2001). We argue below that it is time to develop a whole-organism network theory, as such theory might enhance how we understand important phenomena such as disease and aging but also the ecology and evolution of organisms. Below, we describe how the study of such physiological regulatory networks (Cohen *et al.* 2012) might take place.

WHAT IS A PHYSIOLOGICAL REGULATORY NETWORK (PRN)?

Recently, we proposed two concepts regarding the interconnectedness of physiological systems: the Integrator Network (IN; Martin *et al.* 2011) and Physiological Regulatory Networks (PRNs; Cohen *et al.* 2012). Both concepts emphasize that extensive connections exist among physiological processes, and that these linkages have many consequences. As an IN is an aspect of a PRN, we focus on PRNs in this chapter. A PRN is the full suite of molecules and their regulatory relationships that mediate homeostasis and thus organismal performance (Figure 9.1; Chapter 8, this volume). Because all regulatory molecules within an individual are in some manner connected, each individual has one *PRN structure* but many *PRN states*. As with other networks, PRNs have *nodes* and *edges* (Figure 9.2; Wagner *et al.* 2007). For simplicity, we define PRN nodes as circulating molecules or their functional complexes, a definition that facilitates discussion of edges below. The precise definition of a PRN node will depend on the goal of a given study and could include: (a) the circulating concentrations of a hormone or other molecule; (b) the receptors and carrier molecules for such substances; (c) the gene response elements for hormone-receptor complexes; or (d) some combination of all three. As in gene networks, we define PRN edges as co-regulatory patterns of nodes (Wagner *et al.* 2007); in other words, a PRN edge is the magnitude and direction of change in other nodes when a particular PRN node is perturbed. There are limitations to these definitions, as there are for all complex concepts attempting to explain links between genotype to phenotype. However, as even knocking out a gene can only tell us that that gene has some impact on a phenotype (Noble 2006), we feel that the PRN concept theory will be useful in the short term. As others have emphasized recently (Wingfield 2013), the way forward is to define PRNs so that they are suited to a given context. For most of what we expect to learn from PRNs, node values are unlikely to be as important as PRN structures and states (Cohen *et al.* 2012).

We define *PRN structure* as the network architecture of all regulatory molecules in an individual including potential edges among nodes. We expect that PRN structures resemble genomes in regards to their impacts on phenotypes: in most ways, they are similar among individuals within populations, but they are unique enough that selection can scrutinize them. Indeed, it is PRN structure that evolves. From system-specific physiology, we know that PRNs are comprised of *subnetworks* with particular roles; examples include cytokine cascades in the immune system, the stress responses, lipid balance mediators, coordinators

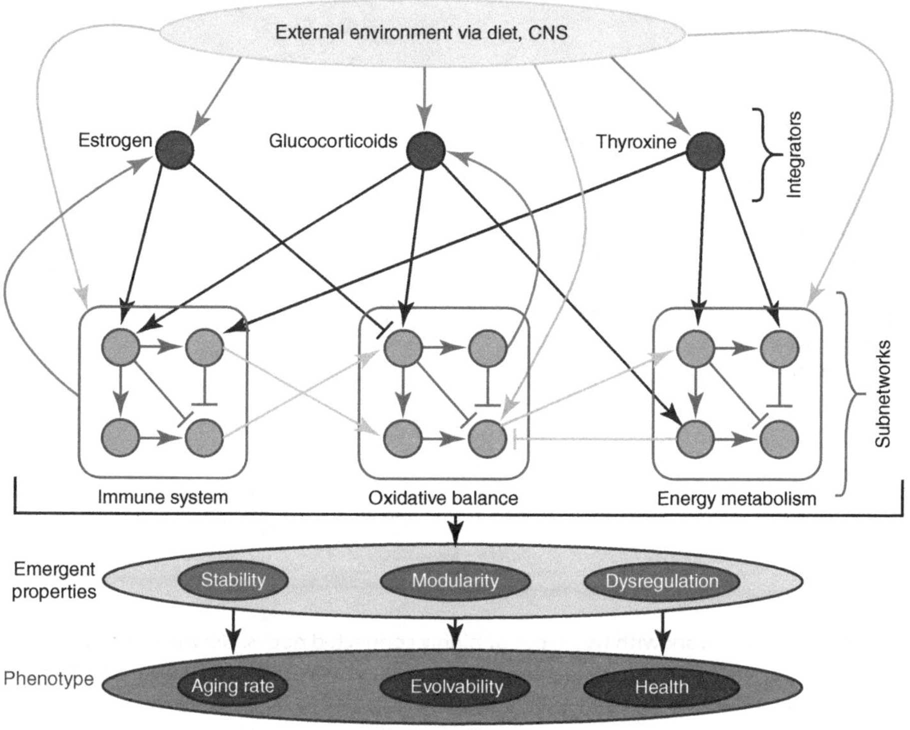

Figure 9.1. A simplified schematic of a physiological regulatory network (PRN). Light grey arrows depict direct effects of the environment on subnetworks, such as content of antioxidants in the diet. Light-blue arrows indicate direct interactions among subnetworks via party hubs (see Figure 9.2), such as immune regulation by dietary antioxidants. Dark grey arrows indicate hierarchical control within PRNs via key nodes (i.e., integrators; see Figure 9.3). Grey arrows indicate environmental regulation of integrators, usually via the central nervous system (CNS). Grey arrows indicate feedback effects of subnetworks on integrators, such as antioxidant effects on glucocorticoids. Emergent, system-level properties of the PRN (grey oval) manifest at different levels, depending on PRN state and structure (see text). Likewise, phenotypic traits (e.g. health and evolvability, respectively; grey oval) arise via individual (state) or species-level (structure) PRN configurations. Connectivity is the number of edges among nodes (total arrows), modularity is the number of possible light-blue arrows relative to all other possible arrows, and hierarchy is the distribution of arrows (edges) among nodes. As edge distributions in bionetworks tend to follow a power law and as integrators appear common in metazoans (see Martin *et al.* 2011), we expect a branching hierarchy in PRNs. *Source*: Adapted from Alan A. Cohen, Lynn B. Martin, John C. Wingfield, Scott R. McWilliams, Jennifer A. Dunne (2012) Physiological regulatory networks: ecological roles and evolutionary constraints. *Trends in Ecology & Evolution*, **28**(8), 428–35. Copyright © 2012, Elsevier. (See plate section for color version.)

of energy metabolism, factors underlying oxidative balance and others. As with other biological networks, nodes and edges are dynamic. We recognize such dynamism as changes in *PRN state*. PRN states shift to achieve stasis or function, including stasis as a function (Chapter 8, this volume). Some such changes are temporary (flexible *sensu* Piersma and Drent 2003), such as when insulin spikes after a meal to drive glucose metabolism. Other

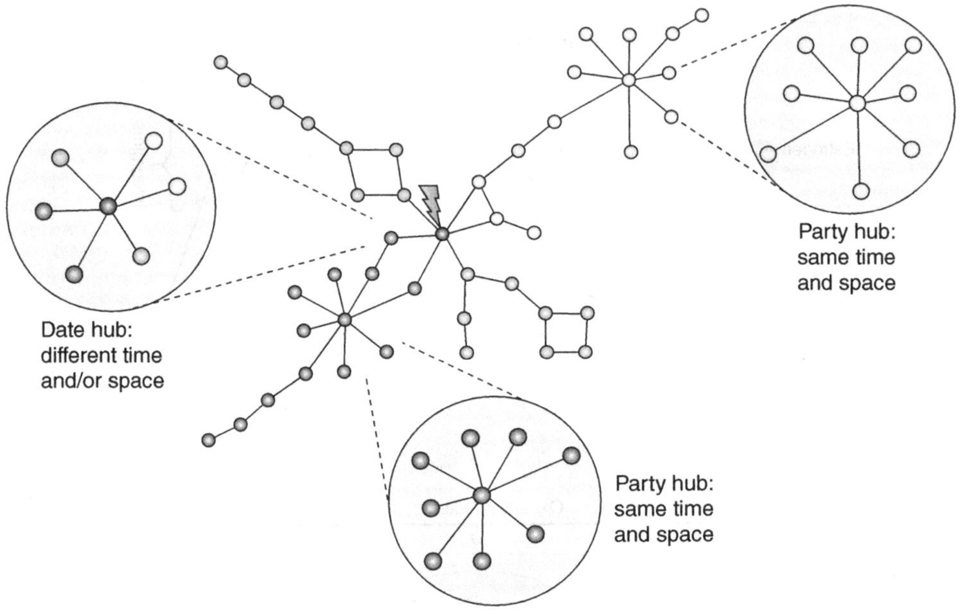

Figure 9.2. A bionetwork with two types of highly connected nodes. Network nodes are colored circles; edges are lines linking nodes. 'Party' hubs (i.e., central blue or yellow nodes in zoomed figure elements) interact simultaneously with many partners whereas 'date' hubs (i.e., red node in figure center; equivalent to integrators in Figure 9.3) interact with many other nodes at many times or locations. Colors depict subnetwork membership. Lightning bolt cartoon depicts particular environmental stimulus inducing this particular network state. *Source*: Adapted from Wagner *et al.* 2007; reprinted with permission. (See plate section for color version.)

changes are enduring, such as alterations to steroid hormone receptors organized in certain developmental windows (i.e., plasticity arising developmentally; Chapters 7, 11, & 12, this volume). Even these enduring changes in PRN state, including transgenerationally inherited epigenetic marks or other parental effects (Martin *et al.* 2011), are characteristics of a particular PRN structure; indeed, PRN structure includes plasticity and flexibility in PRN states. We expect that the evolved function of PRNs is to adjust the whole phenotype cohesively within boundaries set by prior selection as well as the environments experienced by individuals (Martin *et al.* 2011; Cohen *et al.* 2012). Across metazoans, the same modules are used to divide cells, replicate and repair DNA, metabolize energy dense substrates, and achieve other life goals (Kirschner & Gerhart 1998). We propose that PRNs facilitate phenotypic variation at the whole-organism level, much as genetic character identity networks orchestrate the development of homologous morphological traits (Wagner 2007).

KNOWN AND EXPECTED CHARACTERISTICS OF PRNs

In systems biology (Stelling *et al.* 2002) and community ecology (Dunne *et al.* 2002), it is the characteristics of networks, not of nodes or edges, that are points of focus. This observation suggests that network thinking in physiology could provide significant new insights.

In general, networks have three measureable characteristics that affect function (Lipson 2007): *connectivity*, *modularity*, and *hierarchy*. Connectivity is the number of edges within a network: a highly connected network has many edges and thus encodes more information than a less connected one. Most bionetworks tend to have intermediate connectivity, and the majority of connections occur among a few nodes. Most nodes have few connections (Milo *et al.* 2002). Such networks are advantageous because they are resilient to (random) change yet sensitive to targeted activation (Fliri *et al.* 2010). These characteristics might even promote the "weak linkages" important for facilitated variation (Kirschner & Gerhart 2010).

Intermediate connectivity also often facilitates modularity, another key characteristic of bionetworks (Von Dassow & Munro 1999). Modularity is in essence the extent to which networks are comprised of subnetworks, although definitions are diverse (Bolker 2000). Modularity in bionetworks tends to be high (Milo *et al.* 2002), perhaps because repeated use of modules enables organisms to construct complex systems from simple building blocks (Lenski *et al.* 2003). Further, compartmentalization reduces the risk of spillover of maladaptive interactions (i.e., negative pleiotropy) out of subnetworks (Wagner & Altenberg 1996; Kirschner & Gerhart 2010). For most biological networks, intermediate modularity is to be expected due to the costs of connectivity (Clune, Mouret & Lipson 2013); costs include the construction of edges or time lags and error risk in the propagation of information (Woods & Wilson 2013). Empirical studies of bacterial evolution and evolutionary simulations have shown clear advantages of modular networks, including more evolvability (Kirschner & Gerhart 2010) and lower costs of maintenance (Kashtan & Alon 2005). Modularity might also be advantageous because recombination of modules can move populations among new adaptive peaks (and out of ecological traps) efficiently (Kashtan & Alon 2005).

A third important characteristic of bionetworks is hierarchy, which can be defined as the variance in the impact of node or edge identity on whole-network performance. Because bionetwork connectivity tends to follow a power law distribution, a few nodes will act as hubs. Most known bionetworks have two types of hubs (Wagner *et al.* 2007): *party hubs*, nodes that are highly-connected *within* subnetworks, and *date hubs*, nodes that are highly-connected *among* subnetworks (Figure 9.2). In gene networks, the extensive epistatic interactions of date hubs have earned some the title of master regulatory genes (Ketterson & Nolan 1999; Segal *et al.* 2003). In food webs, some taxa act as date hubs (i.e., keystone species), strongly influencing the performance of entire communities (Mills *et al.* 1993).

Currently, PRN connectivity, modularity, and hierarchy are not well characterized, mostly because knowledge about edges tends to be biased to within-subnetwork connections. However, if PRNs mirror other bionetworks, we expect intermediate modularity and connectivity and a preponderance of party and date hubs. The prediction about intermediate modularity is reinforced by of the successes of traditional physiology; modularity must be reasonably high because single-system approaches to physiology have been productive in the past. Indeed, for most physiological processes, certain molecules are major mediators of functional outcomes, and they are locally regulated via negative and positive feedback. Within these subnetworks, party hubs likely coordinate the actions of various effector nodes (Medzhitov 2008), which alter cell functions directly (e.g., receptor ligation) and other physiological processes indirectly (e.g., interference with information transfer, oxidative damage, resource usurpation).

In terms of network hierarchy, we are less confident that a few molecules dominate network behavior, however, in many animals, especially vertebrates, a few PRN nodes seem to coordinate subnetworks toward similar goals (Chrousos 1998). Such date hubs (e.g., certain cytokines, growth hormones, the glucocorticoids and other steroids) have been termed integrators (Figure 9.3; Martin *et al.* 2011). Interleukin-1β, for instance, a major coordinator of inflammatory responses (Chapter 11, this volume), also instigates stress responses (e.g., the release of glucocorticoids) and affects various other life processes via impacts on luteinizing hormones, androgens and estrogens, and even growth hormones (Adelman & Martin 2009). The extensive effects of such integrators led many researchers to focus on the constraints these factors impose on organismal functions (Ketterson & Nolan 1999; McGlothlin & Ketterson 2008); however, as we emphasize below, these integrators might also provide viable context-dependent plasticity and flexibility to phenotypes.

ORGANISMAL IMPLICATIONS OF PRNs: STABILITY AND RESILIENCY

PRNs will be challenging and expensive to characterize, but a potential payoff is the insight provided into PRN *stability* and *resiliency*. PRN stability and/or resiliency may underlie ill-defined yet critical organismal phenomena such as health, disease, and stress. Since at least the late 1800s, concepts trying to distinguish health from disease have struggled with the complexity and dynamism inherent in most biological processes. Indeed, to maintain stability of one physiological process (homeostasis) and thus function at the organismal level, another system must change (Chapter 8, this volume). Perhaps the best modern example of this conundrum is stress: as Hans Selye famously claimed, "everyone knows what stress is but nobody really knows." We have made some progress in disentangling pattern (stress) from process (stress responses and their impacts on other systems) (McEwen & Wingfield 2003; Romero *et al.* 2009; Schulkin 2010), and incorporating classic approaches in other fields (e.g., quantitative genetics) could improve our understanding (Chapter 8, this volume). However, we might still fall short of meaningfully integrating stress into ecology and evolutionary biology – because what is most relevant to organisms is not levels of hormones or even the naked stability of one physiological system. Rather, what amorphous concepts like stress seek to capture are the complex connections between physiology and organismal performance, which are the integrated, organismal traits on which selection likely acts (Ghalambor *et al.*, Chapters 1 & 8, this volume).

An example of the potential value of PRN thinking is that it may reveal stress as an emergent property of PRN instability, or a perturbation of PRN state beyond which network compensation may or may not be possible. This perspective is supported by work in systems biology indicating that complex traits are emergent outcomes of several processes at multiple biological levels. Nerve impulse propagation (Deisseroth *et al.* 1996), circadian rhythmicity (Foster & Kreitzman 2005), and heart pacemaker actions (Noble & Noble 1984) cannot be explained bottom-up from genes; they emerge from bi-directional interactions among many molecules, cells and tissues (Noble 2006). For example, heart pacemaker activity occurs as heart muscle voltage affects fluxes of sodium through protein channels in neurons; clamping this voltage, which is itself instigated by ion fluxes, prevents pacemaker oscillations (Figure 9.4a). Traditional physiological thinking assumes that genes produce proteins, which coordinate tissue function. However, heart pacemaker cycles are middle-out processes (Figure 9.5); for the pacemaker to function, genes, proteins, and cells work together, so that causality cannot be attributed to a single biological level.

To determine whether stress is an emergent property of PRN instability, we could measure PRN connectivity and hierarchy across many environmental contexts. We expect that particular PRN states will covary with individual performance in the same way that species composition influences community productivity. For example, reductions in biodiversity or over-representation by certain species (i.e., introduction of nonnative species) reduces

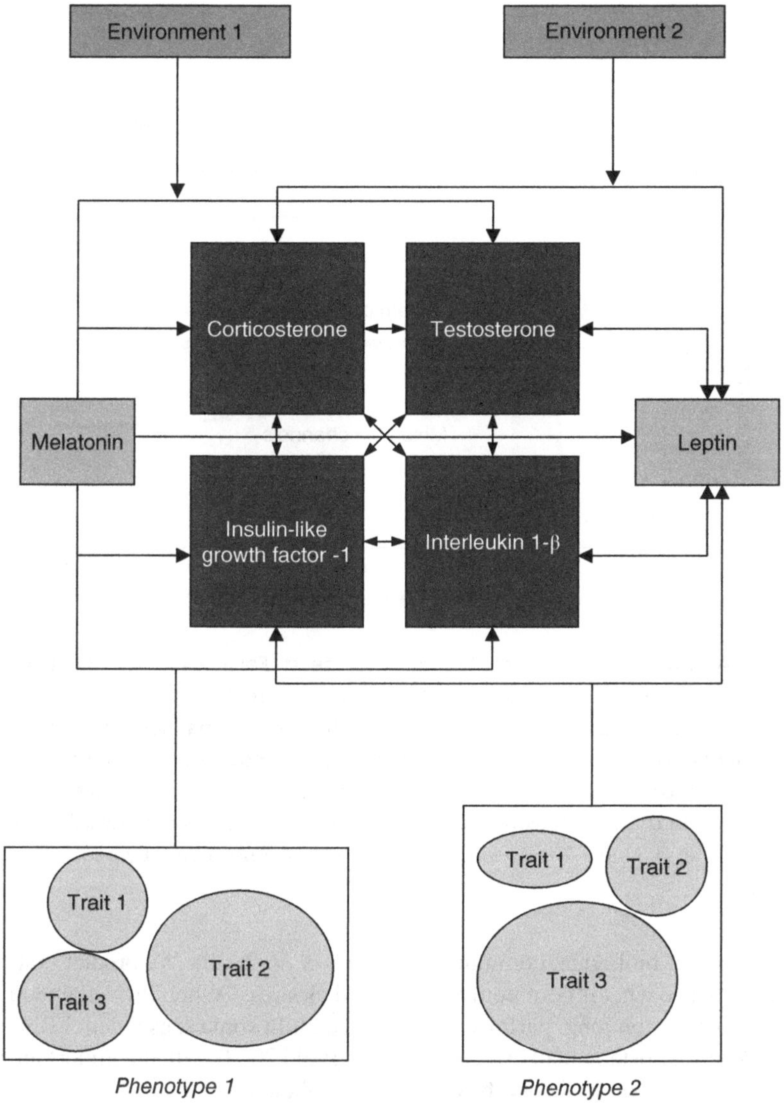

Figure 9.3. Integrators encode present and past environment to produce individual phenotypic variation from one genotype. Dark grey boxes represent integrators (i.e., nodes acting as PRN date hubs) that regulate performance such as coping with stressors (corticosterone), reproduction (testosterone), growth (IGF-1), and interactions with parasites (IL-1β). Arrows between nodes depict edges of network, the configuration of which determines the specific and coordinated phenotypic response. Two integrators are set apart from the main network as their influences on PRNs may be unique. Melatonin impacts, but is not impacted by, other integrators; leptin, insulin and functionally similar molecules may have more labile connection to other date hubs, thus acting as mediators of resource-based trade-offs. The green boxes depict two different environments (e.g., conspecific density, latitude), and hence determine the specific phenotypic outcome (boxes with light grey circles) of the same genotype. Only a few integrators are depicted here, for clarity. (See plate section for color version.)

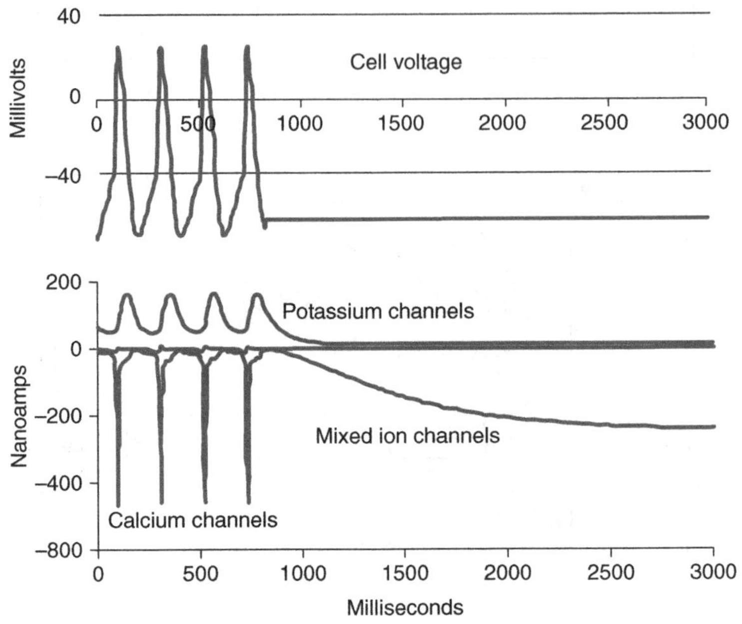

Figure 9.4. (A) Model of pacemaker rhythm in the heart. For the first four beats the model is allowed to run normally and generates rhythm similar to a real heart. Then the feedback from cell voltage to protein channels is interrupted. All the protein channel oscillations then cease and slowly change to steady constant values. The diagram shows the causal loop involved. Protein channels carry current that changes the cell voltage (upward arrow) while the cell voltage changes the protein channels (downward arrow). In the model, this downward arrow was broken. *Source*: Noble, D. (2006) *The Music of Life: Biology beyond Genes*. By permission of Oxford University Press.

the performance of biological communities (Mack *et al.* 2000). We predict that particular PRN states underlie what is commonly recognized as stress. When PRN connectivity takes an unstable state for too long, performance is reduced. In some sense then, PRN instability equates to Type II allostatic load (McEwen & Wingfield 2003) with the important exception that PRN instability can, in theory, be quantified explicitly.

PRN stability as a determinant of stress is a complex concept, so consider as an example glucocorticoid (GC)-mediated immune regulation. Acute stressors tend to stimulate the immune system while chronic stressors tend to suppress it. Whereas the former could be explained by traditional thinking about glucocorticoids as mediators of emergency life history states (Wingfield *et al.* 1998), the latter is harder to reconcile. PRN theory accommodates the distinct effects of GCs on performance, however, going so far as to suggest that a reduction in immunity in the face of sustained high GCs may be a stable PRN state (Dhabhar 2009; Martin 2009). In other words, even immune suppression could be beneficial because it could enable an individual to free up resources for other life processes and/or damp the risk of autoimmunity as novel self-antigens are released in times of adversity (Martin 2009). Stress, disease, allostatic load, and related phenomena only occur when a PRN state takes a form outside the PRN structure accommodated by past selection (i.e., nonadaptive phenotypic plasticity (Ghalambor *et al.* 2007)).

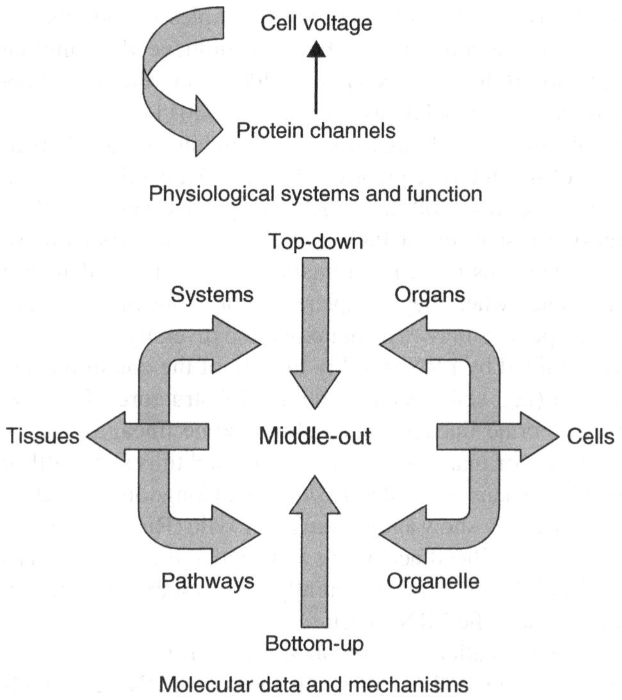

Figure 9.5. Relationship between the bottom-up, top-down, and middle-out approaches to reconstructing organs and systems of the body. This diagram envisages the middle level being cells or tissues, but any biological level can be a middle starting point. The essential concept is that, since all levels can be the starting point for a causal chain. There is not privileged level in biological systems that 'dictates' the rest. *Source*: Noble, D. (2006) *The Music of Life: Biology beyond Genes.* By permission of Oxford University Press.

In spite of the challenges of describing PRN states, investigations of PRN stability likely will provide us with new, actionable definitions of stress, health and disease. Indeed, most current concepts (e.g., frailty (Bortz 2002)) are difficult (if not impossible) to measure, and few include the capacity of individuals to recover function or maintain viability (Sterling 2012). Another advantage of PRN theory is that it accommodates common phenomena such as predictive adaptive responses (PARs), which are major plastic alterations to phenotype mediated by early-life experiences. According to the PAR concept, early-life conditions are used as cues for deciding which future phenotypic strategy will pay the greatest dividends (Gluckman *et al.* 2005). In natural systems, there are few examples of genuinely adaptive PARs (Monaghan 2008), but a growing literature hints that alternative PRN states are involved (Martin *et al.* 2011). A good example is epigenetic alterations of glucocorticoid regulation (Chapters 7 & 12, this volume). High maternal care in domesticated rats reduces methylation of the glucocorticoid receptor promoter, and offspring inheriting the methylated promoter cope better with adversity as adults (Weaver *et al.* 2004). In PRN terms, early-life experience may alter the risk of PRN states becoming unstable by changing the connectivity among glucocorticoid regulators and other PRN nodes. Indeed, developmental plasticity in PRN states is implicated in many human diseases (Gluckman *et al.* 2009). Examples include the hygiene hypothesis for allergies (Yazdanbakhsh *et al.* 2002), obesity as influenced by alteration to insulin-like growth factor regulation (Gluckman & Hanson 2008), and the role of apolipoprotein E in Alzheimer's disease and atherosclerosis (Trotter *et al.* 2011).

A final potential value of studying PRNs is insight into aging (Chapter 16, this volume). There are hundreds of mechanistic theories of aging (Medvedev 1990), and no consensus view has emerged (Kirkwood 2005). A PRN perspective suggests that aging may stem partly from increasing instability of PRNs over a lifetime (Woods & Wilson 2013). The many molecular mechanisms (e.g., inflammation and oxidative damage) related to aging only instigate senescence when they escape regulatory control (e.g., Magalhães & Church 2006). This same perspective may explain taxonomic diversity in aging. In the PRN model, rate of aging is determined by PRN stability in light of the conditions that one's ancestors routinely encountered (i.e., stable states within PRN structure). Because PRNs are likely conserved within vertebrate lineages (see below), some lineages (e.g., turtles) may have relatively resilient PRNs or many stable PRN states and thus show little evidence of aging within their realized lifespans (Congdon *et al.* 2001; Congdon *et al.* 2003). Other lineages (e.g., mammals) consistently show aging even in the wild (Ricklefs 1998). In these lineages, PRN resiliency may be sacrificed according to selective pressures experienced by populations. In other words, PRN states may inherently be less stable in some taxa than others due to differences in taxon-specific PRN structure.

Perhaps the greatest challenge in reconciling the role of PRNs in stress, aging, and organismal performance generally will be to measure PRNs efficiently and accurately. For instance, to use PRN stability as a diagnostic for stress requires that we capture the dynamism in nodes and edges. Just as populations of organisms can stabilize at different equilibria, PRNs probably can take different stable states. Furthermore, certain PRN nodes are likely to exhibit high redundancy, such as the molecules that control infections, heal tissues, or repair DNA with different efficiency and efficacy. PRN instability and/or resiliency may be difficult to quantify because changes at one node may be compensated elsewhere in networks. All these challenges notwithstanding, recent work suggests that the task of linking PRN state to health may not be so daunting after all. Mahalanobis distance, a measure of multivariate statistical distance, can be applied to suites of biomarkers to measure physiological dysregulation. It was found to differ between healthy and sick individuals, both in shorebirds and humans. Moreover, the predictive value of the measure is relatively robust to changes in which biomarkers are used for the calculation (Cohen *et al.* 2013; Milot *et al.* in review).

ECOLOGICAL AND EVOLUTIONARY IMPLICATIONS OF PRNs: CONSTRAINTS AND EVOLVABILITY

PRN theory seeks to showcase mechanisms that link genotypes to phenotypes and thus the traits on which selection acts (Ghalambor *et al.* 2007). The machinery that transduces environmental and genetic variation into phenotypic variation mostly remains a black box (West-Eberhard 2003). How will studying PRNs open this box? We argue that the concept of the trade-off is a good first step. Adaptationists routinely (and often rightly) assume that physiological processes (subnetworks) are coordinated such that the costs and benefits of life functions are reasonably balanced (Chapters 10, 11, & 18, this volume). The simplicity of the "trade-off" concept underlies its dominance in evolutionary biology since at least the 1960s (Williams 1966) and evolutionary (symmorphosis; Weibel *et al.* 1991) and ecophysiology more recently (Sibly & Calow 1986). In other words, adaptive changes in one PRN subnetwork could have maladaptive effects in others, causing PRNs to evolve slowly. Whereas there are not yet documented examples of this effect, there are several reasons to anticipate them. For example, different pathogens are controlled by different immune responses (Graham 2002) (e.g., T-helper 1 cells for intra-cellular and T-helper 2 cells for extra-cellular parasites) and these responses routinely trade-off within individuals. Infection with a particular pathogen probably pushes the PRN into one state and away from another. Depending on the preponderance of parasites in an area, vulnerability to micro- or macroparasites could persist longer than would be expected if T-helper cells, and the PRN subnetworks that coordinate them, evolved independently.

Although PRN-based constraints may be common, PRN traits and outcomes, such as compartmentalization and developmental canalization, can turn apparent constraints into double-edged swords (Kirschner & Gerhart 2010). In other words, genetic and developmental architectures impose constraints *and* facilitate variation within safe bounds (Newman 2010), a property called evolvability (Kirschner & Gerhart 1998). Therefore, the more modular a PRN is, the easier it should be for selection to adjust one subnetwork without affecting others. Connections among subnetworks also help optimize regulatory decision-making. For example, the decision of an organism to shift to a breeding state could involve many factors, including time of year, temperature, age, energy reserves, prior experience, current infections, etc. (Chapter 15, this volume). In principle, the more information an organism incorporates into its regulatory algorithm, the more precisely it can tune its biology to current conditions. However, a species that incorporated many environmental factors into timing its breeding would be unable to adapt quickly to shifting conditions; a species with few modules (or a greater propensity to link and decouple them) could match environments quickly but perhaps with low precision. Such species differences in physiological epistasis might alter how genetic epistasis is understood and require new lines of enquiry into interactions between epistasis at two levels. The role of genetic epistasis in comparatively simple physiological differences such as oxygen transport via alteration in haemoglobin structure (Natarajan *et al.* 2013) may be easier to resolve than the role of genetic epistasis in PRNs.

OPEN ISSUES

There is much to learn about PRNs, most importantly what forms of connectivity, modularity and hierarchy exist in natural systems. For most taxa we understand physiological interconnectedness at only rudimentary levels (Wingfield *et al.* 2013); targeted comparative work (i.e., comparisons of physiological networks among congeners with different life histories) could be especially illuminating, revealing for example the most and least labile PRN elements. Another issue, which assumes that PRNs are hierarchical, is to determine what other date hubs exist in metazoans. The authors' vertebrate bias leads us to ignore most other multicellular life, but the concept should transfer easily across taxa. A third issue to resolve is how PRNs interface with the central nervous system. Most tissues are run through with neurons (Glaser & Kiecolt-Glaser 2005), so perhaps the brain and PRNs are intimately linked because of the advantages of using information encoded over multiple time scales to match phenotypes to environments. Brains continually learn from, and elicit responses to, the environment but over rapid time scales. By contrast, the muscles, skeletons, organs, and immune systems of individuals do most of their learning early in life and become comparatively inflexible thereafter (Turner 2009). Although there is still much work to be done regarding brain-PRN links (Schulkin 2003), it is important to recall that Ivan Pavlov won the Nobel prize for his discovery that the brain facilitates anticipatory matching of organ systems (Sterling 2012).

CONCLUSION

In spite of the historical practice (and success) of single-systems physiology, we propose that developing a whole-organism physiological theory remains an important and unrealized goal. The theory should be conceived in an ecological and evolutionary context. The implicit goal of any physiological process is to maximize the fitness of the individual (Zera & Harshman 2001; Ricklefs & Wikelski 2002), but this idea deserves more explicit attention (Chapter 5, this volume). Even in systems biology, which has revealed the importance of emergent properties in physiology, investigations have not been undertaken for whole organisms. Without question, PRN research will be challenging, but as prices decrease for high-throughput biology, data will become increasingly plentiful. There are also many useful statistical and analytical approaches available already for bionetworks including graph theory (GT), structural equation modeling (SEM), vector field editing (VFE), and control theory (CT; Martin *et al.* 2011). Some of these (VFE) require large datasets, and others (GT, CT) require nearly complete maps of networks. It is important to recognize though that the first mapped ecological networks were simple compared to the complex networks of 1500+ species that are studied today. We advocate starting small, focusing first on putative hubs in ecologically unique species (i.e., those living in extreme environments) and/or contexts (i.e., native versus invasive species). The traditional concept of the reaction norm (Chapter 2, this volume) will also be valuable to these pursuits, especially given recent extensions of these approaches in ways that are valuable to organismal biology (Chapter 4, this volume). Most importantly, a PRN perspective will generate new questions. For instance, can we see a shift in the PRN subnetworks controlled by a given date hub in a certain clade, and what does this tell us about the physiological evolution and historical environment of that clade? What aspects of PRNs are conserved, and what aspects are evolutionarily labile? We close by agreeing with Dover and Darwin (2000):

> "We don't have a theory of interactions, and until we do we don't have a theory of development or evolution."
>
> – *(Dover & Darwin 2000).*

GLOSSARY

Alternative (stable) PRN states – different PRN configurations that produce adaptive and/or functional outcomes for a given species in a given environment

Connectivity – the number of edges within a network

Dysregulation – PRN instability or lack of resiliency

Edge – a linkage (regulatory relationship) between two nodes

Emergent (system-level) property – a property of a PRN not deducible from individual nodes or edges

Epigenetic – alteration of relationships between genes and phenotypes not involving variation in DNA sequences

Epistasis – interaction among nodes in a network

Facilitated variation – a theory proposing how viable, diverse forms of phenotypic variation can arise via little genetic variation

Flexibility – transient or long-term, but typically reversible, variation in phenotype within individuals

Hierarchy – the manner in which network function is regulated; predominance among nodes mediated by patterns of connectivity

Homeostasis – the stablest PRN states, which optimize organismal function in light of internal and external conditions and evolutionary history

Hub – a node connected to many other nodes, often divided into party (many simultaneous partners, mediating within-module interactions) and date (partner with other nodes at different places/times, which integrate between modules) forms

Integrator – a highly-connected node; a date hub

Module – a relatively autonomous, but highly interconnected part of a PRN; a subnetwork

Modularity – the distribution of edges within networks

Node – circulating molecules or their functional complexes

Physiological regulatory network (PRN) – the network of molecules (nodes) and their regulatory relationships (edges) that enable organismal performance

Plasticity – the capacity for one genotype to mediate multiple phenotypes contingent on environment

PRN state – a particular PRN configuration

PRN structure – the network architecture of all regulatory molecules in an individual including potential edges among nodes

Resiliency – the ability to recover or maintain a stable PRN state

Stability – PRN states accommodated by past selection; stable PRN equilibria

Subnetwork – a highly-connected, and thus largely independent, part of a PRN responsible for a particular organismal function; a module.

REFERENCES

Adelman, J. and Martin, L. (2009) Vertebrate sickness behavior: an adaptive and integrated neuroendocrine immune response. *Integrative and Comparative Biology*, **49**, 202–14.

Ader, R. and Cohen, N. (1975) Behaviorally conditioned immunosuppression. *Psychosomatic Medicine*, **37**, 333–40.

Bolker, J.A. (2000) Modularity in development and why it matters to evo-devo. *American Zoologist*, **40**, 770–6.

Bortz, W.M. (2002) A Conceptual Framework of Frailty A Review. *The Journals of Gerontology Series A: Biological Sciences and Medical Sciences*, **57**, M283–M288.

Chrousos, G.P. (1998) Stressors, stress, and neuroendocrine integration of the adaptive response: the 1997 Hans Selye Memorial Lecture. *Annals of the New York Academy of Sciences*, **851**, 311–35.

Clune, J., Mouret, J.-B. and Lipson, H. (2013) The evolutionary origins of modularity. *Proceedings of the Royal Society B: Biological Sciences*, **280**.

Cohen, A.A., Martin, L.B., Wingfield, J.C., McWilliams, S.R. & Dunne, J.A. (2012) Physiological regulatory networks: ecological roles and evolutionary constraints. *Trends in Ecology & Evolution*, **28**(8), 428–35.

Cohen, A.A., Milot, E., Yong, J. *et al.* (2013) A novel statistical approach shows evidence for multi-system physiological dysregulation during aging. *Mechanisms of Ageing and Development*, **134**, 110–17.

Congdon, J., Nagle, R., Kinney, O. and van Loben Sels, R. (2001) Hypotheses of aging in a long-lived vertebrate, Blanding's turtle (*Emydoidea blandingii*). *Experimental Gerontology*, **36**, 813–27.

Congdon, J.D., Nagle, R.D., Kinney, O.M. *et al.* (2003) Testing hypotheses of aging in long-lived painted turtles (*Chrysemys picta*). *Experimental Gerontology*, **38**, 765–72.

Deisseroth, K., Bito, H. and Tsien, R.W. (1996) Signaling from synapse to nucleus: postsynaptic CREB phosphorylation during multiple forms of hippocampal synaptic plasticity. *Neuron*, **16**, 89–101.

Dhabhar, F.S. (2009) A hassle a day may keep the pathogens away: The fight-or-flight stress response and the augmentation of immune function. *Integrative and Comparative Biology*, **49**, 215–36.

Dover, G.A. & Darwin, C. (2000) *Dear Mr. Darwin: Letters on the evolution of life and human nature*. University of California Press.

Dunne, J.A., Williams, R.J. and Martinez, N.D. (2002) Network structure and biodiversity loss in food webs: robustness increases with connectance. *Ecology Letters*, **5**, 558–67.

Fliri, A.F., Loging, W.T. and Volkmann, R.A. (2010) Cause-effect relationships in medicine: a protein network perspective. *Trends in Pharmacological Sciences*, **31**, 547–55.

Folstad, I. and Karter, A.J. (1992) Parasites, bright males, and the immunocompetence handicap. *American Naturalist*, **139**, 603–22.

Foster, R.G. and Kreitzman, L. (2005) *Rhythms of Life: The Biological Clocks that Control the Daily Lives of Every Living Thing*, Yale University Press.

Ghalambor, C.K., McKay, J.K., Carroll, S.P. and Reznick, D.N. (2007) Adaptive versus non-adaptive phenotypic plasticity and the potential for contemporary adaptation in new environments. *Functional Ecology*, **21**, 394–407.

Glaser, R. (2005) Stress-associated immune dysregulation and its importance for human health: a personal history of psychoneuroimmunology. *Brain Behav Immun*, **19**, 3–11.

Glaser, R. and Kiecolt-Glaser, J.K. (2005) Stress-induced immune dysfunction: implications for health. *Nat Rev Immunol*, **5**, 243–51.

Gluckman, P., Beedle, A. and Hanson, M. (2009) *Principles of Evolutionary Medicine*, Oxford University Press.

Gluckman, P. and Hanson, M. (2008) Developmental and epigenetic pathways to obesity: an evolutionary-developmental perspective. *International Journal of Obesity*, **32**, S62–S71.

Gluckman, P.D., Hanson, M.A. and Spencer, H.G. (2005) Predictive adaptive responses and human evolution. *Trends in Ecology & Evolution*, **20**, 527–33.

Graham, A.L. (2002) When T-helper cells don't help: Immunopathology during concomitant infection. *Quarterly Review of Biology*, **77**, 409–34.

Jones, O.R., Scheuerlein, A., Gómez, S.R. *et al.* (2014) Diversity of ageing across the tree of life. *Nature* **505**, 169–73. doi:10.1038/nature12789.

Kashtan, N. and Alon, U. (2005) Spontaneous evolution of modularity and network motifs. *Proceedings of the National Academy of Sciences of the United States of America*, **102**, 13773–8.

Ketterson, E.D. and Nolan, V. (1999) Adaptation, exaptation, and constraint: A hormonal perspective. *American Naturalist*, **154**, S4–S25.

Kirkwood, T.B. (2005) Understanding the odd science of aging. *Cell*, **120**, 437–47.

Kirschner, M. and Gerhart, J. (1998) Evolvability. *Proceedings of the National Academy of Sciences of the United States of America*, **95**, 8420–7.

Kirschner, M.W. and Gerhart, J.C. (2010) Facilitated variation, in *Evolution: The Extended Synthesis* (eds M. Pigliucci and G.B. Muller), MIT Press, Cambridge, MA, pp. 253–80.

Lenski, R.E., Ofria, C., Pennock, R.T. and Adami, C. (2003) The evolutionary origin of complex features. *Nature*, **423**, 139–44.

Lessells, C.M. (2008) Neuroendocrine control of life histories: what do we need to know to understand the evolution of phenotypic plasticity? *Philosophical Transactions of the Royal Society B-Biological Sciences*, **363**, 1589–98.

Lipson, H. (2007) Principles of modularity, regularity, and hierarchy for scalable systems. *Journal of Biological Physics and Chemistry*, **7**, 125.

Mack, R.N., Simberloff, D., Lonsdale, W.M. *et al.* (2000) Biotic invasions: Causes, epidemiology, global consequences, and control. *Ecological Applications*, **10**, 689–710.

Magalhães, J.P.d. and Church, G.M. (2006) Cells discover fire: Employing reactive oxygen species in developmentand consequences for aging. *Experimental Gerontology*, **41**, 1–10.

Martin, L.B. (2009) Stress and immunity in wild vertebrates: timing is everything. *General and Comparative Endocrinology*, **163**, 70–6.

Martin, L.B., Liebl, A.L., Trotter, J.H. *et al.* (2011) Integrator networks: Illuminating the black box linking genotype and phenotype. *Integrative and Comparative Biology*, **51**, 514–27.

Martin, L.B., Weil, Z.M. and Nelson, R.J. (2008) Seasonal changes in vertebrate immune activity: mediation by physiological trade-offs. *Philosophical Transactions of the Royal Society of London B Biological Sciences*, **363**, 321–39.

McEwen, B.S. and Wingfield, J.C. (2003) The concept of allostasis in biology and biomedicine. *Hormones and Behavior*, **43**, 2–15.

McGlothlin, J.W. and Ketterson, E.D. (2008) Hormone-mediated suites as adaptations and evolutionary constraints. *Philosophical Transactions of the Royal Society B-Biological Sciences*, **363**, 1611–20.

Medvedev, Z.A. (1990) An attempt at a rational classification of theories of ageing. *Biological Reviews*, **65**, 375–98.

Medzhitov, R. (2008) Origin and physiological roles of inflammation. *Nature*, **454**, 428–35.

Mills, L.S., Soule, M.E. and Doak, D.F. (1993) The keystone-species concept in ecology and conservation. *Bioscience*, **43**, 219–24.

Milo, R., Shen-Orr, S., Itzkovitz, S. *et al.* (2002) Network motifs: Simple building blocks of complex networks. *Science*, **298**, 824–7.

Milot, E., Cohen, A.A., Vezina, F. *et al.* (2014) A novel, integrative method for measuring body condition in ecological studies based on physiological dysregulation. *Methods in Ecology and Evolution*, **5**(2) 146–55.

Monaghan, P. (2008) Early growth conditions, phenotypic development and environmental change. *Philosophical Transactions of the Royal Society B-Biological Sciences*, **363**, 1635–45.

Natarajan, C., Inoguchi, N., Weber, R.E. *et al.* (2013) Epistasis among adaptive mutations in deer mouse hemoglobin. *Science*, **340**, 1324–7.

Newman, S.A. (2010) Dynamic patterning modules, in *Evolution: The Extended Synthesis* (eds M. Pigliucci and G.B. Muller), MIT Press, Cambridge, MA, pp. 281–306.

Noble, D. (2006) *The Music of Life: Biology beyond Genes*, Oxford University Press, New York.

Noble, D. & Noble, S. (1984) A model of sino-atrial node electrical activity based on a modification of the DiFrancesco–Noble (1984) equations. *Proceedings of the Royal Society of London. Series B, Biological Sciences*, 295–304.

Ricklefs, R.E. (1998) Evolutionary theories of aging: confirmation of a fundamental prediction, with implications for the genetic basis and evolution of life span. *The American Naturalist*, **152**, 24–44.

Ricklefs, R.E. and Wikelski, M. (2002) The physiology/life-history nexus. *Trends in Ecology & Evolution*, **17**, 462–8.

Romero, L.M., Dickens, M.J. and Cyr, N.E. (2009) The reactive scope model – A new model integrating homeostasis, allostasis, and stress. *Hormones and Behavior*, **55**, 375–89.

Schulkin, J. (2003) Allostasis: a neural behavioral perspective. *Hormones and behavior*, **43.1**, 21–7.

Schulkin, J. (2011) Social allostasis: anticipatory regulation of the internal milieu. *Frontiers in Evolutionary Neuroscience*, **2**, 111.

Segal, E., Shapira, M., Regev, A. *et al.* (2003) Module networks: identifying regulatory modules and their condition-specific regulators from gene expression data. *Nature Genetics*, **34**, 166–76.

Segerstrom, S.C. (2007) Stress, energy, and immunity: An ecological view. *Current Directions in Psychological Science*, **16**, 326–30.

Sibly, R.M. and Calow, P. (1986) *Physiological Ecology of Animals: An Evolutionary Approach*, Blackwell Scientific Publications Oxford.

Stelling, J.R., Klamt, S., Bettenbrock, K. *et al.* (2002) Metabolic network structure determines key aspects of functionality and regulation. *Nature*, **420**, 190–3.

Sterling, P. (2012) Allostasis: a model of predictive regulation. *Physiology & Behavior*, **106**, 5–15.

Tilman, D., Reich, P.B., Knops, J. *et al.* (2001) Diversity and productivity in a long-term grassland experiment. *Science*, **294**, 843–5.

Trotter, J.H., Liebl, A.L., Weeber, E.J. and Martin, L.B. (2011) Linking ecological immunology and evolutionary medicine: the case for apolipoprotein E. *Functional Ecology*, **25**, 40–7.

Turner, J.S. (2009) *The Tinkerer's Accomplice: How Design Emerges from Life Itself*, Harvard University Press.

Von Dassow, G. and Munro, E. (1999) Modularity in animal development and evolution: elements of a conceptual framework for EvoDevo. *The Journal of Experimental Zoology*, **285**, 307–25.

Wagner, G.P. (2007) The developmental genetics of homology. *Nature Reviews Genetics*, **8**, 473–479.

Wagner, G.P. and Altenberg, L. (1996) Perspective: Complex adaptations and the evolution of evolvability. *Evolution*, **50**, 967–76.

Wagner, G.P., Pavličev, M. and Cheverud, J.M. (2007) The road to modularity. *Nature Reviews Genetics*, **8**, 921–31.

Weaver, I.C.G., Cervoni, N., Champagne, F.A. *et al.* (2004) Epigenetic programming by maternal behavior. *Nature Neuroscience*, **7**, 847–54.

Weibel, E.R., Taylor, C.R. and Hoppeler, H. (1991) The concept of symmorphosis – a testable hypothesis of structure-function relationship. *Proceedings of the National Academy of Sciences of the United States of America*, **88**, 10357–61.

West-Eberhard, M. (2003) *Developmental Plasticity and Evolution*, Oxford University Press, Oxford.

Williams, G.C. (1966) *Adaptation and Natural Selection*, Princeton University Press, Princeton, N.J.

Wingfield, J.C. (2013) The comparative biology of environmental stress: behavioural endocrinology and variation in ability to cope with novel, changing environments. *Animal Behaviour*, **85** (5), 1127–33.

Wingfield, J.C., Maney, D.L., Breuner, C.W. *et al.* (1998) Ecological bases of hormone-behavior interactions: The "emergency life history stage." *American Zoologist*, **38**, 191–206.

Woods, H.A. and Wilson, J.K. (2013) An information hypothesis for the evolution of homeostasis. *Trends in Ecology & Evolution*, **28**, 283–9.

Yazdanbakhsh, M., Kremsner, P.G. and van Ree, R. (2002) Immunology – Allergy, parasites, and the hygiene hypothesis. *Science*, **296**, 490–4.

Zera, A.J. and Harshman, L.G. (2001) The physiology of life history trade-offs in animals. *Annual Review of Ecology and Systematics*, **32**, 95–126.

10

INTEGRATING COSTS OF REPRODUCTION BETWEEN THE SEXES

Robert M. Cox

Department of Biology, University of Virginia, Charlottesville, VA, USA

INTRODUCTION

Reproduction and survival are the two fundamental components of Darwinian fitness, but traits that improve reproductive success often do so at the expense of survival, and vice versa. This can create a negative correlation, or tradeoff, between fitness components. This concept of a tradeoff can also be extended to describe negative correlations between the underlying traits and processes that influence fitness. Tradeoffs are structured by costs of reproduction, which can be measured as decrements to fitness (survival and future reproductive success), reductions in performance (locomotion, immune function, cellular maintenance), or expenditures of limited resources (time, energy, nutrients). The theoretical and empirical literature on costs of reproduction is vast, and has been iteratively synthesized by many excellent reviews and insightful perspectives (Partridge & Harvey 1985; Reznick 1985; Bell & Koufopanou 1986; Stearns 1989; Partridge 1992; Reznick 1992; Stearns 1992; Roff 2002; Roff & Fairbairn 2007). With the growth of integrative organismal biology, these reviews have increasingly focused on the mechanistic underpinnings of reproductive tradeoffs (Finch & Rose 1995; Rose & Bradley 1998; Zera & Harshman 2001; Partridge *et al.* 2005; Williams 2005; Harshman & Zera 2007; Flatt 2011; Flatt & Heyland 2011). The goal of this chapter is not to review this expansive literature, but to call attention to an important feature that has been largely absent from most perspectives: empirical and theoretical integration between the sexes.

Costs and tradeoffs shape our thinking in many areas of integrative biology (e.g., Chapters 14, 16, & 18, this volume), and those associated with reproduction are particularly

Integrative Organismal Biology, First Edition.
Edited by Lynn B. Martin, Cameron K. Ghalambor, and H. Arthur Woods.

important to theories of life-history evolution and sexual selection. However, conceptual integration between these two areas of evolutionary research has been hindered by a tendency to study reproductive tradeoffs separately in each sex. Males and females are inherently predisposed to different reproductive strategies due to anisogamy, the fundamental difference in the size and number of gametes that defines separate sexes. Despite this inherent asymmetry, average investment in reproduction should be equivalent in males and females because each sex receives equal genetic benefits from reproduction (i.e., each offspring has a mother and a father; Fisher 1930; Queller 1997). Although males and females each experience costs of reproduction, these costs have traditionally been considered in relative isolation from one another (but see e.g., Queller 1997; Bonduriansky *et al.* 2008; Chu & Lee 2012).

From a theoretical perspective, the historical focus on costs incurred by females was motivated in part by the initial framing of life-history theory in demographic and population-dynamic models centered on population growth as a key fitness parameter (e.g., Cole 1954; Gadgil & Bossert 1970). Because males do not typically provide energy and nutrients to eggs and embryos, they do not contribute directly to population growth (the two-fold cost of males; Maynard Smith 1978), so their reproductive costs are less relevant from this perspective. Instead, costs of reproduction in males have been explored primarily within the theoretical framework of sexual selection, as potential explanations for the maintenance of honest signaling in sexually selected traits (Zahavi 1975; Folstad & Karter 1992).

From an empirical perspective, the major challenges associated with comparing reproductive costs between the sexes are that these costs are measured in different currencies, occur at different times, and encompass different processes and regulatory axes in each sex. Despite these challenges, some empirical studies have directly compared costs of reproduction between males and females (Kotiaho & Simmons 2003; Penn & Smith 2007; Hoffman *et al.* 2008; Christe *et al.* 2012). However, these studies have yet to be synthesized into an integrative theory for life-history evolution that encompasses both sexes (Queller 1997; Rolff 2002; Bonduriansky *et al.* 2008). A first step in this process is to identify a framework in which we can begin to directly compare the costs of reproduction between the sexes.

INTEGRATING COSTS OF REPRODUCTION THROUGH COMMON CURRENCIES

There are at least two currencies that should be of general importance for comparing reproductive tradeoffs between the sexes: energy and fitness. Energy is a useful currency because can be rigorously quantified in absolute units that are directly comparable between sexes, across seasons, among species, and over a range of biological scales (Chapter 14, this volume). Moreover, energy is a common denominator for many aspects of reproductive investment, including production costs (i.e., energy deposited in new tissue) associated with making gametes and offspring; respiratory costs (i.e., metabolic expenditure) required for the synthesis and maintenance of reproductive tissues and for the expression of reproductive behaviors; and acquisition costs due to decreased foraging opportunities or limited abdominal space for food (Weeks 1996). Finally, many of the tradeoffs that link current reproduction to survival and future reproduction are thought to involve processes that require energy, such as growth and self-maintenance (e.g., Chapters 14, 16, & 18, this volume).

Although energy provides a useful unifying currency for comparison, the nature of energetic costs of reproduction often differs dramatically between the sexes (Queller 1997; Hoffman *et al.* 2008). Females typically incur large production costs from provisioning eggs and embryos. Quantifying this cost is fairly straightforward for any reproductive episode if the number, size, and energetic content of eggs or embryos can be obtained. By contrast, quantifying the production cost associated with spermatogenesis is difficult, and the extent to which this cost is even comparable to that associated with egg production is a matter of debate (Olsson *et al.* 1997; Hayward & Gillooly 2011). In many species, the primary energetic costs associated with male reproduction occur in the form of respiratory expenditure due to increased activity, courtship, mate-searching, combat, and territory defense. Females also incur respiratory costs, but largely from other sources, such as from biosynthesis of eggs or embryos and the maintenance of active reproductive tissues. Thus, even when comparing males and females in the common currency of energy expenditure, we are often confronted with a sex difference in the relative contributions of production versus respiratory costs, each requiring separate methods to quantify. The situation becomes more complicated when incorporating sex-specific costs of reduced energy intake, as can occur in females when the volume of a clutch or litter limits the abdominal space available for food (Weeks 1996), or in males when prolonged territory defense or courtship limits the time available for foraging.

An additional layer of complexity can arise from temporal dissociation between the major costs of reproduction in each sex. Even in the absence of true sexual bimaturation, the sexes often incur their primary reproductive costs at different points in the reproductive season. Among vertebrates, males often incur costs early in the season, during courtship, mate searching and territory defense, whereas females incur costs later in the season, during vitellogenesis, pregnancy, and offspring provisioning (Hoffman *et al.* 2008). Perhaps not surprisingly, those aspects of reproduction for which we know the most about sex-specific costs are those in which both the timing and the currency of the cost are aligned in males and females. One example comes from the large experimental literature on effects of brood manipulations (egg removals and additions) in birds, in which the nature of parental investment (egg incubation, offspring provisioning) and its associated costs (loss of body mass, increased parasite loads, reduced survival) can be measured over the same intervals and with the same procedures, thus permitting comparison between the sexes (Hõrak 2003; Knowles *et al.* 2009; Christe *et al.* 2012). However, brood manipulations are also limited because they consider only the latter stages of reproductive investment, provide only partial estimates of the total cost of reproduction, and may lack generality beyond the relatively small proportion of organisms that provide parental care.

Comparisons based solely on energetics are also complicated by the myriad ways in which individuals can adjust both acquisition and allocation of energy to compensate for costs (Chapter 14, this volume), raising the daunting question of whether the entire energy budget of an organism must be quantified to provide a realistic comparison of reproductive costs between sexes. This concern is exacerbated by the likely possibility that energy allocation and acquisition rules vary not just between sexes, but also among individuals or genotypes within each sex, in response to environmental factors, and as a function of individual condition (van Noordwijk & Jong 1986; Pettifor *et al.* 2001). As with other aspects of life-history evolution, comparisons of the energetic costs of reproduction may therefore require explicit consideration of genotype-by-environment or individual-by-environment reaction norms (Chapters 2 & 3, this volume), phenotypic plasticity (Chapter 5, this volume), and individual variation (Chapter 4, this volume).

Individual variation is central to the measurement of fitness and, with respect to evolution, fitness is the only currency that matters (Stearns 1989). This is the rationale for a "fitness-based framework" of reproductive costs (Edwards & Chapman 2011) and an essential component in any attempt to compare the evolutionary dynamics of reproductive tradeoffs between sexes. Survival is usually estimable using identical methods in either sex (e.g., mark-recapture), and overall sex differences in survival associated with reproduction have thus been inferred in some systems (e.g., Kotiaho & Simmons 2003; Penn & Smith 2007; Hoffman *et al.* 2008). However, it is important to know not only the sex difference in mean survival as it relates to reproduction, but also to quantify the survival costs *and* reproductive benefits associated with individual variation in reproductive investment. This is because the evolutionary trajectory of the life history and other phenotypic traits ultimately depends upon the net effects of these costs and benefits on total lifetime fitness, and because individual variation is the raw material for natural and sexual selection (Chapter 2, this volume). Unfortunately, few studies of wild populations have simultaneously measured selection on both sexes through even a single component of fitness (reviewed by Cox & Calsbeek 2009), and even fewer have integrated survival and reproductive success to describe the complete dynamics of sex-specific selection acting through lifetime fitness (e.g., Badyaev & Martin 2000; Preziosi & Fairbairn 2000; McGlothlin *et al.* 2005). Quantifying net fitness in both sexes is important not only for assessing tradeoffs within each sex, but also for determining whether these tradeoffs interact to constrain the evolution of optimal life histories in either sex, as predicted in the presence of intralocus sexual conflict.

INTRALOCUS SEXUAL CONFLICT AND LIFE-HISTORY EVOLUTION

Intralocus sexual conflict occurs because males and females share an autosomal genome, which can constrain their ability to evolve independently in response to sex-specific selection (Rice & Chippindale 2001; Bonduriansky & Chenoweth 2009; Cox & Calsbeek 2009). When considering the evolution of reproductive investment, we often overlook this genetic constraint because the reproductive functions of males and females are so divergent that we can realistically treat most aspects of reproduction as if they map to separate loci in each sex, thus apparently falling outside the conceptual framework of "intralocus" sexual conflict (Figure 10.1). For example, the loci that underlie variation in sperm production and courtship behavior in males are likely different from those that influence clutch size and egg provisioning in females. Hence, we tend to implicitly treat reproductive traits as though their evolution is shaped only by the fitness costs incurred by the sex in which they are expressed (Figure 10.1a).

The oversight in this approach is that, even if the genetic basis of reproductive investment is largely separate in each sex, males and females are still constrained by between-sex genetic correlations for other shared traits that are linked to reproduction via tradeoffs (Figure 10.1b,c). Consider the classic life-history approach of asking how females should optimize their current reproductive investment to maximize lifetime reproductive success in light of tradeoffs with survival and future reproduction (reviewed by Stearns 1977). Lande (1982) first demonstrated the importance of augmenting this optimality approach with dynamic evolutionary models incorporating genetic correlations between life-history traits, while separately developing the theoretical and analytical foundations for assessing genetic correlations between the sexes (Lande 1980, 1987). However, a dynamic approach to life-history tradeoffs must include genetic correlations both among traits *and* between sexes. This explicit integration of sexual conflict into life-history theory is relatively new, but may be essential for understanding life-history evolution (Rolff *et al.* 2005; Wedell *et al.* 2006; Bonduriansky *et al.* 2008; Berg & Maklakov 2012).

The empirical data that best support this view come from quantitative-genetic studies estimating the between-sex genetic correlation (r_{mf}) for fitness itself, rather than for individual phenotypic traits. If this correlation is strong and positive, then genomes that confer high fitness in one sex will tend to confer high fitness in the other sex. However, an increasing number of studies show that r_{mf} is often negative, such that sires with high

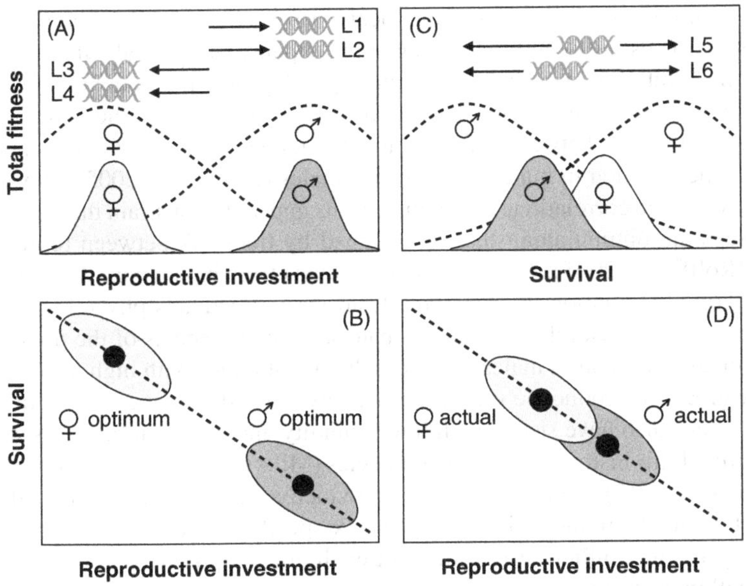

Figure 10.1. Simplified conceptual diagram of the role of intralocus sexual conflict in shaping life-history tradeoffs. Panel A illustrates hypothetical sex-specific fitness surfaces (dashed lines) for reproductive investment. DNA icons represent loci (L1, L2 ...) influencing this life-history phenotype. Shaded areas represent phenotypic distributions for this trait in each sex, which we might expect to be highly dimorphic, given the sex-specific fitness surfaces and the fact that the genetic basis for reproductive investment in males (e.g., L1 influences spermatogenesis; L2 influences investment in courtship) is likely very different from that in females (e.g., L3 influences egg size; L4 influences clutch size). Panel B assumes a classic tradeoff between survival and reproduction (dashed line) and illustrates the sex-specific phenotypic means (black dots) and variances (shaded ellipses) we might expect given the fitness surfaces above in A. Panel C is analogous to A, but illustrates a situation in which the genetic basis for traits influencing survival (e.g., L5 influences immune function, L6 influences energy storage) is shared between sexes, resulting in intralocus conflict that constrains evolution toward sex-specific fitness optima. Panel D illustrates the discrepancy between optimal life-history phenotypes in panel B and actual levels of reproductive investment and survival that might be observed in each sex due to intralocus conflict arising from strong between-sex genetic correlations at L5 and L6. The heuristic framework for sexual conflict is modified from Cox and Calsbeek (2009).

fitness produce daughters with low fitness, and dams with high fitness produce sons with low fitness (Chippindale *et al.* 2001; Fedorka & Mousseau 2004; Brommer *et al.* 2007; Foerster *et al.* 2007). This should dramatically limit the potential for "good-genes" mate choice and could also weaken other forms of sexual selection, and has therefore led to a surge of recent research and critical reappraisal of sexual selection theory (Pischedda & Chippindale 2006; Brommer *et al.* 2007; Cox & Calsbeek 2010a). Our view of life-history evolution could be similarly enriched by considering that the costs and benefits experienced by females are not the sole factors that shape the evolution of the female life history: genetic correlations with males may often prevent females from attaining what would otherwise appear to be the optimal solutions to tradeoffs between reproduction and survival (Figure 10.1d). Reciprocally, genetic correlations with females may constrain the evolution of an optimal life history in males (Berg & Maklakov 2012).

As an example of this process, male mealworm beetles (*Tenebrio molitor*) maximize their fitness by investing heavily in mating at the expense of survival, whereas females are limited in the number of eggs they can produce per day, and are thought to maximize their reproductive success via longevity. This creates a sex difference in the optimal solution to the life-history tradeoff between reproduction and longevity, which favors reduced allocation to immune function in males, relative to females (Rolff *et al.* 2005). However, strong between-sex genetic correlations for immune traits appear to constrain the sex-specific evolution of separate optima along the axes defined by tradeoffs between reproduction and longevity (Rolff *et al.* 2005). Another example comes from bank voles (*Myodes glareolus*), in which heritable variation in circulating testosterone mediates physiological (immunocompetence) and behavioral (dominance, mate searching) aspects of the tradeoff between reproduction and survival in males. Whereas dominant males with high testosterone levels experience greater reproductive success, especially when dominance is rare in the population, these males also have sisters with low fecundity, indicating intralocus sexual conflict over the optimal life-history and mating strategies (Mills *et al.* 2009; Mokkonen *et al.* 2011; Mokkonen *et al.* 2012). Thus, even when reproductive strategies are mediated by physiological factors that facilitate a high degree of sex-specific expression, such as testosterone, underlying genomic conflict can constrain the evolution of optimal solutions to reproductive tradeoffs within each sex.

IMMUNOCOMPETENCE HANDICAPS AND SEX-SPECIFIC COSTS

Similar integration in the framework of energetics could aid in the conceptual unification of other hypotheses that traditionally fall outside the classic life-history perspective on costs of reproduction. One example is the immunocompetence handicap hypothesis (ICHH, Folstad & Karter 1992), which proposes that the sex steroid testosterone stimulates the development and expression of sexually selected signals (a form of reproductive investment that leads to a fitness benefit) while also suppressing the immune system (a potential fitness cost due to increased morbidity or mortality from parasites and pathogens). As in the case of intralocus sexual conflict, this hypothesis has been explored primarily from the perspective of sexual selection on males, as an example of a handicap mechanism that maintains honest signaling in epigamic traits. The original formulation of the ICHH recognized the importance of nutritional status for the expression of sexual signals, immune response, and testosterone production, but did not invoke a direct role for energetic tradeoffs, requiring only that the expression of the sexual signal is negatively coupled to immune function by testosterone to provide an honest indicator of genetic resistance to parasites and pathogens (Folstad & Karter 1992). Subsequent work has largely assumed, either implicitly or explicitly, that this negative coupling of trait expression and immune function reflects an adaptive role of testosterone as a mediator of energy allocation between reproduction and self-maintenance (Wedekind & Folstad 1994; Sheldon & Verlhust 1996; Westneat & Birkhead 1998; Rolff 2002; Owen-Ashley *et al.* 2004; Alonso-Alvarez *et al.* 2007). Without this adaptive framework involving energy allocation, there is no obvious reason why testosterone should be linked to both trait expression and immune function, and sexual selection would presumably favor any mutant "cheaters" in which this hormonal coupling is broken (Wedekind & Folstad 1994).

If the immunosuppressive effects of testosterone derive largely from its role in coordinating energy allocation between immune function and reproductive investment, then the tradeoffs structuring the ICHH should be generalizable beyond testosterone (e.g., to invertebrates lacking testosterone) and even beyond males (Rolff 2002; McKean & Lazzaro 2011).

Yet, historical focus on sexual selection as a process and testosterone as a mechanism has created the perception that compromised immune function and increased susceptibility to parasites are reproductive costs borne largely by males. From an energy-allocation perspective, this is somewhat at odds with the common view that females incur substantial energetic costs of reproduction. The expanding field of eco-immunology (Chapter 11, this volume) has begun to balance this perception by demonstrating that reproductive females often suffer reduced immune function and increased parasite loads (e.g., Ardia *et al.* 2003; French *et al.* 2007; Cox *et al.* 2010; Box 10.1; McKean & Lazzaro 2011), presumably as a consequence of energy allocation tradeoffs (Sheldon & Verlhust 1996; Martin *et al.* 2003, 2011a). The fact that compromised immune function and increased parasite or pathogen loads occur as costs of reproduction in both sexes may help to explain why simplistic tests for male biases in these parameters tend to yield equivocal results (Roberts *et al.* 2004). Whereas males exhibit higher parasite loads and lower immunocompetence than females in some taxa (e.g., mammals, Moore & Wilson 2002; Nunn *et al.* 2009), the reverse is true in others (e.g., birds, McCurdy *et al.* 1998). Explaining such differences requires a framework that explicitly integrates costs of reproduction in both sexes.

One example of such a framework comes from Rolff's (2002) generalization of the ICHH to predict that males should typically exhibit lower immunocompetence than females not because of any inherent immunosuppressive effects of testosterone, but because Bateman's principle (i.e., reproductive success increases as a function of mating success to a greater degree in males than in females) predisposes them to greater investment in current reproduction. By contrast, females should generally maximize their fitness by investing more in longevity (also see Bonduriansky *et al.* 2008; Nunn *et al.* 2009). This framework is appealing because it views reproductive tradeoffs in each sex holistically in the context of life-history strategies, rather than focusing on one sex or the other based on taxon-specific or sex-specific mechanisms, such as testosterone. It can also be used to derive specific predictions about the conditions under which we expect to see sex differences in particular costs of reproduction. Does Bateman's principle predispose males to greater investment in current reproduction at the expense of self-maintenance, relative to females? Presumably, the answer will vary among species and will depend on the strength of the Bateman gradient in males (and females), as well as the location of the optimal life history for each sex along the continuum between investment in current reproduction versus investment in survival and future reproduction. Once the problem is cast in this framework, other questions follow. Do observed sex differences in immune function and parasite load reflect actual sex differences in the extent to which reproductive investment trades off with immunocompetence and self-maintenance, or are they often the result of cross-sectional sampling at points in the reproductive season when one sex or the other is investing more heavily in reproduction? Are males (or females) differentially predisposed to incur energetic costs related to self-maintenance because other sex-specific mortality costs, such as increased susceptibility to predators, supersede the effects of lowered immune function on survival?

INTEGRATING SEX-SPECIFIC PROCESSES THROUGH SHARED REGULATORY AXES

From an empirical standpoint, comparing the costs of reproduction between sexes calls attention to the importance not only of measuring costs in common currencies, but also of focusing on underlying mechanisms that are shared between the sexes. In vertebrates, for example, focusing on sex steroids as mediators of reproductive tradeoffs with immune function makes it difficult to directly compare mechanisms between males and females, as illustrated above in the case of the ICHH. The same is true for any perspective that focuses on tissues (e.g., testes, ovaries), processes (e.g., spermatogenesis, vitellogenesis), and behaviors (e.g., courtship, oviposition) that are highly sexually differentiated (Figure 10.2b,c). A more general approach would be to focus on physiological regulatory networks (Cohen *et al.* 2012; Chapter 9, this volume) and hormonal integrators (Martin *et al.* 2011b; Chapter 11, this volume) that are shared between the sexes (Figure 10.2d).

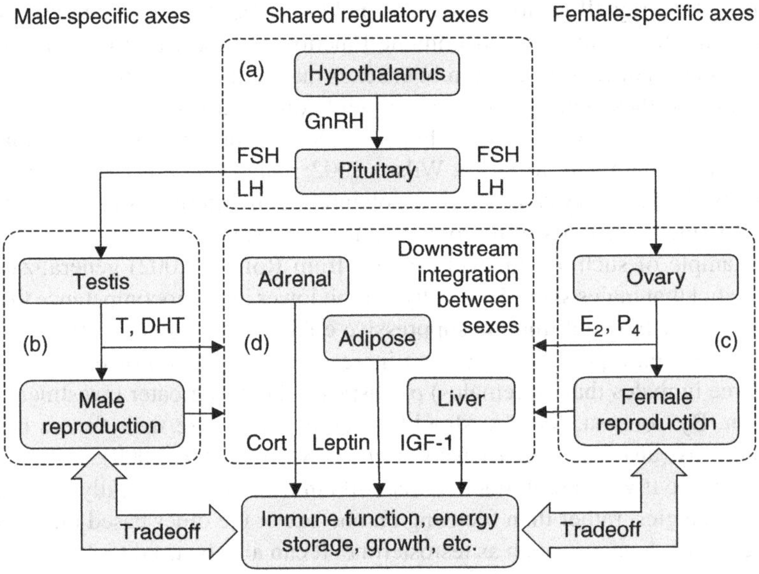

Figure 10.2. Simplified diagram of some pathways in the vertebrate endocrine system, illustrating how sex-specific regulatory axes feed into shared regulatory axes, which provide downstream points of integration through which the costs of reproduction can be studied in a common framework between sexes. Upstream components of the hypothalamic-pituitary-gonadal (HPG) axis are shared between the sexes (module A), but are probably not good candidates for manipulating and/or comparing the costs of reproduction between sexes because similar levels of upstream signaling molecules (GnRH = gonadotropin-releasing hormone; FSH = follicle-stimulating hormone; LH = lutenizing hormone) likely have sex-specific impacts on downstream levels of reproductive investment. Downstream components of the HPG axis (modules B and C) are sexually differentiated in terms of tissues (testes, ovaries), endocrine products (T = testosterone; DHT = 5α-dihydrotestosterone; E_2 = estradiol; P_4 = progesterone), and the aspects of reproductive investment they regulate (e.g., spermatogenesis; oogenesis). However, downstream from the HPG axis, these sex-specific pathways are translated back into shared regulatory axes and tissues (module D) that are influenced directly by gonadal steroids and/or indirectly by other aspects of reproductive investment. Examples include adrenal tissue, which produces corticosteroids (Cort), adipose tissue, which produces leptin, and the liver, which produces insulin-like growth factor 1 (IGF-1). These and many other endocrine tissues and messengers modulate many of the shared maintenance, growth, and survival processes that are involved in tradeoffs with sex-specific aspects of reproductive investment, making them strong candidates for the comparison of male and female costs in a common regulatory framework. Note that many details have been omitted for clarity, such as negative feedback loops, additional connections between various axes, and additional hormones, enzymes, metabolites and binding globulins involved in these pathways.

For example, rather than focusing on sex-biased endocrine messengers such as androgens and estrogens, recent work has begun to focus on hormonal integrators such as leptin, which is secreted by adipose tissue and coordinates energy allocation between reproduction and immune function in both males and females (Alonso-Alvarez *et al.* 2007; French *et al.* 2011). Direct and indirect effects of gonadal sex steroids and reproductive processes on the secretion of signaling molecules such as leptin can be conceptualized as "edges" (Chapter 9, this volume), or points of integration at which sex-specific "subnetworks" of reproductive investment are translated into downstream regulatory networks that are shared between the sexes (Figure 10.2d). Similar approaches can be applied to other regulatory networks that are shared between the sexes, such as the somatotrophic axis or the hypothalamic-pituitary-adrenal axis (Chapter 12, this volume), each of which interacts with sex steroids to coordinate energy allocation between reproduction and processes such as growth and self-maintenance (Figure 10.2d).

Even when focusing on underlying mechanisms that are shared between the sexes, we are still left with the problem of how to directly compare two fundamentally different processes of reproduction. This problem can be put into perspective by considering how we might design an experiment to manipulate reproductive investment and test whether males and females differ in their costs of reproduction. Manipulations that permit measurement of both the costs and benefits of altered reproductive investment are generally either sex-specific, as in the case of testosterone supplementation in males (Reed *et al.* 2006; Hau & Wingfield 2011) and follicular ablation or stimulation in females (Landwer 1994; Sinervo & DeNardo 1996; Kalmbach *et al.* 2004), or they involve only a portion of total reproductive investment, as in the case of clutch- or brood-size manipulations (Koivula *et al.* 2003; Knowles *et al.* 2009). Manipulating the shared, upstream regulatory aspects of the hypothalamic-pituitary-gonadal (HPG) axis (e.g., gonadotropins that influence both spermatogenesis and oogenesis) is unlikely to solve the problem because there is no guarantee that identical manipulations (e.g., stimulating the secretion, supplementing the circulation, or blocking the action of gonadotropins) will alter downstream reproductive investment to the same degree in each sex (Figure 10.2a).

Given these limitations, the simplest approach may often be to infer the costs of reproduction in each sex from comparisons of males and females in which reproductive function is completely ablated. In this regard, gonadectomy and sterilization are potentially useful because they permit estimates of the total costs of reproduction summed across multiple components of investment. Box 10.1 provides a brief summary of how this approach has recently been used to quantify reproductive costs in female brown anole lizards (*Anolis sagrei*). In principle, there is no reason why castration should be any less informative as a manipulation of total reproductive investment in males. However, castration is still used largely in conjunction with replacement of exogenous androgens to mechanistically identify effects of specific hormones, rather than as a holistic manipulation of reproductive investment (Cox *et al.* 2005; Cox & John-Alder 2007).

Box 10.1. Experimentally Assessing Costs of Reproduction in Brown Anoles

The brown anole (*Anolis sagrei*) is a small lizard in which females iteratively lay single-egg clutches at roughly 10-d intervals throughout a breeding season that spans six months or more. Annual survival is low (e.g., 5–15% of adults survive to a second reproductive season) and population densities are among the highest recorded for any lizard, which facilitates large-scale phenotypic manipulations

(*continued*)

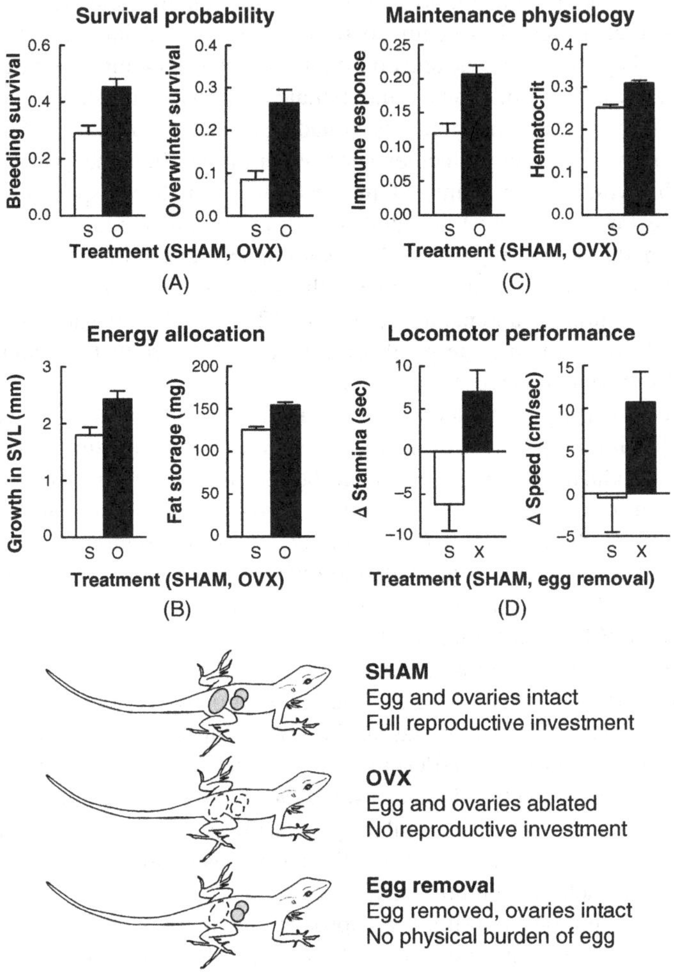

Figure 10.3. Experimental evidence for costs of reproduction in wild female brown anoles (*Anolis sagrei*). Illustrations depict three treatment groups: (1) SHAM, which received a surgical control but no manipulation of the ovaries, leaving them fully reproductive over the subsequent breeding season; (2) OVX, which received a bilateral ovariectomy, rendering them nonreproductive; and (3) Egg removal, in which the single-egg clutch was ablated to eliminate the physical burden of carrying an egg. Reproduction imposes costs with respect to (A) probability of survival over the breeding season (May–September) and subsequent overwinter period (September–May); (B) energy allocation, as inferred from treatment effects on growth in snout-vent length (SVL) over the breeding season, and on the wet mass of abdominal fat bodies; and (C) maintenance physiology, as inferred from treatment effects on localized swelling response to a novel antigen (phytohemagglutinin, PHA) and on hematocrit (proportion of red blood cells to total blood volume). The physical burden of an egg also reduces (D) locomotor performance, as inferred from changes between pre- and post-treatment measures of stamina (time to exhaustion on an inclined, motorized treadmill) and sprint speed (maximal speed on an inclined dowel). Data are summarized from three years of study involving hundreds of individual animals per treatment (Cox and Calsbeek 2010b; Cox *et al.* 2010).

and direct estimates of natural selection (Cox & Calsbeek 2010c). Following the elimination of reproduction via surgical ovariectomy, survival increases significantly over both the breeding and nonbreeding seasons (Figure 10.3), such that inter-annual survival to a second reproductive season increases threefold. This shifts females from a nearly annual life history to one in which a quarter or more of the adult population survives to a second year (Cox & Calsbeek 2010b). Elimination of reproduction also significantly increases growth, locomotor performance, energy storage, immune function, hematocrit, and parasite tolerance (Cox & Calsbeek 2010b; Cox *et al.* 2010; Figure 10.3). Removal of a single ovary maintains endogenous steroid production, reduces reproductive investment by roughly half, and produces intermediate costs with respect to growth, energy storage, and survival. Whether the energetic cost of reproduction is directly responsible for reducing survival can be tested by simultaneously manipulating both reproductive investment (via ovariactomy) and energy storage (via removal of fat bodies), thereby decoupling reproduction from its downstream effects on energy storage (Cox *et al.* 2014). In principle, similar manipulations of reproductive function can be achieved by castration in males, thereby facilitating a direct comparison of costs of reproduction in each sex.

A clear disadvantage to sterilization or gonadectomy is that it prevents an accounting of the balance between survival costs and reproductive benefits because it eliminates reproduction. Gonadectomy also removes the primary source of sex steroids and prostaglandins that coordinate numerous behavioral and physiological processes in addition to gamete production, which can complicate interpretations about the specific processes and pathways that shape the costs of reproduction. However, in the broadest sense, most of these behavioral and physiological effects can be viewed as components of reproductive investment, strengthening the argument for this approach to estimating the total costs of reproduction in each sex. Nonetheless, given the emerging view that many reproductive tradeoffs may be structured by endocrine signaling pathways rather than proximate energy allocation tradeoffs *per se* (Williams 2005; Edwards & Chapman 2011), it will often be desirable to discern between these mechanisms.

SUMMARY

Costs of reproduction structure important evolutionary tradeoffs in both sexes, but we still lack a general framework for comparing these sex-specific costs and integrating them into a holistic view of life-history evolution. In part, this is due to the historical separation of research on life-history evolution, which has focused primarily on females, versus sexual selection, which has focused primarily on males. Empirically, a major challenge to this integration arises because the primary costs of reproduction are often measured in different currencies, occur at different times, and encompass different regulatory processes in each sex. A first step toward integration between sexes is the measurement of reproductive costs in common currencies, particularly energy and fitness. The emerging field of intralocus sexual conflict provides an example of how genetic correlations between the sexes can constrain the evolution of optimal solutions to reproductive tradeoffs within each sex, emphasizing the importance of expressing reproductive tradeoffs in units of fitness. The immunocompetence handicap hypothesis illustrates the potential limitations of viewing costs of reproduction through mechanisms that are sex-specific, especially when the energetic tradeoffs that often structure such costs are generalizable to either sex. In this and other cases, a more informative framework could be established by focusing on downstream regulatory axes that are shared between the sexes, and by employing experimental manipulations that abolish the sex-specific aspects of reproductive investment. Taken together, these ideas form the basis of a more general framework for integrating theory on life-history evolution and sexual selection by focusing explicitly on the interaction of reproductive tradeoffs in both sexes.

ACKNOWLEDGMENTS

Many of the ideas presented here grew out of conversations and collaborations with Ryan Calsbeek and Henry John-Alder. Thanks to Malcolm Augat, Edmund Brodie III, Cameron Ghalambor, Amanda Hanninen, Ariel Kahrl, Megan Kobiela, Scott MacDougall-Shackleton, Marty Martin, Aaron Reedy, Brian Sanderson, Corlett Wood, and Art Woods for valuable comments on earlier versions of this chapter.

REFERENCES

Alonso-Alvarez, C., Bertrand, S. and Sorci, G. (2007) Energetic reserves, leptin and testosterone: a refinement of the immunocompetence handicap hypothesis. *Biology Letters*, **3**, 271–4.

Ardia, D.R., Schat, K.A. and Winkler, D.W. (2003) Reproductive effort reduces long-term immune function in breeding tree-swallows (*Tachycineta bicolor*). *Proceedings of the Royal Society of London B*, **270**, 1679–83.

Badyaev, A.V. and Martin, T.E. (2000) Sexual dimorphism in relation to current selection in the house finch. *Evolution*, **54**, 987–97.

Bell, G. and Koufopanou, V. (1986) The cost of reproduction, in *Oxford Surveys in Evolutionary Biology* (eds R. Dawkins and M. Ridley), Oxford University Press, Oxford, pp. 83–131.

Berg, E.C. and Maklakov, A.A. (2012) Sexes suffer from suboptimal lifespan because of genetic conflict in a seed beetle. *Proceedings of the Royal Society B: Biological Sciences*, **279**, 4296–4302.

Bonduriansky, R. and Chenoweth, S.F. (2009) Intralocus sexual conflict. *Trends in Ecology and Evolution*, **24**, 280–8.

Bonduriansky, R., Maklakov, A., Zajitschek, F. and Brooks, R. (2008) Sexual selection, sexual conflict and the evolution of ageing and life span. *Functional Ecology*, **22**, 443–53.

Brommer, J.E., Kirkpatrick, M., Qvarnström, A. and Gustafsson, L. (2007) The intersexual genetic correlation for lifetime fitness in the wild and its implications for sexual selection. *PLoS ONE*, **2**, e744.

Chippindale, A.K., Gibson, J.R. and Rice, W.F. (2001) Negative genetic correlation for adult fitness between sexes reveals ontogenetic conflict in *Drosophila*. *Proceedings of the National Academy of Science, USA*, **98**, 1671–5.

Christe, P., Glaizot, O., Strepparava, N. *et al.* (2012) Twofold cost of reproduction: an increase in parental effort leads to higher malarial parasitaemia and to a decrease in resistance to oxidative stress. *Proceedings of the Royal Society B: Biological Sciences*, **279**, 1142–9.

Chu, C.Y.C. and Lee, R.D. (2012) Sexual dimorphism and sexual selection: a unified economic analysis. *Theoretical Population Biology*, **82**, 335–63.

Cohen, A.A., Martin, L.B., Wingfield, J.C. *et al.* (2012) Physiological regulatory networks: ecological roles and evolutionary constraints. *Trends in Ecology & Evolution*, **27**, 428–35.

Cole, L.C. (1954) The population consequences of life history phenomena. *Quarterly Review of Biology*, **29**, 103–37.

Cox, R.M. and Calsbeek, R. (2009) Sexually antagonistic selection, sexual dimorphism, and the resolution of intralocus sexual conflict. *The American Naturalist*, **173**, 176–87.

Cox, R.M. and Calsbeek, R. (2010a) Cryptic sex-ratio bias provides indirect genetic benefits despite sexual conflict. *Science*, **328**, 92–4.

Cox, R.M. and Calsbeek, R. (2010b) Severe costs of reproduction persist in *Anolis* lizards despite the evolution of a single-egg clutch. *Evolution*, **64**, 1321–30.

Cox, R.M. and Calsbeek, R. (2010c) Sex-specific selection and intraspecific variation in sexual size dimorphism. *Evolution*, **64**, 798–809.

Cox, R.M. and John-Alder, H.B. (2007) Increased mite parasitism as a cost of testosterone in male striped plateau lizards, *Sceloporus virgatus*. *Functional Ecology*, **21**, 327–34.

Cox, R.M., Lovern, M.B., and Calsbeek, R. (2014) Experimentally decoupling reproductive investment from energy storage to test the functional basis of a life-history trade-off. *Journal of Animal Ecology*, **83**, 888–98.

Cox, R.M., Parker, E.U., Cheney, D.M. *et al.* (2010) Experimental evidence for physiological costs underlying the trade-off between reproduction and survival. *Functional Ecology*, **24**, 1262–9.

Cox, R.M., Skelly, S.L. and John-Alder, H.B. (2005) Testosterone inhibits growth in juvenile male eastern fence lizards (*Sceloporus undulatus*): Implications for energy allocation and sexual size dimorphism. *Physiological and Biochemical Zoology*, **78**, 531–45.

Edwards, D.A. and Chapman, T. (2011) Mechanisms underlying reproductive trade-offs: costs of reproduction, in *Mechanisms of Life History Evolution* (eds T. Flatt and A. Heyland), Oxford University Press, Oxford, pp. 137–52.

Fedorka, K.M. and Mousseau, T.A. (2004) Female mating bias results in conflicting sex-specific offspring fitness. *Nature*, **429**, 65–7.

Finch, C.E. and Rose, M.R. (1995) Hormones and the physiological architecture of life history evolution. *The Quarterly Review of Biology*, **70**, 1–52.

Fisher, R.A. (1930) *The Genetical Theory of Natural Selection*, Clarendon Press, Oxford.

Flatt, T. (2011) Survival costs of reproduction in *Drosophila*. *Experimental Gerontology*, **46**, 369–75.

Flatt, T. and Heyland, A. (2011) *Mechanisms of Life History Evolution: The Genetics and Physiology of Life History Traits and Trade-Offs: The Genetics and Physiology of Life History Traits and Trade-Offs*, Oxford University Press, Oxford.

Foerster, K., Coulson, T., Sheldon, B.C. *et al.* (2007) Sexually antagonistic genetic variation for fitness in red deer. *Nature*, **447**, 1107–11.

Folstad, I. and Karter, A.J. (1992) Parasites, bright males and the immunocompetence handicap hypothesis. *American Naturalist*, **139**, 603–22.

French, S.S., DeNardo, D.F. and Moore, M.C. (2007) Trade-offs between the reproductive and immune systems: facultative responses to resources or obligate responses to reproduction? *The American Naturalist*, **170**, 79–89.

French, S.S., Denise Dearing, M. and Demas, G.E. (2011) Leptin as a physiological mediator of energetic trade-offs in ecoimmunology: implications for disease. *Integrative and Comparative Biology*, **51**, 505–13.

Gadgil, M. and Bossert, W.H. (1970) Life historical consequences of natural selection. *The American Naturalist*, **104**, 1–24.

Harshman, L.G. and Zera, A.J. (2007) The cost of reproduction: the devil in the details. *Trends in Ecology and Evolution*, **22**, 80–6.

Hau, M. and Wingfield, J.C. (2011) Hormonally-regulated trade-offs: Evolutionary variability and phenotypic plasticity in testosterone signalling pathways, in *Mechanisms of Life History Evolution* (eds T. Flatt and A. Heyland), Oxford University Press, Oxford, pp. 349–61.

Hayward, A. and Gillooly, J.F. (2011) The cost of sex: quantifying energetic investment in gamete production by males and females. *PLoS ONE*, **6**, e16557.

Hoffman, C.L., Ruiz-Lambides, A.V., Davila, E. *et al.* (2008) Sex differences in survival costs of reproduction in a promiscuous primate. *Behavioral Ecology and Sociobiology*, **62**, 1711–18.

Hõrak, P. (2003) When to pay the cost of reproduction? A brood size manipulation experiment in great tits (*Parus major*). *Behavioral Ecology and Sociobiology*, **54**, 105–12.

Kalmbach, E., Griffiths, R., Crane, J.E. and Furness, R.W. (2004) Effects of experimentally increased egg production on female body condition and laying dates in the great skua *Stercorarius skua*. *Journal of Avian Biology*, **35**, 501–14.

Knowles, S.C.L., Nakagawa, S. and Sheldon, B.C. (2009) Elevated reproductive effort increases blood parasitaemia and decreases immune function in birds: a meta-regression approach. *Functional Ecology*, **23**, 405–15.

Koivula, M., Koskela, E., Mappes, T. and Oksanen, T.A. (2003) Cost of reproduction in the wild: manipulation of reproductive effort in the bank vole. *Ecology*, **84**, 398–405.

Kotiaho, J.S. and Simmons, L.W. (2003) Longevity cost of reproduction for males but no longevity cost of mating or courtship for females in the male-dimorphic dung beetle Onthophagus binodis. *Journal of Insect Physiology*, **49**, 817–22.

Lande, R. (1980) Sexual dimorphism, sexual selection, and adaptation in polygenic characters. *Evolution*, **34**, 292–307.

Lande, R. (1987) Genetic correlations between the sexes in the evolution of sexual dimorphism and mating preferences, in *Sexual Selection: Testing the Alternatives* (eds J.W. Bradbury and M.B. Andersson), John Wiley & Sons, Chichester, pp. 83–94.

Landwer, A. (1994) Manipulation of egg production reveals costs of reproduction in the tree lizard (*Urosaurus ornatus*). *Oecologia*, **100**, 243–9.

Martin, L.B., Hawley, D.M. and Ardia, D.R. (2011a) An introduction to ecological immunology. *Functional Ecology*, **25**, 1–4.

Martin, L.B., Liebl, A.L., Trotter, J.H. *et al.* (2011b) Integrator networks: Illuminating the black box linking genotype and phenotype. *Integrative and Comparative Biology*, **51**, 514–27.

Martin, L.B., Scheuerlein, A. and Wikelski, M. (2003) Immune activity elevates energy expenditure of house sparrows: a link between direct and indirect costs? *Proceedings of the Royal Society, London B: Biological Sciences*, **270**, 153–8.

Maynard Smith, J. (1978) *The Evolution of Sex*, Cambridge University Press, Cambridge.

McCurdy, D.G., Shutler, D., Mullie, A. and Forbes, M.R. (1998) Sex-biased parasitism of avian hosts: relations to blood parasite taxon and mating system. *Oikos*, **82**, 303–12.

McGlothlin, J.W., Parker, P.G., Nolan, V. Jr. and Ketterson, E.D. (2005) Correlational selection leads to genetic integration of body size and an attractive plumage trait in dark-eyed juncos. *Evolution*, **59**, 658–71.

McKean, K.A. and Lazzaro, B.P. (2011) The costs of immunity and the evolution of immunological defense mechanisms, in *Mechanisms of Life History Evolution* (eds T. Flatt and A. Heyland), Oxford University Press, Oxford, pp. 299–310.

Mills, S.C., Grapputo, A., Jokinen, I. *et al.* (2009) Testosterone-mediated effects on fitness-related phenotypic traits and fitness. *The American Naturalist*, **173**, 475–87.

Mokkonen, M., Kokko, H., Koskela, E. *et al.* (2011) Negative frequency-dependent selection of sexually antagonistic alleles in myodes glareolus. *Science*, **334**, 972–4.

Mokkonen, M., Koskela, E., Mappes, T. and Mills, S.C. (2012) Sexual antagonism for testosterone maintains multiple mating behaviour. *Journal of Animal Ecology*, **81**, 277–83.

Moore, S.L. and Wilson, K. (2002) Parasites as a viability cost of sexual selection in natural populations of mammals. *Science*, **297**, 2015–18.

Nunn, C.L., Lindenfors, P., Pursall, E.R. and Rolff, J. (2009) On sexual dimorphism in immune function. *Philosophical Transactions of the Royal Society B: Biological Sciences*, **364**, 61–9.

Olsson, M., Madsen, T. and Shine, R. (1997) Is sperm really so cheap? Costs of reproduction in male adders, Vipera berus. *Proceedings of the Royal Society B: Biological Sciences*, **264**, 455–9.

Owen-Ashley, N.T., Hasselquist, D. and Wingfield, J.C. (2004) Androgens and the immunocompetence handicap hypothesis: unraveling direct and indirect pathways of immunosuppression in song sparrows. *The American Naturalist*, **164**, 490–505.

Partridge, L. (1992) Measuring reproductive costs. *Trends in Ecology and Evolution*, **7**, 99–100.

Partridge, L., Gems, D. and Withers, D.J. (2005) Sex and Death: What is the connection? *Cell*, **120**, 461–72.

Partridge, L. and Harvey, P.H. (1985) Costs of reproduction. *Nature*, **316**, 20–1.

Penn, D.J. and Smith, K.R. (2007) Differential fitness costs of reproduction between the sexes. *Proceedings of the National Academy of Sciences*, **104**, 553–8.

Pettifor, R.A., Perrins, C.M. and McCleery, R.H. (2001) The individual optimization of fitness: variation in reproductive output, including clutch size, mean nestling mass and offspring recruitment, in manipulated broods of great tits *Parus major*. *Journal of Animal Ecology*, **70**, 62–79.

Pischedda, A. and Chippindale, A.K. (2006) Intralocus sexual conflict diminishes the benefits of sexual selection. *Public Library of Science Biology*, **4**, 1–5.

Preziosi, R.F. and Fairbairn, D.J. (2000) Lifetime selection in adult body size and components of body size in a waterstrider: opposing selection and maintenance of sexual size dimorphism. *Evolution*, **54**, 558–66.

Queller, D.C. (1997) Why do females care more than males? *Proceedings of the Royal Society B: Biological Sciences*, **264**, 1555–7.

Reed, W.L., Clark, M.E., Parker, P.G. *et al.* (2006) Physiological effects on demography: a long term experimental study of testosterone's effects on fitness. *The American Naturalist*, **167**, 667–83.

Reznick, D. (1985) Costs of reproduction: an evaluation of the empirical evidence. *Oikos*, **44**, 257–67.

Reznick, D. (1992) Measuring the costs of reproduction. *Trends in Ecology and Evolution*, **7**, 42–5.

Rice, W.R. and Chippindale, A.K. (2001) Intersexual ontogenetic conflict. *Journal of Evolutionary Biology*, **14**, 865–93.

Roberts, M.L., Buchanan, K.L. and Evans, M.R. (2004) Testing the immunocompetence handicap hypothesis: a review of the evidence. *Animal Behaviour*, **68**, 227–39.

Roff, D.A. (2002) *Life History Evolution*, Sinauer Associates Inc., Boston.

Roff, D.A. and Fairbairn, D.J. (2007) The evolution of trade-offs: where are we? *Journal of Evolutionary Biology*, **20**, 433–7.

Rolff, J. (2002) Bateman's principle and immunity. *Proceedings of the Royal Society of London. Series B: Biological Sciences*, **269**, 867–72.

Rolff, J., Armitage, S.A.O. & Coltman, D.W. (2005) Genetic constraints and sexual dimorphism in immune defense. *Evolution*, 1844–50.

Rose, M.R. and Bradley, T.J. (1998) Evolutionary physiology of the cost of reproduction. *Oikos*, **83**, 443–51.

Sheldon, B.C. and Verlhust, S. (1996) Ecological immunology: costly parasite defences and trade-offs in evolutionary ecology. *Trends in Ecology and Evolution*, **11**, 317–21.

Sinervo, B. and DeNardo, D.F. (1996) Costs of reproduction in the wild: path analysis of natural selection and experimental tests of causation. *Evolution*, **50**, 1299–1313.

Stearns, S.C. (1977) The evolution of life history traits: A critique of the theory and a review of the data. *Annual Review of Ecology and Systematics*, **8**, 145–71.

Stearns, S.C. (1989) Trade-offs in life-history evolution. *Functional Ecology*, **3**, 259–68.

Stearns, S.C. (1992) *The Evolution of Life Histories*, Oxford University Press, Oxford.

van Noordwijk, A.J. and Jong, G.d. (1986) Acquisition and allocation of resources: Their influence on variation in life history tactics. *The American Naturalist*, **128**, 137–42.

Wedekind, C. and Folstad, I. (1994) Adaptive or nonadaptive immunosuppression by sex hormones? *The American Naturalist*, **143**, 936–8.

Wedell, N., Kvarnemo, C., Lessells, C.M. and Tregenza, T. (2006) Sexual conflict and life histories. *Animal Behaviour*, **71**, 999–1011.

Weeks, S.C. (1996) The hidden cost of reproduction: reduced food intake caused by spatial constraints in the body cavity. *Oikos*, **75**, 345–9.

Westneat, D.F. and Birkhead, T.R. (1998) Alternative hypotheses linking the immune system and mate choice for good genes. *Proceedings of the Royal Society of London. Series B: Biological Sciences*, **265**, 1065–73.

Williams, T.D. (2005) Mechanisms underlying the costs of egg production. *BioScience*, **55**, 39–48.

Zahavi, A. (1975) Mate selection – a selection for a handicap. *Journal of Theoretical Biology*, **53**, 205–14.

Zera, A.J. and Harshman, L.G. (2001) The physiology of life history trade-offs in animals. *Annual Review of Ecology and Systematics*, **32**, 95–126.

11

IMMUNE SYSTEMS: LINKING ORGANISMS, POPULATIONS, AND EVOLUTION THROUGH DISEASE

James S. Adelman

Department of Biological Sciences, Virginia Polytechnic Institute and State University, Blacksburg, VA, USA

INTRODUCTION

At one point or another, almost every organism will face infection by parasites.[1] Despite this near universality, immune responses to infection vary markedly across individuals, sexes, populations, and species (Sheldon & Verhulst 1996; Norris & Evans 2000; Martin *et al.* 2008; Lazzaro & Little 2009). Because immune defenses play a crucial role in combating parasites, understanding variation in these organismal traits is integral to understanding higher-level biological processes such as disease transmission and host-parasite co-evolution (Roy & Kirchner 2000; Miller *et al.* 2006; Lazzaro & Little 2009; Little *et al.* 2010; Long & Boots 2011). Stated in the terms of this volume's overarching themes, variation in immune defense lies squarely at the interface between organism, ecology, and evolution (Ghalambor *et al.* 2014).

At the most basic level, the function of an immune system appears straightforward: to detect and neutralize threats posed by infecting parasites or damaged host tissues (Matzinger 1994). However, like any physiological process, immune defense incurs costs, including the utilization of limited resources and the potential to damage a host's own

[1]Here and throughout this chapter, I use the term "parasite" to include any organism that lives all or part of its life on or within another organism (its host), resulting in negative effects on that host. This encompasses both microparasites (viruses, bacteria, and protozoa) and macroparasites (helminths and other metazoa) (Dobson & Hudson 1986; Combes 2005; Boots *et al.* 2009).

Integrative Organismal Biology, First Edition.
Edited by Lynn B. Martin, Cameron K. Ghalambor, and H. Arthur Woods.

tissues (Sheldon & Verhulst 1996; Råberg *et al.* 1998; Lochmiller & Deerenberg 2000; Graham *et al.* 2005). Therefore, effective immune responses must be balanced against other fitness-enhancing traits, through coordination with diverse physiological systems. For example, wild vertebrates often show reduced immune responses during the breeding season, consistent with a need to allocate limited resources to reproduction (Martin *et al.* 2008). Such coordination is doubly critical because parasites can strike in almost any tissue type, requiring precise orchestration between immune and other cell types. For example, in mice experimentally infected with the intestinal helminth *Trichuris muris*, specific cytokines, signaling molecules secreted by immune cells, induce rapid proliferation of epithelial cells within the digestive tract, effectively dislodging parasites and facilitating their excretion (Cliffe *et al.* 2005). Both of these examples highlight an important caveat when considering variation in immune responses: these defenses do not act in a vacuum, but can interact with other host traits in a bidirectional manner.

In recent decades, the nascent field of ecological immunology has sought to understand how such interactions and trade-offs with other physiological systems can drive variation in immune defenses (Sheldon & Verhulst 1996; Viney *et al.* 2005; Lee 2006; Martin *et al.* 2008; Demas *et al.* 2012; Adelman *et al.* 2013a). However, few studies have focused on the consequences of this variation for parasite transmission or evolution, and even fewer have addressed how these effects on epidemiological dynamics will alter future selection pressures on immune defenses. In this chapter, I briefly review current theories regarding the drivers of immune heterogeneity, focusing on differences at the individual level. I then explore how theoretical and empirical studies have begun to link this variation with rates of parasite transmission and evolution. Finally, by highlighting gaps between current theoretical and empirical studies of immune variation and infectious disease dynamics, I propose avenues for future research.

IMMUNE SYSTEMS: A BRIEF OVERVIEW

Examining specific causes and consequences of variation in immune defenses requires some initial discussion of immune systems more generally. Historically, the term "immune" referred to complete protection from a particular disease or parasite. With increasing research into the processes underlying such protection, "immune" came to encompass a broad array of mechanisms that defend against parasites, many of which also protect against other noxious agents (such as foreign toxins or malfunctioning host tissues and molecules) (Matzinger 1994; OED 2013). Throughout this chapter, I will refer to several classes of such mechanisms: (1) sensors, those mechanisms that detect threats; (2) mediators, those mechanisms that signal detection to other immune cells and physiological systems; and (3) effectors, those mechanisms that kill or clear problems directly (Medzhitov 2008). The mechanisms in each class are numerous and diverse, often with redundancies within an individual (Nish & Medzhitov 2011) and analogies or homologies across species (Nurnberger *et al.* 2004; Schmid-Hempel 2005; Spoel & Dong 2012). Given this diversity, and in the interest of highlighting general principles of immune systems, I do not attempt an exhaustive review of immune mechanisms in any specific group or groups of organisms. Rather, when discussing examples of defenses in particular hosts, I will refer to the classifications of immune mechanisms put forward by Schmid-Hempel and Ebert (2003) (Figure 11.1). Because their classification relies upon general descriptions of immune defenses, rather than specific molecular definitions or strict divisions by cell type, it facilitates comparisons across broad taxa. Such breadth will be essential to uncovering

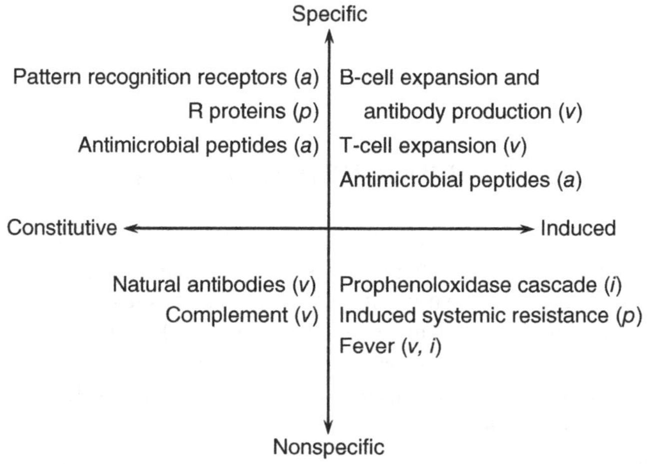

Figure 11.1. Immune defenses can be classified along two axes: (1) constitutive vs. induced and (2) specific vs. nonspecific. Here, common sensor and effector mechanisms from vertebrates (*v*), invertebrates (*i*), plants (*p*), or all (*a*) are placed in appropriate quadrants along these axes. Redrawn from Figure 1 in Schmid-Hempel, P., and D. Ebert (2003) On the evolutionary ecology of specific immune defence. *Trends in Ecology & Evolution*, **18**, 27–32. Copyright © Elsevier, 2003.

generalizable links among immunology, organismal biology, and population-level disease dynamics. The Schmid-Hempel and Ebert system organizes immune defenses along two axes: (1) constitutive *vs.* induced and (2) specific *vs.* nonspecific. Constitutive defenses are present and readily active in the body, even in the absence of infection. Induced defenses, by contrast, require some activation event, and often involve a delay before they can combat parasites or other dangers effectively. Specific defenses are targeted exclusively against individual threats, parasites, or molecular components thereof. Specific defenses have long been known to exhibit immune memory: upon encountering the same parasite for a second time, both the rate and magnitude of these responses is increased (Janeway *et al.* 2005). Nonspecific defenses do not exhibit the same degree of precision, responding to and combating broad classes of threats. Still, these defenses can exhibit a degree of immune memory: after initial infection by one type of parasite, nonspecific responses to different types of parasites can either increase or decrease in magnitude (West & Heagy 2002; Spoel & Dong 2012). Figure 11.1 places commonly studied sensor and effector mechanisms from vertebrates, invertebrates, and plants along these axes.

Traditionally, immunology has focused on mechanisms that kill or remove parasites, also known as mechanisms of "resistance." However, hosts can mitigate the effects of infection in another way: "tolerance," or the ability to minimize per-parasite reductions to fitness during infection, rather than reducing parasite numbers (Caldwell *et al.* 1958; Simms 2000; Råberg *et al.* 2009). Although tolerance has been documented in a number of taxa, from plants to animals, the mechanisms underlying this trait are largely unknown (Råberg *et al.* 2009; Sears *et al.* 2011). Recent work has suggested that among vertebrates, tolerance could be achieved by favoring specific rather than nonspecific induced responses (Råberg *et al.* 2009; Sears *et al.* 2011; Adelman *et al.* 2013b). In particular, inflammatory responses, such as the production of free-radicals, can help destroy numerous types of parasites, but can also cause severe damage to a host's own tissues, potentially reducing overall fitness (see references to immunopathology in the following section). Favoring highly targeted, specific defenses, such as antibody responses, could result in less damage, reducing fitness losses during infection.

PROPOSED DIVERS OF IMMUNE HETEROGENEITY

As Figure 11.1 indicates, hosts have evolved a broad array of defenses against parasitism. Given that most of Earth's species are parasites (during at least one part of their life cycles), such a breadth of defense is not surprising (May 1988; Thompson 1994; Windsor 1998). However, with the threat of parasitism so high, why hasn't selection swept immune defenses to maximal levels, leaving little variation? Answering this question has been a principal goal of the emerging field of ecological immunology and several nonexclusive hypotheses seek to explain this apparent paradox.

First, immune defenses can incur costs in terms of energy, amino acids, micronutrients, and time (Hart 1988; Klasing 2004; Demas *et al.* 2012). If any of these resources is scarce, organisms must trade-off the utilization of immune defenses against the expression of other costly, fitness-enhancing traits (Lochmiller & Deerenberg 2000; Demas *et al.* 2012). Substantial evidence suggests that immune defenses incur direct costs in terms of these resources, although these costs may vary significantly across species or environments (Yirmiya *et al.* 1995; Aubert *et al.* 1997; Svensson *et al.* 1998; Martin *et al.* 2003; Demas *et al.* 2012). Additionally, a number of studies have shown that increased activation of immune defenses results in decreased reproduction or growth, and vice versa, further supporting the idea that resource-based costs and trade-offs could help drive immune heterogeneity (Demas *et al.* 2012). Although the above studies focus on resistance mechanisms, tolerance can also incur life history costs, including reduced reproductive investment (Simms & Triplett 1994).

Second, whereas maximal levels of immune defenses may kill parasites most effectively, these can cause substantial immunopathology, or damage to a host's own tissues (Graham *et al.* 2005; Råberg *et al.* 2009). Overactive immune responses have been known to cause severe disease in humans for some time, but recent evidence shows that immunopathology also occurs in other species (Graham *et al.* 2005; Sadd & Siva-Jothy 2006; Graham *et al.* 2010). Costs associated with such self-damage will be twofold: (1) reductions in current function of damaged tissue and (2) resource costs needed to repair this damage.

Third, an organism's current and prior infection will also contribute to immune heterogeneity at multiple time scales. First, and most obviously, current infection will activate a given subset of immune defenses, causing currently infected individuals to display different immune phenotypes than uninfected individuals. More subtly, infection with one type of parasite can bias concurrent responses to different types of parasites. For example, mammalian immune responses to helminth worms can down-regulate responses to microparasites (Jackson *et al.* 2009). Such altered immune responses can result from normal constraints and interactions within the immune system itself and from direct attempts by parasites to modulate host immunity (Schmid-Hempel 2008; Jackson *et al.* 2009). Additionally, immune memory of past infections often increases the rate and magnitude of induced responses to subsequent exposure with the same parasite (Janeway *et al.* 2005). Interestingly, prior infection with one type of parasite can influence specific responses to another type of parasite (as distantly related as nematode worms and malarial protozoa), when mechanisms of immune memory allow for cross-reactivity (Fairlie-Clarke *et al.* 2010).

In some species, immune memory can span generations, with offspring of immune-challenged parents showing heightened induced defenses (Sadd *et al.* 2005; Roth *et al.* 2009; Spoel & Dong 2012). For instance, in bumble bees, offspring from queens previously injected with heat-killed bacteria show more pronounced antimicrobial activity than offspring from control queens (Sadd *et al.* 2005). Finally, at the evolutionary timescale, strong epidemics can select for specific immune phenotypes, resulting in temporal and geographic heterogeneity across populations. One example of such patterns involves major histocompatibility complex (MHC) proteins, which display foreign particles to immune effector- and mediator-cells in vertebrates (Edwards & Hedrick 1998). Because mutations in MHC alleles can alter resistance to particular parasites, the frequency and diversity of MHC alleles often change after epidemics; repeated epidemics of different parasites can maintain variation in these alleles at the level of the population or meta-population (Edwards & Hedrick 1998; Aguilar *et al.* 2004; Hawley & Fleischer 2012).

Finally, another possibility, which has received less attention in ecological immunology, is that the structure of physiological regulatory networks may constrain immune defenses below maximal levels (Cohen *et al.* 2012; Martin & Cohen 2014). As in the example of mice clearing intestinal parasites (Cliffe *et al.* 2005), immune defenses can require the recruitment of diverse tissue types, all of which must perform myriad other functions. Because selection does not act on any single function or tissue type alone, but on the whole-organism (Ghalambor *et al.* 2014), the structure of communication networks among these tissues will not be organized to maximize any single function, but must enable robust homeostasis of the whole organism (Cohen *et al.* 2007; Martin & Cohen 2014; Woods & Wilson 2014). So, although a maximal immune response may enable rapid clearance of parasites, this maximal response would likely have consequences for a diversity of tissue types within a physiological network, some of which would shift their other functions (or ability to interact with one another) away from optimal. As such, network structure, selected over long time periods, may help mediate costs to the entire organism, constraining immune responses while maintaining overall homeostasis. The idea that evolved signaling mechanisms can promote trade-offs among specific traits represents some overlap with the first proposed driver of immune variation (current costs and resource-based trade-offs). However, this possibility has been discussed more often in the context of individual signaling molecules (Ketterson *et al.* 2005; Muehlenbein & Bribiescas 2005; Hau 2007), rather than entire regulatory networks.

CONSEQUENCES OF IMMUNE HETEROGENEITY: THEORETICAL STUDIES

Although empirical studies have made considerable progress in understanding the causes of immune heterogeneity, we know less about what this variation means for parasite replication, spread, and evolution. Theoretical studies can examine the consequences of immune heterogeneity on infectious disease dynamics, both within individual hosts and at the population level. In doing so, such studies can highlight gaps in empirical knowledge that will be critical to linking individual variation in immunity with population-level processes.

Theoretical studies of disease dynamics (including those that incorporate immune heterogeneity) fall into two broad categories: within-host models and between-host models. Within-host models generally incorporate greater immunological detail than between-host models, focusing on how the timing of and interaction between immune defenses predict traits such as parasite load, time to parasite clearance, probability of host survival, and duration of immune memory (Alizon & van Baalen 2008). Between-host models, in contrast, generally include fewer details of host defenses, focusing instead on the rates at which hosts transition between three stages of infection: susceptible to infection, currently infected, and recovered from infection (S, I, and R, respectively, see Box 11.1) (Anderson & May 1979). Because differences in host defenses should influence transitions among all of these stages (Box 11.1), between-host models can examine the consequences of immune heterogeneity for population-level processes, including parasite spread and evolution.

Box 11.1. Immunological Heterogeneity and Models of Disease Dynamics

Including immune variation in mathematical models can help researchers understand infectious disease dynamics at both ecological and evolutionary timescales. Here I consider incorporating immune variation into a population model with directly transmitted parasites (Anderson & May 1979). Models of vector-borne parasites (e.g., malaria is transmitted between hosts by mosquitoes) can also accommodate such variation, with the complication that both host and vector immune systems will vary.

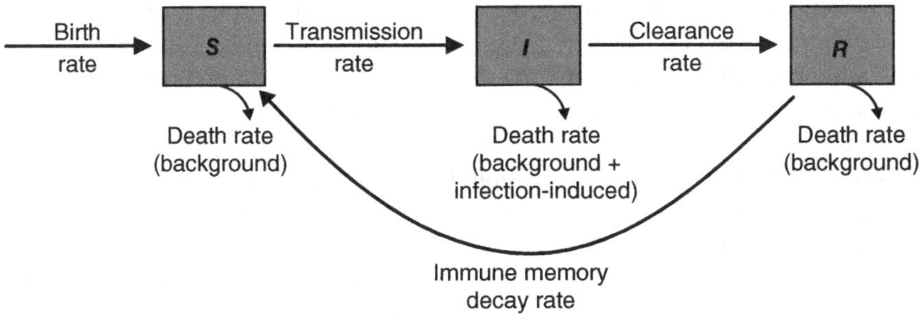

Individual hosts within a population can be classified as susceptible to infection (*S*), currently infected (*I*), or recovered (*R*). The rates at which hosts transition among these categories are the parasite transmission rate, parasite clearance rate, and immune memory decay rate, with hosts entering or leaving the population through birth and death.

The effects of variation in the memory decay rate are most straightforward. *R* individuals that mount less effective memory responses will reenter the *S* class more rapidly, increasing the abundance of the *S* individuals, and thus a parasite's opportunities to persist.

Immune variation will also affect recovery rates in an intuitive manner: *I* individuals mounting more effective responses will clear infections more rapidly, reducing a parasite's ability to spread. Although this effect is not surprising, complex dynamics can arise when immune responses themselves contribute to infection-induced mortality via immunopathology (Day *et al.* 2007; Best *et al.* 2012).

Parsing the effects of immune heterogeneity on parasite transmission rate requires breaking transmission into two components: 1) the contact rate between *S* and *I* individuals and 2) the probability that an *S* individual becomes infected after contacting an *I* individual (Dobson 1995). Contact rates will depend on the behavior of both *S* and *I* individuals. Because immune responses often promote sickness behaviors, e.g., reduced locomotion, *I* hosts mounting more pronounced responses will likely move less, reducing contact rates (Adelman & Martin 2009; Hawley *et al.* 2011). Immune variation within both *S* and *I* classes will also affect the probability of infection, once contact has occurred. *S* individuals with robust defenses are less likely to become infected, while *I* individuals mounting more robust immune responses will have lower parasite burdens, also reducing the likelihood of transmission.

Although the above analysis only considers resistance (the ability to kill parasites), variation in tolerance (the ability to minimize fitness effects of parasites) can also have dramatic effects on disease dynamics. For example, more tolerant individuals will likely remain in the *I* class longer, potentially accumulating higher parasite loads and increasing rates of transmission to *S* individuals.

Inter-individual variation in all of the above parameters can be incorporated using random values, drawn from empirical estimates. Additionally, variation can be partitioned among different classes of individuals, using variables such as sex, age, or reproductive status. Incorporating such realistic variation can improve the fit of disease models to empirical data (Lloyd-Smith *et al.* 2005). Moreover, examining how models perform across a range of immune parameters can yield insight into the types of empirical data most needed to understand epidemiological dynamics.

Hurtado's (2012) treatment of within-host parasite dynamics provides an excellent framework in which to examine the utility of such models (Figure 11.2). Although designed to capture the dynamics of a specific disease system, *Mycoplasma gallisepticum* (MG) infection in house finches (*Haemorhous mexicanus*), the methods and assumptions of this model can apply to a variety of systems. The core model (Figure 11.2A) considers immune defenses along the axes of Figure 11.1, with specific defenses induced directly by the parasites and indirectly by nonspecific host defenses. Such dynamics are typical of vertebrate immune systems, with initial nonspecific defenses (e.g., macrophages and dendritic cells) that help activate specific effectors (e.g., B- and T-cells) by presenting parasite-derived molecules and producing widely circulating signaling molecules (e.g., cytokines) (Janeway *et al.* 2005).

By incorporating these details, this within-host model can test how heterogeneity in immune activation and signaling alters both the magnitude and duration of parasite load during infection. For example, variation in how quickly nonspecific immune defenses are down-regulated after their initial induction ("decay rate," Figure 11.2A) can be a crucial determinant of variation in parasite load (Figure 11.2B) (Hurtado 2012). Importantly, testing such variation generates theoretical outputs that closely match the heterogeneity observed in empirical studies of this disease (Figure 11.2C). Because both transmission rates and infection-induced mortality will scale with parasite load, such models provide an important tool to examine how variation in particular aspects of host immunity should affect population-level disease processes.

Whereas within-host models can quantify how immune heterogeneity affects epidemiologically important parameters at the individual level (e.g., parasite load, pathology), between-host models can examine such effects at the population level and across evolutionary time. Although different modeling frameworks can facilitate such inquiries, those allowing hosts and parasites to coevolve in the presence of ecological feedbacks (e.g., density dependent birth or death rates) are most comparable to real biological systems (reviewed in Boots *et al.* 2009). For example, Best and colleagues (2009) examine how immune defenses can alter the evolution of parasite virulence (the likelihood that a parasite kills its hosts) and vice versa. Given the complexity of such coevolutionary scenarios, this model places specific limits on the influence of immune variation. First, no recovered class exists in this model (see Box 11.1), removing any influence of immune memory. Second, the model constrains how both hosts and parasites alter the rate of parasite transmission:

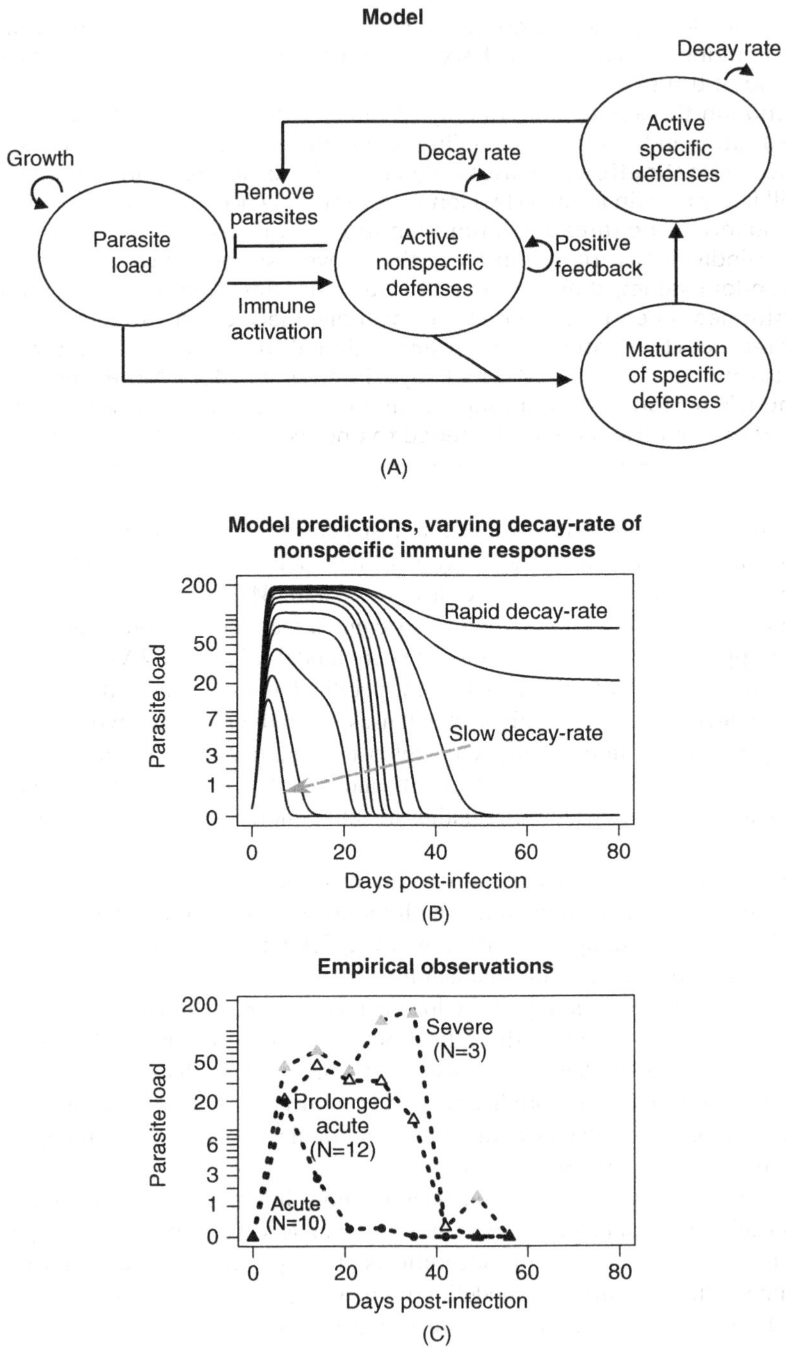

Figure 11.2. A within-host model of *Mycoplasma gallisepticum* infection in house finches (A) shows how modeling heterogeneity in interactions among immune defenses can closely match variation in parasite loads observed in empirical studies (B, C). In (A), arrows indicate positive relationships, blunt lines indicate negative relationships. In (C), data from experimental infections show that animals can be classified into three categories based upon the intensity and duration of their parasite load. *Source*: Redrawn from Figures 2 and 11 of Hurtado, P.J. (2012) Within-host dynamics of mycoplasma infections: Conjunctivitis in wild passerine birds. *Journal of Theoretical Biology*, **306**: 73–92. Copyright © 2012, Elsevier.

host immune defenses alter the probability that a susceptible host becomes infected, but once infected, a host's ability to transmit the parasite is controlled solely by the parasite. Importantly, the model assumes costs for both hosts and parasites: hosts must trade-off defenses against fecundity while parasites must trade-off transmissibility against virulence (the probability of killing the host in this model).

By assuming that small, heritable mutations arise rarely in both host and parasite, Best and colleagues (2009) show that a number of coevolutionary outcomes are possible. For instance, hosts and parasites can evolve to intermediate levels of immune defense and virulence, respectively, with only moderate heterogeneity in each. In contrast, hosts or parasites can evolve bimodal heterogeneity in terms of immune defense or virulence. For example, half of the host population can exhibit extremely high levels of immune defense, with the other half exhibiting very low defenses (and no individuals in between). With such bimodal heterogeneity in host immune defenses, the parasites tend to evolve higher virulence. Critically, these results are highly sensitive to the shapes of the trade-off curves for both the parasite and the host (Best *et al.* 2009) (Figure 11.3). This final result highlights a recurring theme in between-host models that incorporate immune defenses: outcomes often depend upon how the costs of an immune response increase with its magnitude (Figure 11.3). This is true whether costs are encoded as life-history trade-offs (based upon resource utilization) or increased mortality due to immunopathology (Day *et al.* 2007; Best *et al.* 2009; Boots *et al.* 2009; Best *et al.* 2012).

While the Best *et al.* model examines the importance of different relationships between the costs of an immune defense and its magnitude, the model does not incorporate different relationships between the benefits of an immune defense and its magnitude. Understanding both of these phenomena, i.e., how the *net* benefit of an immune response varies with its magnitude, should also have important consequences for the dynamics of host-parasite systems. As few studies have explicitly addressed this issue, I will return to this topic in the final section of this chapter.

Both the Best *et al.* and Hurtado models focus on issues associated with immune mechanisms of resistance, omitting tolerance (the host's ability to minimize per-parasite reductions in fitness). However, incorporating variation in tolerance can have important consequences for disease dynamics. For example, tolerance mechanisms can maintain host fitness during infection by either reducing the risk of mortality or by maintaining fecundity. The evolutionary dynamics of many between-host models can change dramatically depending upon which of these mechanisms hosts employ: tolerance mechanisms that maintain host fitness by reducing mortality will increase parasite prevalence, whereas tolerance mechanisms that maintain host fitness by increasing fecundity will not (reviewed in Boots *et al.* 2009).

Figure 11.3. The relationship between the magnitude of an immune response and its associated costs could follow linear, accelerating, or decelerating curves. These different possibilities can have dramatic consequences for host-parasite population dynamics and coevolution (see Best *et al.* 2009; Boots *et al.* 2009). However, empirical tests have yet to determine when and why each type of curve will exist.

Taken together, results from these and other within- and between-host models illustrate several important points about theory's role in understanding the consequences of immune heterogeneity. First and most generally, these models can make quantitative predictions about how immune variation will alter infectious disease dynamics, from parasite spread to evolution. Second and perhaps most importantly, because these predictions are based upon the assumptions each model makes, mechanistic details are critical. For example, in Best and colleagues' model, the specific relationship between the cost of an immune response and its magnitude greatly alters both host and parasite evolution. In addition, different models have shown that other mechanistic assumptions can affect model outcomes in similarly dramatic ways. For instance, using a model that assumes complete host recovery from infection and subsequent lifelong immunity, Pugliese (2011) has shown that population-level heterogeneity in immune defenses can select for decreased parasite virulence. This result contrasts with findings from Best and colleagues' model, which does not allow for host recovery or lifelong immunity. The discrepancy between these two models highlights a final point about mechanistic details and theoretical models: by examining how different assumptions influence model outcomes, theory can focus future empirical work on critical details of immune systems. From these between-host examples, details of immune memory, its potential to decay, and how varying degrees of memory may incur different levels of costs (for example increased energy use), all seem particularly relevant. Additionally, from the Hurtado within-host model, variation in the down-regulation of nonspecific immune responses may be critical in determining eventual parasite load. This pattern points toward studying particular mechanisms, including immunological signaling molecules such as interleukin-1β and interleukin-10, which increase and decrease the expression of nonspecific defenses, respectively (Vilček 1998; Giansanti *et al.* 2006). So, while theory can explore the potential consequences of variation in immune function, it can also encourage specific avenues for empirical study that will have the largest impacts upon population-level disease processes.

CONSEQUENCES OF IMMUNE HETEROGENEITY: EMPIRICAL STUDIES

The examples above highlight how theoretical models can explore the consequences of immune heterogeneity and guide new empirical studies of this variation. In addition to this theoretical work, several empirical studies have begun to examine the consequences of immune variation for infection outcomes, parasite transmission, and parasite evolution. Although such experimental work is less abundant than similarly-themed theory, the techniques used thus far provide important frameworks for exploring these links in the future.

One promising method for understanding the impact of immune heterogeneity on epidemiology involves experimentally increasing or decreasing immune responses. For example, recent experiments on rodent malaria have blocked immune signaling components to examine how immune differences alter parasite load and transmission potential. Long and colleagues (2008) have shown that blocking the action of tumor necrosis factor α (TNF-α), a signaling molecule that promotes inflammatory immune responses, increases both parasite load and transmission to mosquito vectors (when hosts are anesthetized) (Figure 11.4). This study represents a crucial step forward in linking differences in immune responses with disease dynamics. Future studies involving both vector-borne and directly transmitted parasites could be expanded to include the entire transmission cycle, from original host (through vectors) on to new hosts.

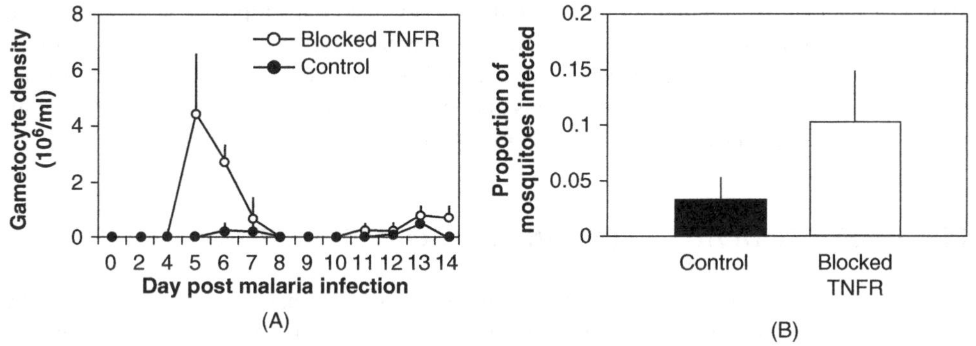

Figure 11.4. In mice, blocking the receptor (TNFR) for the pro-inflammatory signaling molecule (TNF-α) leads to increased levels of transmissible *Plasmodium chabaudi chabaudi* (gametocytes) in the host (A) and an increased probability of transmitting these parasites to mosquito vectors (B). *Source*: Redrawn from Figures 2 and 3 in Long, G.H., B.H.K. Chan, J.E. Allen, A.F. Read, and A.L. Graham (2008) Blockade of TNF receptor 1 reduces disease severity but increases parasite transmission during *Plasmodium chabaudi chabaudi* infection. *International Journal for Parasitology*, **38**, 1073–81. Copyright © Elsevier, 2008.

Another promising technique for linking immune variability with infectious disease dynamics capitalizes upon variability in parasite resistance across host populations or strains. In several experiments, Kubinak and colleagues (Kubinak *et al.* 2012; Kubinak & Potts 2013) have examined how different immune genotypes influence evolution of the Friend virus complex, whose primary pathology involves gross enlargement of the spleen. Specifically, the authors use mice from several inbred genetic strains, each of which possesses a unique combination of MHC alleles and a different level of initial resistance against the virus (recall that MHC genes enable host cells to present parasite-derived molecules to immune cells, making them critical players in inducing immune responses). Initial hosts from all genotypes are experimentally infected with the Friend virus complex, then during infection, virus is isolated from these infected hosts and experimentally passed to new, naïve hosts. This process is then repeated for ten rounds within each host strain. Utilizing this protocol, the authors have shown that after ten rounds of infection, the virus evolves two interesting and related features. First, viruses from higher resistance hosts replicate two times faster in these hosts than in other host genotypes, whereas viruses from lower resistance hosts replicate equally well across host genotypes. Second, viruses from higher resistance hosts evolve to a lower level of virulence than do viruses from lower resistance hosts (virulence being measured as the average pathology observed in subsequent experimental infections performed across all host genotypes) (Kubinak & Potts 2013). Taken together, these results suggest that variation in host immune defenses can select for both varying levels of parasite specialization and virulence. As with the example of murine malaria, this system represents an important step forward in understanding how heterogeneity in host defenses, here genetically encoded, can alter the evolution of a parasite. However, as above, future experiments could be expanded to include more realistic biological conditions, such as natural transmission among hosts.

Although currently limited, examples such as these illustrate that experiments can test the effects of immune heterogeneity on parasite transmission and evolution. Moreover, similar to the theoretical studies outlined in the previous section, empirical work shows that variation in specific mechanisms of defense can have pronounced impacts on epidemiological processes. Taken together these theoretical and empirical data suggest that not only are individual immune variation and infectious disease dynamics linked, but that understanding these links requires knowledge of specific mechanisms at the organismal level.

BRIDGING THE DIVIDE BETWEEN THEORY AND EXPERIMENT

As indicated above, both theoretical and empirical studies have begun to explore the consequences of immune heterogeneity for population-level processes, including parasite transmission and evolution. However, a substantial gap exists between the problems raised by theoretical studies and those addressed by experiments. Fortunately, there is substantial interest in bridging this gap (Alizon & van Baalen 2008; Hawley & Altizer 2010; Day *et al.* 2011). For example, a number of theoretical studies have begun embedding detailed within-host dynamics into between-host models (reviewed in Mideo *et al.* 2008). While such models can yield very powerful predictions, they are currently less common than simpler between-host models, likely because of the level of immune detail required to parameterize them accurately (Day *et al.* 2011). In order to realize the full potential of such models, including accurate predictions of how parasites will evolve under different conditions, empiricists have several clear tasks ahead.

First, empiricists must quantify the links among immune variation, parasite load, and parasite transmission directly. The work of Long and colleagues (2008) illustrates how this can be accomplished in vector transmitted diseases (Figure 11.4), but directly transmitted parasites will require different methods. One of the simplest designs would introduce infected individuals into groups of immune-deficient or wild-type conspecifics and track parasite spread. Depending upon the types of parasite used, this work may be constrained by logistical and ethical concerns. However, numerous species exist in which such dynamics can be studied, including the house finch-*Mycoplasma* system (Figure 11.2), in which immune signaling could potentially be manipulated as in Long *et al.* (2008) (Adelman *et al.* 2013b). Invertebrates, such as *Drosophila*, offer exciting possibilities for such studies, as animals with specific immune mutations are commercially available. Such systems would also facilitate expanded tests of the evolution of parasite virulence, including more natural modes of parasite transmission than have been previously utilized.

Additionally, empiricists must estimate how the costs and benefits of an immune response vary with the magnitude of that response (Figure 11.3). Whereas theoretical studies have shown that the relationships between the cost of an immune response and its magnitude can be critical to host-parasite coevolution, this cost is only part of the equation. The degree to which an organism employs a particular immune defense must depend more on the net benefit of that defense than simply its cost. Determining how the net benefits of an immune response vary with its magnitude could be achieved by combining experimental infections with immune manipulation. Ideally, several treatment groups would be infected with the same initial dose of a parasite. Concurrently, immune responses would be augmented or diminished within each group to yield a range of response magnitudes. Antibodies that disrupt immune signaling pathways, like those used in Long and colleagues' study (Figure 11.4), provide one potential mechanism, though the particular reagents used in that study did not produce dose-dependent effects (Long *et al.* 2008). Benefits could then be assessed in terms of pathogen clearance rates or reductions pathology (normalized to the lowest immune response level). Costs could be measured in terms of resource utilization (e.g., energy expended, amino acids utilized, etc.) and subtracted from benefits to yield a net benefit curve across this range of immune response. An alternative approach could incorporate numerous cost and benefit metrics into a principal components analysis, first assigning negative values to metrics of costs and positive values to metrics of benefits. In this construction, costs and benefits would load in opposite directions onto the first principal component, forming a new composite variable. This new variable could then be used as a single metric of net benefit and analyzed across the experimentally altered levels of immune activation.

Finally, empiricists must expand efforts to quantify interactions within the immune system and among immune defenses and other physiological processes. Because immune defenses can influence diverse tissue types, these defenses play a pivotal role in integrating information about a host's parasitic environment into whole-organism homeostatic networks (Husband 1995). Moreover, communication between the immune system and other physiological systems is bidirectional, enabling modulation of immune responses based upon external and internal states (e.g., ambient temperature, breeding season, social setting) (Martin *et al.* 2008; Adelman & Martin 2009). As the prior sections of this chapter have shown, the ways in which specific immune mechanisms vary will impact infectious disease dynamics. For instance, simply altering the timing of communication within the immune system can have profound effects on parasite loads and transmission potential (Pilyugin & Antia 2000, Wick & Self 2000, Hurtado 2012). The dynamic communication among immune defenses and other physiological systems could have similarly dramatic effects, potentially augmenting or diminishing immune variation, though these vast networks of interactions are currently poorly understood (Martin & Cohen 2014). Empirical data on these links could lend insight into how immune responses will vary under different social or environmental conditions, issues that will be critical to building the most realistic and predictive theoretical models.

Although such endeavors seem daunting, many of the necessary immunological tools exist, particularly for laboratory model species (Long *et al.* 2008; Shapira & Hacohen 2011). Moreover, with ecological immunology shifting its focus toward mechanisms that defend against ecologically relevant parasites (Adamo 2004; Owen & Clayton 2007; Graham *et al.* 2010), and theoretical models of parasite transmission incorporating detailed within-host processes (Alizon & van Baalen 2008), we are poised to make great strides in understanding how individual immune variation both shapes and responds to ecological and coevolutionary processes.

ACKNOWLEDGMENTS

I am very grateful to the editors of this volume and to Paul Hurtado, Cassandra Nuñez, Sahnzi Moyers, Kevin Matson, and one anonymous reviewer for their insightful comments on earlier versions of this manuscript.

REFERENCES

Adamo, S.A. (2004) How should behavioural ecologists interpret measurements of immunity? *Animal Behaviour*, **68**, 1443–9.

Adelman, J.S., Ardia, D.R. and Schat, K.A. (2013a) Ecoimmunology, in *Avian Immunology* (eds K.A. Schat, B. Kaspers and P. Kaiser), Academic Press, San Diego, pp. 391–411.

Adelman, J.S., Kirkpatrick, L., Grodio, J.L. and Hawley, D.M. (2013b) House finch populations differ in early inflammatory signaling and pathogen tolerance at the peak of *Mycoplasma gallisepticum* infection. *The American Naturalist*, **181**, 674–89.

Adelman, J.S. and Martin, L.B. (2009) Vertebrate sickness behaviors: Adaptive and integrated neuroendocrine immune responses. *Integrative and Comparative Biology*, **49**, 202–14.

Aguilar, A., Roemer, G., Debenham, S. *et al.* (2004) High MHC diversity maintained by balancing selection in an otherwise genetically monomorphic mammal. *Proceedings of the National Academy of Sciences of the United States of America*, **101**, 3490–4.

Alizon, S. and van Baalen, M. (2008) Acute or chronic? Within-host models with immune dynamics, infection outcome, and parasite evolution. *American Naturalist*, **172**, E244–E256.

Anderson, R.M. and May, R.M. (1979) Population biology of infectious diseases: Part I. *Nature*, **280**, 361–7.

Aubert, A., Goodall, G., Dantzer, R. and Gheusi, G. (1997) Differential effects of lipopolysaccharide on pup retrieving and nest building in lactating mice. *Brain Behavior and Immunity*, **11**, 107–18.

Best, A., Long, G., White, A. and Boots, M. (2012) The implications of immunopathology for parasite evolution. *Proceedings of the Royal Society B-Biological Sciences*, **279**, 3234–40.

Best, A., White, A. and Boots, M. (2009) The implications of coevolutionary dynamics to host-parasite interactions. *American Naturalist*, **173**, 779–91.

Boots, M., Best, A., Miller, M.R. and White, A. (2009) The role of ecological feedbacks in the evolution of host defence: what does theory tell us? *Philosophical Transactions of the Royal Society B-Biological Sciences*, **364**, 27–36.

Caldwell, R.M., Schafer, J.F., Compton, L.E. and Patterson, F.L. (1958) Tolerance to cereal leaf rusts. *Science*, **128**, 714–15.

Cliffe, L.J., Humphreys, N.E., Lane, T.E. *et al.* (2005) Accelerated intestinal epithelial cell turnover: A new mechanism of parasite expulsion. *Science*, **308**, 1463–5.

Cohen, A., Klasing, K. and Ricklefs, R. (2007) Measuring circulating antioxidants in wild birds. *Comparative Biochemistry and Physiology B – Biochemistry & Molecular Biology*, **147**, 110–21.

Cohen, A.A., Martin, L.B., Wingfield, J.C. *et al.* (2012) Physiological regulatory networks: ecological roles and evolutionary constraints. *Trends in Ecology & Evolution*, **27**, 428–35.

Combes, C. (2005) *The Art of Being a Parasite*, University of Chicago Press, Chicago.

Day, T., Alizon, S. and Mideo, N. (2011) Bridging scales in the evolution of infectious disease life histories: theory. *Evolution*, **65**, 3448–61.

Day, T., Graham, A.L. and Read, A.F. (2007) Evolution of parasite virulence when host responses cause disease. *Proceedings of the Royal Society B: Biological Sciences*, **274**, 2685–92.

Demas, G., Greives, T., Chester, E. and French, S. (2012) The energetics of immunity: mechanisms mediating trade-offs in ecoimmunology, in *Ecoimmunology* (eds G.E. Demas and R.J. Nelson), Oxford University Press, Oxford, pp. 259–96.

Dobson, A. (1995) The ecology and epidemiology of Rinderpest Virus in Serengeti and Ngorongoro Conservation Area, in *Serengeti II. Dynamics, Management, and Conservation of an Ecosystem* (eds A.R.E. Sinclair and P. Arcese), University of Chicago Press, Chicago, pp. 485–505.

Dobson, A.P. and Hudson, P.J. (1986) Parasites, disease, and the structure of ecological communities. *Trends in Ecology & Evolution*, **1**, 11–15.

Edwards, S.V. and Hedrick, P.W. (1998) Evolution and ecology of MHC molecules: from genomics to sexual selection. *Trends in Ecology & Evolution*, **13**, 305–11.

Fairlie-Clarke, K.J., Lamb, T.J., Langhorne, J. *et al.* (2010) Antibody isotype analysis of malaria-nematode co-infection: problems and solutions associated with cross-reactivity. *BMC Immunology*, **11**.

Ghalambor, C.K., Martin, L.B. and Woods, H.A. (2014) The conceptual links between organismal biology and evolutionary ecology: plasticity, integration and a theory of the individual, in *Integrative Organismal Biology* (eds L.B. Martin, C.K. Ghalambor and H.A. Woods), John Wiley & Sons, Inc., Hoboken.

Giansanti, F., Giardi, M.F. and Botti, D. (2006) Avian cytokines – an overview. *Current Pharmaceutical Design*, **12**, 3083–99.

Graham, A.L., Allen, J.E. and Read, A.F. (2005) Evolutionary causes and consequences of immunopathology. *Annual Review of Ecology, Evolution and Systematics*, **36**, 373–97.

Graham, A.L., Hayward, A.D., Watt, K.A. *et al.* (2010) Fitness correlates of heritable variation in antibody responsiveness in a wild mammal. *Science*, **330**, 662–665.

Hart, B.L. (1988) Biological basis of the behavior of sick animals. *Neuroscience and Biobehavioral Reviews*, **12**, 123–37.

Hau, M. (2007) Regulation of male traits by testosterone: implications for the evolution of vertebrate life histories. *Bioessays*, **29**, 133–44.

Hawley, D.M. and Altizer, S.M. (2011) Disease ecology meets ecological immunology: understanding the links between organismal immunity and infection dynamics in natural populations. *Functional Ecology*, **251**, 48–60.

Hawley, D.M., Etienne, R.S., Ezenwa, V.O. and Jolles, A.E. (2011) Does animal behavior underlie covariation between hosts' exposure to infectious agents and susceptibility to infection? Implications for disease dynamics. *Integrative and Comparative Biology*, **51**, 528–39.

Hawley, D.M. and Fleischer, R.C. (2012) Contrasting epidemic histories reveal pathogen-mediated balancing selection on class II MHC diversity in a wild songbird. *PLoS ONE*, **7**, e30222.

Hurtado, P.J. (2012) Within-host dynamics of mycoplasma infections: Conjunctivitis in wild passerine birds. *Journal of Theoretical Biology*, **306**, 73–92.

Husband, A.J. (1995) The immune system and integrated homeostasis. *Immunol Cell Biol*, **73**, 377–82.

Jackson, J.A., Friberg, I.M., Little, S. and Bradley, J.E. (2009) Review series on helminths, immune modulation and the hygiene hypothesis: Immunity against helminths and immunological phenomena in modern human populations: coevolutionary legacies? *Immunology*, **126**, 18–27.

Janeway, C.A., Travers, P., Walport, M. and Capra, J.D. (2005) *Immunobiology: The Immune System in Health and Disease*, 6th edn, Garland, New York.

Ketterson, E.D., Nolan, V. and Sandell, M. (2005) Testosterone in females: Mediator of adaptive traits, constraint on sexual dimorphism, or both? *American Naturalist*, **166**, S85–S98.

Klasing, K.C. (2004) The costs of immunity. *Acta Zoologica Sinica*, **50**, 961–9.

Kubinak, J.L. and Potts, W.K. (2013) Host resistance influences patterns of experimental viral adaptation and virulence evolution. *Virulence*, **4**, 410–18.

Kubinak, J.L., Ruff, J.S., Hyzer, C.W. *et al.* (2012) Experimental viral evolution to specific host MHC genotypes reveals fitness and virulence trade-offs in alternative MHC types. *Proceedings of the National Academy of Sciences of the United States of America*, **109**, 3422–7.

Lazzaro, B.P. and Little, T.J. (2009) Immunity in a variable world. *Philosophical Transactions of The Royal Society B: Biological Sciences*, **364**, 15–26.

Lee, K.A. (2006) Linking immune defenses and life history at the levels of the individual and the species. *Integrative and Comparative Biology*, **46**, 1000–15.

Little, T.J., Shuker, D.M., Colegrave, N. *et al.* (2010) The coevolution of virulence: tolerance in perspective. *PLoS Pathogens*, **6**, e1001006.

Lloyd-Smith, J.O., Schreiber, S.J., Kopp, P.E. and Getz, W.M. (2005) Superspreading and the effect of individual variation on disease emergence. *Nature*, **438**, 355–9.

Lochmiller, R.L. and Deerenberg, C. (2000) Trade-offs in evolutionary immunology: just what is the cost of immunity? *Oikos*, **88**, 87–98.

Long, G.H. and Boots, M. (2011) How can immunopathology shape the evolution of parasite virulence? *Trends in Parasitology*, **27**, 300–5.

Long, G.H., Chan, B.H.K., Allen, J.E. *et al.* (2008) Blockade of TNF receptor 1 reduces disease severity but increases parasite transmission during *Plasmodium chabaudi chabaudi* infection. *International Journal for Parasitology*, **38**, 1073–81.

Martin, L.B. and Cohen, A. (2014) Physiological regulatory networks: the orchestra of life?, in *Integrative Organismal Biology* (eds L.B. Martin, C.K. Ghalambor and H.A. Woods), John Wiley & Sons Inc., Hoboken.

Martin, L.B., Scheuerlein, A. and Wikelski, M. (2003) Immune activity elevates energy expenditure of house sparrows: a link between direct and indirect costs? *Proceedings of the Royal Society of London Series B-Biological Sciences*, **270**, 153–8.

Martin, L.B., Weil, Z.M. and Nelson, R.J. (2008) Seasonal changes in vertebrate immune activity: mediation by physiological trade-offs. *Philosophical Transactions of the Royal Society B-Biological Sciences*, **363**, 321–39.

Matzinger, P. (1994) Tolerance, danger, and the extended family. *Annual Review of Immunology*, **12**, 991–1045.

May, R.M. (1988) How many species are there on earth? *Science*, **241**, 1441–9.

Medzhitov, R. (2008) Origin and physiological roles of inflammation. *Nature*, **454**, 428–35.

Mideo, N., Alizon, S. and Day, T. (2008) Linking within- and between-host dynamics in the evolutionary epidemiology of infectious diseases. *Trends in Ecology & Evolution*, **23**, 511–17.

Miller, M.R., White, A. and Boots, M. (2006) The evolution of parasites in response to tolerance in their hosts: The good, the bad, and apparent commensalism. *Evolution*, **60**, 945–56.

Muehlenbein, M.P. and Bribiescas, R.G. (2005) Testosterone-mediated immune functions and male life histories. *American Journal of Human Biology*, **17**, 527–58.

Nish, S. and Medzhitov, R. (2011) Host defense pathways: Role of redundancy and compensation in infectious disease phenotypes. *Immunity*, **34**, 629–36.

Norris, K. and Evans, M.R. (2000) Ecological immunology: life history trade-offs and immune defense in birds. *Behavioral Ecology*, **11**, 19–26.

Nurnberger, T., Brunner, F., Kemmerling, B. and Piater, L. (2004) Innate immunity in plants and animals: striking similarities and obvious differences. *Immunological Reviews*, **198**, 249–66.

OED (2013) "immune, adj. and n." In *Oxford English Dictionary*. Oxford University Press.

Owen, J.P. and Clayton, D.H. (2007) Where are the parasites in the PHA response? *Trends in Ecology & Evolution*, **22**, 228–9.

Pilyugin, S.S. and Antia, R. (2000) Modeling immune responses with handling time. *Bulletin of Mathematical Biology*, **62**, 869–90.

Pugliese, A. (2011) The role of host population heterogeneity in the evolution of virulence. *Journal of Biological Dynamics*, **5**, 104–19.

Råberg, L., Graham, A.L. and Read, A.F. (2009) Decomposing health: tolerance and resistance to parasites in animals. *Philosophical Transactions of the Royal Society B: Biological Sciences*, **364**, 37–49.

Råberg, L., Grahn, M., Hasselquist, D. and Svensson, E. (1998) On the adaptive significance of stress-induced immunosuppression. *Proceedings of the Royal Society of London Series B-Biological Sciences*, **265**, 1637–41.

Roth, O., Sadd, B.M., Schmid-Hempel, P. and Kurtz, J. (2009) Strain-specific priming of resistance in the red flour beetle, Tribolium castaneum. *Proceedings of the Royal Society B-Biological Sciences*, **276**, 145–51.

Roy, B.A. and Kirchner, J.W. (2000) Evolutionary dynamics of pathogen resistance and tolerance. *Evolution*, **54**, 51–63.

Sadd, B.M., Kleinlogel, Y., Schmid-Hempel, R. and Schmid-Hempel, P. (2005) Trans-generational immune priming in a social insect. *Biology Letters*, **1**, 386–8.

Sadd, B.M. and Siva-Jothy, M.T. (2006) Self-harm caused by an insect's innate immunity. *Proceedings of the Royal Society B-Biological Sciences*, **273**, 2571–4.

Schmid-Hempel, P. (2005) Evolutionary ecology of insect immune defenses. *Annual Review of Entomology*, **50**, 529–51.

Schmid-Hempel, P. (2008) Parasite immune evasion: a momentous molecular war. *Trends in Ecology & Evolution*, **23**, 318.

Schmid-Hempel, P. and Ebert, D. (2003) On the evolutionary ecology of specific immune defence. *Trends in Ecology & Evolution*, **18**, 27–32.

Sears, B.F., Rohr, J.R., Allen, J.E. and Martin, L.B. (2011) The economy of inflammation: when is less more? *Trends in Parasitology*, **27**, 382–7.

Shapira, S.D. and Hacohen, N. (2011) Systems biology approaches to dissect mammalian innate immunity. *Current Opinion in Immunology*, **23**, 71–7.

Sheldon, B.C. and Verhulst, S. (1996) Ecological immunology: Costly parasite defences and trade-offs in evolutionary ecology. *Trends in Ecology and Evolution*, **11**, 317–21.

Simms, E.L. (2000) Defining tolerance as a norm of reaction. *Evolutionary Ecology*, **14**, 563–70.

Simms, E.L. and Triplett, J. (1994) Costs and benefits of plant responses to disease: resistance and tolerance. *Evolution*, **48**, 1973–85.

Spoel, S.H. and Dong, X. (2012) How do plants achieve immunity? Defence without specialized immune cells. *Nature Reviews: Immunology*, **12**, 89–100.

Svensson, E., Raberg, L., Koch, C. and Hasselquist, D. (1998) Energetic stress, immunosuppression and the costs of an antibody response. *Functional Ecology*, **12**, 912–19.

Thompson, J.N. (1994) *The Coevolutionary Process*, University of Chicago Press, Chicago.

Vilček, J. (1998) The Cytokines: An overview, in *The Cytokine Handbook* (ed A. Thomson), Academic Press, San Diego, CA.

Viney, M.E., Riley, E.M. and Buchanan, K.L. (2005) Optimal immune responses: immunocompetence revisited. *Trends in Ecology & Evolution*, **20**, 665–9.

West, M.A. and Heagy, W. (2002) Endotoxin tolerance: A review. *Critical Care Medicine*, **30**, S64–S73.

Wick, D. and Self, S.G. (2000) Early HIV infection in vivo: branching-process model for studying timing of immune responses and drug therapy. *Mathematical Biosciences*, **165**, 115–34.

Windsor, D.A. (1998) Most of the species on Earth are parasites. *International Journal for Parasitology*, **28**, 1939–41.

Woods, H.A. and Wilson, J.K. (2014) An elephant in the fog: unifying concepts of physiological stasis and change, in *Integrative Organismal Biology* (eds L.B. Martin, C.K. Ghalambor and H.A. Woods), John Wiley & Sons, Inc., Hoboken.

Yirmiya, R., Avitsur, R., Donchin, O. and Cohen, E. (1995) Interleukin-1 inhibits sexual behavior in female but not in male rats. *Brain Behavior and Immunity*, **9**, 220–33.

12

DEVELOPMENTAL PLASTICITY OF INDIVIDUAL VARIATION IN STRESS RESPONSES

Haruka Wada

Department of Biological Sciences, Auburn University, Auburn, AL, USA

INTRODUCTION

How an individual responds to physical, physiological, and psychological challenges is fundamental to life. Animals have evolved both specific and nonspecific organismal responses to escape from and adjust to new environments and overcome challenges. These responses, called stress responses (Box 12.1), vary greatly among species and populations as well as between individuals (Guimont & Wynne-Edwards 2006; Schrandt & Lema 2011). They are generally repeatable within individuals across time (Wada *et al.* 2008; Rensel & Schoech 2011), but the origins and consequences of such individual differences are not well understood. The external environment is constantly in flux, and internal physiological demands also change with physical activity and life history stages. Thus the first question to answer is under what conditions a change in environment becomes a stressor. Interestingly, even in identical environments, individuals at the same life history stage, e.g., lactation, hibernation, or migration, show different stress responses. This leads to the second question; where does the individual variation come from?

Box 12.1. Definition of Components of Stress

The word "stress" has been used in everyday language as well as in range of disciplines from engineering to medicine, causing confusion and debate over the definitions. Stress at times refers to a stimulus, yet it also can refer to a state, or consequences of such stimuli or states (see Romero *et al.* 2009; Koolhaas *et al.* 2011

(*continued*)

Integrative Organismal Biology, First Edition.
Edited by Lynn B. Martin, Cameron K. Ghalambor, and H. Arthur Woods.

for discussion). In most cases, stress tends to refer to something undesirable and harmful. In this chapter, I try to avoid references to stress at all, instead emphasizing stressors, stress responses and components and consequences thereof. I define stress responses as modifications in physiology and behavior that attempt to return the body to equilibrium (i.e., rheostasis, allostasis, and predictive and reactive homeostasis). I define a stressor as an external or internal stimulus that initiates a stress response. As the body's response to stressors evolved to adjust to new or evolutionarily rare environments, stress responses can have both adaptive and harmful outcomes, depending on the duration and magnitude of the stressors as well as physiological state of the animal and its past experiences, as each phenotype is composed of gene (G), environment (E), G × E, as well as irreversible developmental plasticity, reversible phenotypic flexibility, and their interactions (see Chapter 8, this volume).

This chapter focuses on variation among individuals – what contributes to this variation and how is it maintained through natural selection. I argue that (1) parental environment directly and indirectly affects physiological set points for the stress responses of offspring, (2) individual variation in organismal responses to stressors is a result of an interaction among genes, developmental plasticity, phenotypic flexibility, and current environment, and (3) parental fitness may ultimately direct inter-individual variation in the stress response among siblings. Throughout the chapter, I emphasize the need to consider multiple dimensions of the stress response from physiological mediator interactions (e.g., hormone-hormone, hormone-cytokine) to developmental background of whole individuals. Through this approach, I propose that we can better understand how stress responses are shaped by natural selection.

WHEN IS A CHANGE IN ENVIRONMENT A STRESSOR?

Traditionally, stress responses have been defined as processes aiming to reestablish homeostasis when it is truly, or only perceived as, compromised (Koolhaas *et al.* 2011) (detailed discussion of stasis and change can be found in Chapter 8 this volume). Modern views on stress responses tend to distinguish routine, predictable changes from unpredictable, uncontrollable stimuli as response thresholds between normal and pathological outcomes of stress responses change with context. Two concepts, allostasis and the reactive scope model, address this perspective (McEwen & Wingfield 2003; Landys *et al.* 2006; Romero *et al.* 2009).

Allostasis Model

Allostasis describes a process of "maintaining stability through change" (Sterling & Eyer 1988). Allostatic states describe altered levels of mediators of stress responses, such as glucocorticoids (GCs), catecholamines, and cytokines, that parallel the change in energetic demands of life history stages (McEwen & Wingfield 2003). At any point in time, an organism is at one of the three physiological states: (A) a basal physiological state needed to maintain life, (B) an elevated baseline state associated with daily and seasonal changes in life history priorities and other predictable demands, and (C) a physiological state entered in response to unpredictable, uncontrollable changes in the environment (Landys *et al.* 2006). Circulating hormones and cytokines track physiological states

A, B, and C, and purportedly support the requisite physiological processes of each state. For instance, some migratory birds fly more than 10 000 km to reach their breeding or wintering destinations. These seasonal, energetically demanding journeys are made possible by morphological, physiological, and behavioral modifications, including regression of the nonessential organs, increased ability to catabolize lipid stores, and behavioral adjustments during flight to conserve energy (Weber 2009). Migratory fattening in preparation for long journeys (at physiological state B) appears to be regulated in part by GCs and thyroid hormones (Cornelius *et al.* 2013). In other words, GCs and thyroid hormones facilitate the establishment of a life-history-specific set point for body mass and fat stores, which could accommodate predictable increases in energetic demands (Cornelius *et al.* 2013). However, these modified hormone levels tend not be sustained outside seasonal stages. Indeed, allostatic load, physiological "wear and tear" via modified stress response mediator levels, can eventually become allostatic overload (i.e., a pathophysiological state) (McEwen & Wingfield 2003).

Reactive Scope Model

A modified version of allostasis, the reactive scope model, incorporates permanent changes in the thresholds that separate normal from pathological states (Romero *et al.* 2009). This model also distinguishes predictive homeostasis – i.e., adjusted levels of physiological mediators in response to predictable changes in life-history (internal) demands – from reactive homeostasis – i.e., altered physiological mediator levels in response to unpredictable changes in the environment (external). Predictive and reactive homeostasis are both well-regulated by healthy individuals (Romero 2009), and they correspond to physiological states B and C in the allostasis model (Landys 2006; Romero 2009). Homeostatic failure refers to a range below a threshold of minimum levels for life, corresponding to below the physiological state A in the allostasis model. Homeostatic overload refers to the range above reactive homeostasis when physiological mediators become harmful to the body. Although the two models have their differences, both have enhanced our understanding of stress-responses. As with other models (summarized in Chapter 8, this volume), they emphasize that physiological set points change with time, stress responses shift from adaptive to detrimental depending on physiological state, and an accumulation of "wear and tear" from repeated exposure to stressors leads to pathology and disease.

ORGANISMAL RESPONSES TO STRESSORS

The above models and the many that came before them have been particularly good at focusing attention on the key elements of stress responses organisms use to accommodate stressors. In general, once an individual perceives a stressor, its body initiates a range of physiological and behavioral responses. Initial responses facilitate an escape from the stressor whereas delayed responses aid in managing or adapting to the stressor, such as recovery from injuries or exposure to pathogens. In this section, I discuss one type of organismal responses to stressors, hormonal responses.

Hormonal Responses

Secretion of two classes of hormones, the catecholamines and GCs, is a hallmark of stress responses. Catecholamines, comprised of noradrenaline and adrenaline (also known as norepinephrine and epinephrine) from the sympathetic nervous system and the adrenal medulla, are secreted within milliseconds to seconds after detection of stressors (Sapolsky *et al.* 2000; Romero & Butler 2007). These hormones act rapidly through membrane-bound adrenergic receptors on the peripheral tissues, effectively coordinating fight-or-flight responses. Catecholamines, for instance, route ATP and other resources to vital tissues necessary for a swift escape and increase arousal, heart rate, blood glucose levels, blood pressure, and visual acuity, as well as dilation of the bronchioles and shunting blood to skeletal muscles (Sapolsky *et al.* 2000; Romero & Butler 2007). In contrast, GCs from the hypothalamic-pituitary-adrenal (HPA) axis are slower to respond, and are mainly directed toward recovery from stressors and/or preparing for future stressors (Johnson *et al.* 2002). Within minutes of detecting stressors, GCs are secreted from the adrenal cortex, redirecting energy and behavior from nonessential toward critical processes for survival including gluconeogenesis, and adjustments of the immune and reproductive systems and of feeding behaviors (Wingfield *et al.* 1998; Sapolsky *et al.* 2000).

Apart from responding to stressors, both catecholamine and GC baseline levels change with age and activity. As catecholamines regulate blood pressure and heart rate, catecholamine levels are higher in the morning compared to during sleep and tend to increase with age and exercise (Sachs *et al.* 1985). GC levels are also at their lowest during the inactive phase and start to rise before dawn with their highest levels reached just after awakening, to stimulate appetite and feeding (Sapolsky *et al.* 2000; Dimitrov *et al.* 2009). Baseline GCs also prepare and mediate the transition between critical life-history stages (Crespi & Denver 2005; Wada 2008). Although these hormones play important roles in adjusting the body to daily, seasonal, and age-related physiological demands, prolonged elevation beyond the normal range leads to allostatic and homeostatic overloads (McEwen & Wingfield 2003; Romero *et al.* 2009). For instance, once catecholamines are elevated for a prolonged period of time, systolic hypertension and cardiovascular hypertrophy can result (Sterling & Eyer 1988). Likewise, prolonged elevation of GCs impairs growth, immune function, and reproduction (Hayes *et al.* 1993; Sapolsky *et al.* 2000; Spencer *et al.* 2003; Wingfield & Sapolsky 2003).

Regulators of the Adrenocortical Responses

The efficacy of any hormone is regulated by the abundance of hormone receptors on the target cell, affinity of those receptors, bioavailability of the hormones, and rates of clearance of the hormone from the body. GCs bind to two types of intracellular receptors: high-affinity, mineralocorticoid receptors (MR) and low-affinity glucocorticoid receptors (GR) (de Kloet *et al.* 1993). Once bound to the intracellular receptors, receptors dimerize and act as transcription factor, binding to glucocorticoid response elements on the DNA to activate or repress gene transcription (Archer *et al.* 1990). Although effects of GCs tend to be slower than catecholamine-mediated outcomes, GCs bind nuclear and membrane-bound receptors, making some of their effects rapid (Tasker *et al.* 2006). Such rapid effects include behavioral modifications, e.g., suppression of reproductive behaviors (Rose *et al.* 1993; Moore & Orchinik 1994). GC secretion is tightly regulated by negative feedback through intracellular MR and GR as well. MRs are important for regulating a circadian change in basal GC secretion (Dallman *et al.* 1994). GRs, particularly in hippocampus, hypothalamus, and amygdala, play important roles in the negative feedback regulation of GCs during stress responses (Sapolsky *et al.* 1984). There is recent evidence that GCs can be locally synthesized in intestines, lymphoid organs, skin, and brain (Taves *et al.* 2011). Local synthesis may be partly responsible for the rapid, nongenomic effects of GCs (Schmidt *et al.* 2008).

Two main factors regulate the amount of GCs reaching receptors. The first, hepatic corticosteroid binding globulin (CBG), is thought to limit the availability of GCs to the tissues (free hormone hypothesis (Ekins 1990; Rosner 1990); see also Schoech *et al.* 2013). Second, 11β-hydroxysteroid dehydrogenease type 2 (11β-HSD-2) inactivates GCs to protect the target cells from effects of GCs (Robson *et al.* 1998). Similar to GR and MR, set points of CBG and 11β-HSDs change with age, season, and experience. The seasonal changes in CBG level parallel that of GC in house sparrow brain, with the highest levels found during breeding seasons (Breuner & Orchinik 2001). It is suspected that high levels of CBG-bound corticosterone during a breeding season allow higher availability of corticosterone to tissues by translocating CBG-corticosterone into cells directly or locally cleaving CBG from corticosterone (Breuner & Orchinik 2001). In mammals, 11β-HSD-2 in placenta reduces the exposure of fetus to maternal GCs (Seckl 2001) and the levels in individuals is at highest during development to minimize the negative GCs on growth (Robson *et al.* 1998).

INTEGRATING DEVELOPMENTAL PLASTICITY

The phenotype is a combination of gene-environment interactions. More specifically, it reflects permanent phenotypic change due to developmental environment or past experience, reversible phenotypic state in response to current environment, which are both influenced by variation in genetic background (see Chapter 8, this volume). Such plasticity may largely explain individual variation in responses to stressors. In species with extensive parental care, the developmental environment is defined mostly by the parents. How much each parent can provide for the young depends on the availability of resources, predictability of the environment, and the decisions parents make about allocating resources between offspring and self (Figures 12.1 and 12.2). Evolutionary history of the population, via parental genetic contribution, interacts with the developmental environment to determine the set points of physiological regulators (e.g., GR, MR, CBG, and 11β-HSD), thus adjusting mediator levels (e.g., GCs). Ultimately, direct and inclusive fitness of parents through offspring with various stress response phenotypes shapes hormonal and nonhormonal stress responses in the population (Figure 12.1).

Irreversible Developmental Plasticity vs. Reversible Phenotypic Flexibility

In the HPA axis, corticosteroid receptor density shows irreversible developmental plasticity, which sets the sensitivity of the individual to future stressors. Three series of experiments in domesticated rodents provide the most compelling evidence that developmental environment determines the set points for GC via its regulators. The first examined the correlation among individual variation in maternal care and offspring behavioral and physiological responses to stressors. Rat mothers varied in the amount of licking and grooming (LG) they exhibited toward their offspring (Liu *et al.* 1997; Liu *et al.* 2000). Adult offspring of high LG mothers showed a reduced HPA response to a stressor, were less fearful (Liu *et al.* 1997; Caldji *et al.* 1998), and had better spatial memory (Liu *et al.* 2000), owing to the lower corticotropin releasing hormone mRNA in the paraventricular nucleus of the hypothalamus, and more modest adrenocorticotropic hormone (ACTH) responses to experimenter handling (Liu *et al.* 1997). Cross-fostering of pups between high and low LG mothers revealed that maternal behavior directly influenced the hippocampal development; pups from either HG or LG mothers were equally impacted by variation in maternal behavior (Liu *et al.* 2000).

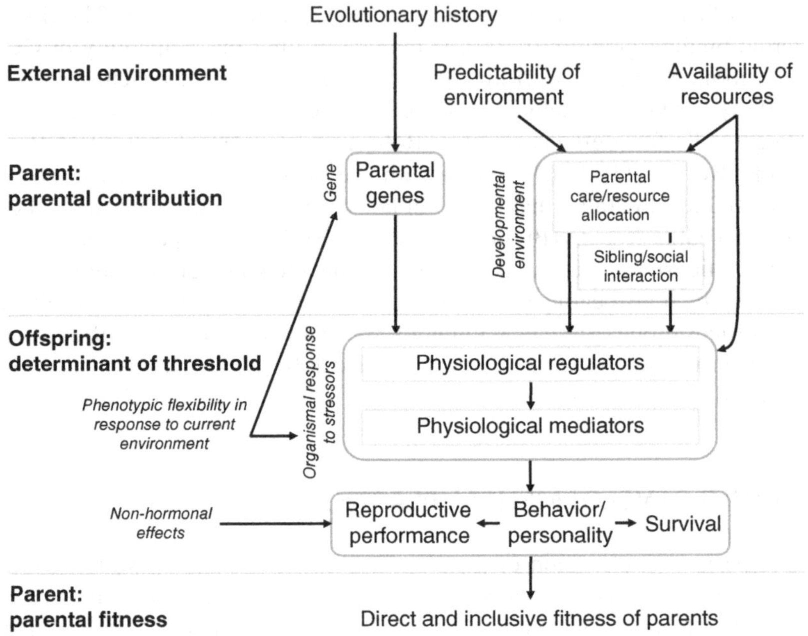

Figure 12.1. Factors influencing offspring mediator levels in species with extensive parental care. Mediator levels (e.g., GCs) are directly controlled by regulators (e.g., glucocorticoid and mineralocorticoid receptors, corticosteroid binding globulins, and 11β-hydroxysteroid dehydrogeneases). Regulator levels are influenced by interactions among genes and the developmental environment, which coarsely determines the set point of the regulators. Current condition/environment fine-tunes these set points. Offspring fitness ties back to parental fitness.

The second series of studies entailed brief daily separation of pups from mothers, which led to hyporesponsive HPA axes and emotional stability of the pups once adults (Meaney 2001). The strength of the negative feedback onto the HPA axis played a key role: early handling up-regulated glucocorticoid receptors in the hippocampus and frontal cortex, exerting stronger negative feedback in handled than in nonhandled rats. Typically, as these rodents age, negative feedback weakens and baseline and stress-induced levels of GC increase (Landfield *et al.* 1978; Sapolsky *et al.* 1983). This weakening is accompanied by a neuron loss in the hippocampus and a deterioration of spatial learning (Meaney *et al.* 1988). Rats handled as neonates slowed down age-related deterioration of hippocampal neurons and spatial learning (Meaney *et al.* 1988).

The third series of experiments exposed rat pups to maternal separation of 1–24 hours. Adults who were exposed to maternal separation as neonates had higher corticotropin releasing hormone expression in the hypothalamus (Plotsky & Meaney 1993), elevated adrenocortical responses to handling (Plotsky & Meaney 1993; Kalinichev *et al.* 2002) and startle responses (Kalinichev *et al.* 2002) compared to adults that experienced brief separation and early handling or no separation or handling at all. Interestingly, artificial tactile stimulation by a paintbrush mimicked maternal licking and either entirely or partially reversed the effects of maternal separation. Offspring with tactile stimulation during the first week of life increased hippocampal GR expression and reduced responses to

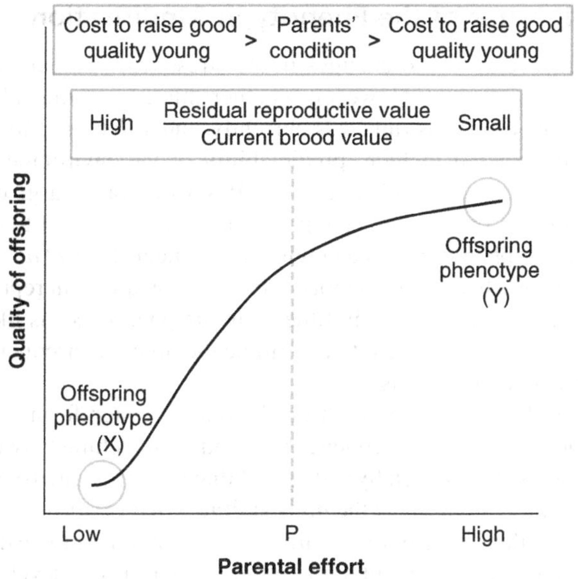

Figure 12.2. Relationship between parental effort and offspring quality. The amount of parental care determines the offspring quality in a sigmoidal fashion (Pianka 1992). The amount of parental care and effort is determined by parents' condition relative to the cost to raise high quality offspring and residual reproductive value (RRV) relative to the current brood value. Two offspring phenotypes are highlighted: low quality offspring phenotype X (when cost to raise good quality young is higher than parents' condition and RRV is relatively high; orange in Figure 12.4) and high quality offspring Y (when cost to raise good quality young is lower than parents' condition and RRV is relatively low; blue in Figure 12.4).

handling (Jutapakdeegul *et al.* 2003). Once female offspring raised with tactile stimulation were allowed to mate, their maternal behavior was indistinguishable from female offspring reared by attentive mothers (Gonzalez *et al.* 2001). These studies show that moderate stressors during development can enable animals to accommodate stressors in adulthood while severe stressor has the opposite effects on the HPA axis function and emotionality.

In addition to enduring changes in HPA setpoints mediated by early-life experiences, reversible, short-term changes in the HPA axis are also possible. Levels of CBG that determine how much GC is biologically available to target cells rapidly change in response to capture and handling stress (Breuner *et al.* 2006) and metabolic needs (Tinnikov 1999; Lynn, Breuner & Wingfield 2003). For instance, in rats, fasting and swimming increased GC and decreased CBG levels within 2 days, elevating free GC that are available to the tissues. Thus, reduction in CBG likely magnifies the bioavailability of GC after GC elevation to match the metabolic demands of fasting and exercise. These and other studies demonstrate that prior experience gives rise to the differences in behavior and sensitivity to future stressors. After developmental environments coarsely set the sensitivity of the HPA axis, more transient changes in HPA set points can ensue in response to other environmental stimuli. Stressors such as exercise and food restriction provoke phenotypic flexibility of set points (Tinnikov 1999; Lynn *et al.* 2003), which reflects the interaction among developmental plasticity and current condition.

Adaptive Significance of the Phenotypic Modification

Overwhelming evidence points to changes in the stress response phenotype being set by the developmental environment. However, whether this environmentally-induced phenotypic change is adaptive or constrained is currently under debate (Monaghan 2008). The key factors being considered include: predictability of the environment, life expectancy of the species, and the selective advantage of HPA variation in a given environment. If environmentally-induced phenotypic change is adaptive, we expect to see a difference in fitness among resulting phenotypes in a single environment (Ellis *et al.* 2006). At the same time, in an environment where each phenotype can employ different reproductive strategies and niche partitioning, the difference in fitness among phenotypes is likely to decrease or diminish altogether. Thus, it is critical to determine the fitness consequences of phenotypes in relation to different environments.

One current focal point for research on the adaptive significance of the phenotypic modification is whether fitness consequences depend on matching between developmental and adult environments. It has been hypothesized that early-life environment programs the phenotype such that individuals have the highest fitness in an environment that matches the environment in which they were born (environmental matching models; reviewed in Monaghan (2008)). When an environmental condition is poor during development, a trade-off arises between allocating limited resources to one organ vs. another (Barker 2004; Bateson *et al.* 2004). This trade-off may initially alleviate mortality risk but can be costly to reproductive performance and adult survival if a change in environment leads to mismatch between what the animal is programmed to expect and what the actual environment turns out to be. This scenario of environmental matching is plausible if the environment is stable across time. Likewise, the environmental match would be more important in short-lived species where environment is less likely to change within developmental stages. However, support for environmental matching models is inconsistent. Marshall & Uller (2007) point out a possible explanation of inconsistencies among developmental studies. They argue that selection pressure acts to maximize parental fitness, rather than offspring fitness (Marshall & Uller 2007; Schwarzkopf & Andrews 2012). If parents can breed in a better circumstance in the future, parents are likely to forgo current breeding, or reduce investment, at the expense of offspring quality and survival (Marshall & Uller 2007). Ultimately, parents with high direct and inclusive fitness have a selective advantage (Figure 12.1), so studies focused on offspring alone could produce mixed results.

Alternatively, individuals born in good environments might be of better quality and have permanent fitness advantages over those born in poor environments (silver-spoon effects (Grafen 1988; Monaghan 2008)). In this model, better environments in adulthood increases fitness regardless of early-life environment, although fitness of individuals born in poor environments never surpasses individuals born in good environments. Varying combinations of silver-spoon and environmental matching models are also possible (outlined in Monaghan 2008); for example, individuals born in good environments generally do better than ones born in poor environments except in very poor adult environments. Another possibility is that while those born in good environment improve fitness as the adult environment improves, those born in poor environments decrease fitness as adult environment improves due to environmental mismatching.

Another hypothesis posits that mothers trigger phenotypic change in the offspring that match maternal condition (maternal matching hypothesis) (Love & Williams 2008; Sheriff & Love 2013). That is, in a poor environment, the energy demand to raise offspring is lowered to match the reduced caring capacity of the mother. Love and colleagues (2005) conducted a series of experiments to test this hypothesis in the European Starlings

(*Sturnus vulgaris*). They found that as maternal body condition declined, plasma GC and GC deposition into the yolk increased. Experimental elevation of yolk GC resulted in sons that were smaller at hatching, had slower growth rates, and higher mortality rates during early stages of nestling compared to sons from vehicle-injected eggs (Love *et al.* 2005; Love & Williams 2008). Most intriguingly, when feather clipping during egg incubation artificially worsened maternal condition in combination with GC or vehicle injection of eggs, return rate of the mothers in the subsequent years was indistinguishable between sham-clipped/vehicle injection and feather-clipped/GC injection (Love & Williams 2008). This indicates that maternal fitness was high if the feeding demands from nestlings matched the body conditions of their mothers.

The above studies suggest that mothers adjusted parental effort based on condition. However, parents may also bias care toward certain offspring to diversify offspring quality within a clutch or brood. A recent study in rodents reveals that within high or low licking and grooming mothers, there is considerable variation in maternal care among siblings (Claessens *et al.* 2011). Two scenarios can explain why mothers differentiated care among siblings. One is that a mother produces more offspring than she can rear successfully. Another is that mothers use a bet-hedging strategy. Whereas a predictable environment enables parents to prepare offspring for the anticipated environment, in an unpredictable environment, parents can maximize their fitness by producing offspring with diverse phenotypes (Marshall & Uller 2007). That is, parents can maintain consistent inclusive fitness if they produce offspring with varied phenotypes so at least some of the offspring survive and reproduce (Figure 12.3). Interestingly, a similar strategy of stabilizing cellular stress occurs at a molecular level. One of the heat shock proteins (Hsp 90) functions to aid folding and maintenance of other proteins in nonstressed cells (see Chapter 1 for details). Without Hsp 90, higher variation in protein expression will manifest, thus Hsp 90 stabilize and minimize the phenotypic variation within the individual.

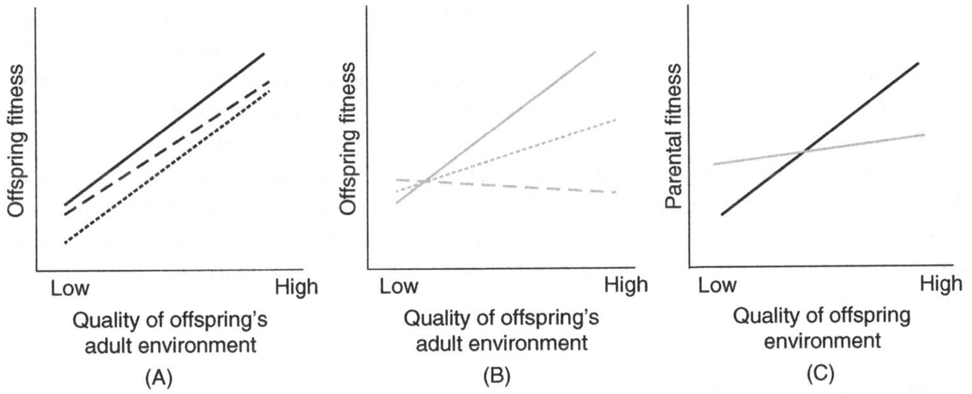

Figure 12.3. Relationship between individual fitness and its adult environment in a predictable (A) and unpredictable environments (B). Three lines in A and B represent 3 distinct relationships between quality of adult environment and fitness for each sibling. In a predictable environment, parents produce offspring that respond to adult environment in a similar way, leading to adaptive phenotypic change (A). However, in an unpredictable environment, parents have better overall fitness if they produce offspring that vary in phenotypic flexibility in response to adult environment, leading to bet hedging by the parents (B). C depicts a resulting parental fitness in A (black line) in B (gray line).

Framework Linking Parental Investment, Physiological Mediators, Regulators, and Offspring Phenotype

When resources are not limited, one may think parents provide good care to all the offspring equally. In reality however, parental care is inconsistent between and within clutches or broods. Studies in rodents showed that the degree of parental care depended on how the parent itself was cared for early in its life. In other cases, the degree of parental care may be attributed to resource availability, attraction of its mate, and future reproductive potential, as life history theory posits that there is a trade-off between current reproduction vs. self-maintenance and future reproduction (Creighton *et al.* 2009). When food availability is low, parental effort and care for the young decreases as parental condition declines (Figure 12.2). Similarly, if parents' residual reproductive value is relatively high compared to the value of the current brood, then parental effort and care for the current brood should be low. As a result, parents give rise to a range of offspring quality with varying performance, including sensitivity to future stressors (Figure 12.4A). The prediction here is that as a result, there will be high variance in physiological set points among offspring with a range of parental care. This is likely established by differential levels of regulators. Because of the interaction between developmental plasticity and phenotypic flexibility, quality and match of the developmental and adult environments may also affect the levels of regulators (Figure 12.1).

In the case of the HPA axis, GR levels are one potential mediator of sensitivity to stressors. GR levels are likely to co-vary with developmental environment and be fine-tuned by current condition. To simplify, only two phenotypes are represented in Figure 12.4. Figure 12.4A shows physiological set points and resulting GC in the low and high quality offspring. Figure 12.4B through D depict possible changes in the regulator set points in different environments according to the environmental matching model (individuals do better when perinatal and adult environments match, Figure 12.4B), silver-spoon model (individuals born in good environments do better than individuals born in poor environments, Figure 12.4C), and the combination of two models (Figure 12.4D) where improvement of the adult environment increases the set points but in a poor adult environment, individuals born in a poor environment have a higher regulator levels than ones born in a good environment (see Monaghan (2008) for the details of each models). These levels of regulator set points translate to the threshold between reactive homeostasis and homeostatic overload. In reality, the reactivity of the HPA axis depends on multiple regulators, such as levels of CBG and 11β-HSDs, thus it is important to consider multiple dimension of the stress responses. These models offer testable predictions for the mediators and regulators of the organismal response to stressors. By relating the regulator and mediator levels to fitness in varying environments will shed light on the adaptive significance of the environmentally-induced plasticity in the stress responses.

FUTURE DIRECTIONS

Individual Variation in Responses to Stressor and Fitness Consequences

There are several recent reviews on the relationship between GCs in adulthood and fitness (Moore & Jessop 2003; Breuner, Patterson & Hahn 2008; Bonier *et al.* 2009a; Breuner

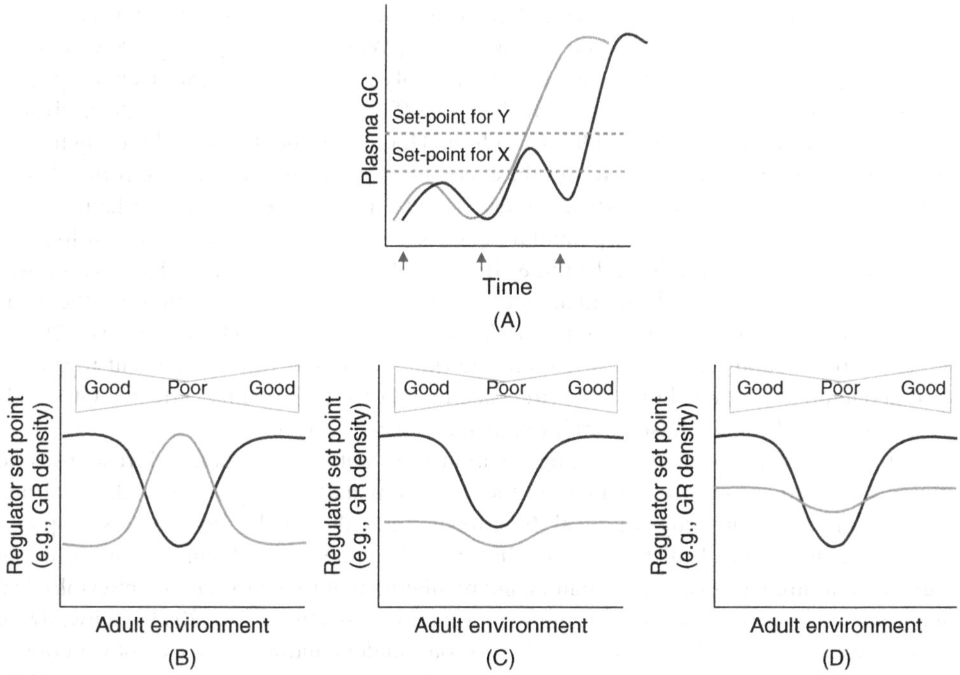

Figure 12.4. Schematic of how parental effort in Figure 12.2 can affect set points of regulators and plasma GC of offspring. To simplify, two offspring phenotypes are represented: low quality offspring phenotype X due to poor developmental environment (orange lines) and high quality offspring phenotype Y due to good developmental environment (blue lines). These phenotypes differ in a set point where if plasma GC surpasses in response to any stimulus (red arrow, e.g., low energy storage), neither behavioral and physiological modification nor negative feedback can restore reactive homeostasis (A). Two lines are off-set to increase visibility. In reality, the threshold between reactive homeostasis and homeostatic overload (in reactive scope model), or between physiological state B & C (in allostasis model) is likely to change with developmental environment, age, and interaction between developmental and adult environment. Figures B through D depict a change in regulator set point across developmental environment according to the environmental matching model (B), silver spoon model (C), and the combination of the two models (D). Gray box represents adult environments. (See plate section for color version.)

2011; Crespi *et al.* 2013). In essence, those papers point out a real dearth of studies that directly measure fitness. For those that did consider fitness, there was equivocal empirical support for relationships between GCs and fitness. I expect that three factors explain some of the discrepancies among studies: 1) the relationship between GCs and fitness-related measures can change across time, age, and sex (Brown *et al.* 2005; Bonier *et al.* 2009b), 2) individuals differ in response to environmental challenges due to varying current condition, such as energy storage and physiological state (Kitaysky *et al.* 2010), and 3) the extent at which baseline GCs change in response to an environmental challenge depends on whether the challenge is within or beyond the predicted setpoint for the life-history stage

(McEwen & Wingfield 2003; Romero, Dickens & Cyr 2009). For example, GCs can suppress reproductive behaviors (Moore & Jessop 2003; Wingfield & Sapolsky 2003; Breuner 2011; Crespi *et al.* 2013), but these effects are probably context-dependent. In environments with limited resources, breeding individuals trade off current reproduction with survival to the future reproduction (Stearns 1992; Creighton, Heflin & Belk 2009). Redirecting behavior away from reproduction by GCs in these conditions may not be adaptive if the chance of future breeding is low. For instance, older parents should invest more in offspring than younger parents as reproductive potential declines with age (terminal investment hypothesis, (Williams 1966; Creighton, Heflin & Belk 2009)). In essence, individuals with lower future reproductive potential should dampen adrenocortical response to minimize the interference of GCs on reproductive activities (Wingfield & Sapolsky 2003; Bokony *et al.* 2009). In fact, experimental reductions of clutch size (i.e., decreased value of current reproduction) and temporal removal of a mate (thus decreased offspring quality) affected GC levels (Lendvai, Giraudeau & Chastel 2007; Lendvai & Chastel 2008).

The next steps to understanding relationships between organismal responses to stressors and fitness are to consider: 1) the developmental history of the individual and 2) interactions among physiological functional units that make up a stress response phenotype. Inclusion of all possible mediators and regulators is daunting but we might start by examining a subset of mediators and regulators that relate to specific physiological functions such as metabolism or immune function (Martin *et al.* 2011; Schwartz & Bronikowski 2011; Cohen *et al.* 2012). As our understanding of these physiological networks progresses, components of the networks and connections among them will be illuminated allowing researchers to select cohesive sets of integrators to examine together.

Epigenetic Modifications by Stressors

Epigenetic modification of HPA setpoints by stressors is a relatively new area of research. Early-life adversity often induces *context-dependent* epigenetic modifications of the HPA axis (Crews 2008). This type of epigenetic modification to the HPA axis may often be important because physiology and behavior can be passed trans-generationally as long as individuals experience the same, particular contexts. For instance, serotonin increases expression of nerve growth factor-inducible factor A (NGFI-A), a transcription factor that binds to the promoter region of the GR gene to initiate GR mRNA expression in hippocampus (Bagot *et al.* 2009). Licking and grooming (LG) as well as simple tactile stimulation increased serotonin activity, NGFI-A expression, and GR expression. High GR expression in offspring of high LG mothers persisted until adulthood via DNA demethylation. The NGFI-A binding site on the GR gene is highly methylated at birth (Weaver *et al.* 2004). LG demethylated the NGFI-A binding site (Weaver *et al.* 2004) and intracerebroventricular infusion of a methyl donor group reversed the effect of maternal care on GR expression (Weaver *et al.* 2005). Although this type of epigenetic modification is not directly inherited, it can be inherited indirectly through behavioral modifications of the parents (Ledon-Rettig *et al.* 2012).

In contrast to context-dependent epigenetic modifications, *germline-dependent* epigenetic modification is not influenced by later environment. It is also heritable, thus this type of modification becomes important in an evolutionary context (Crews 2008; Ledon-Rettig *et al.* 2012). For instance in rats, a single exposure to an endocrine disrupting chemical, the fungicide vinclozolin, during gestation reduced sperm count, motility, and attractiveness of male offspring for generations (Crews 2008). It also affected how individuals respond to chronic stressors: young males 3 generations after the single exposure to vinclozolin reduced metabolic activity in hippocampus, a brain region that exerts negative feedback on the HPA axis, in response to chronic stressors (Crews *et al.* 2012). Ancestral exposure of vinclozolin heightened anxiety in stressed individuals while nonexposed individuals lowered anxiety in response to chronic stressors. These alterations in brain and behavior suggest germ-line epigenetic modifications in stress response-related tissues.

Heritable, epigenetic modification should act like genetic variation within a population (Ledon-Rettig *et al.* 2012). However, epigenetic change is more reversible than genetic change and may be enforced by subsequent environment. Thus epigenetic change is beneficial in fluctuating, unpredictable environments and may minimize the adversity imposed by mismatch between environment and phenotype (Ledon-Rettig *et al.* 2012).

Multidimensional Connections among Physiological Mediators

This chapter mainly focused on the environmentally-induced changes in the HPA axis that gives rise to the individual variation in hormonal responses to stressors. However, the components of the HPA axis represent only a fraction of the myriad physiological mediators that allow animals to adapt to new environments. For example, the immune response is an integral part of the organismal stress response and mediates recovery and preparation for additional stressors by initiating inflammatory response and redistributing immune cells to where they are needed. The immune system has reciprocal relationships with the nervous system and the HPA axis to orchestrate the organismal stress responses. For instance, psychological stressors, such as open field test, restraint, or electric footshock, elevate catecholamines, which increase transcription and plasma levels of pro-inflammatory cytokines, interleukin (IL)-1β (Grebe *et al.* 2010) and IL-6 prior to a rise in GC (Zhou *et al.* 1993; Grebe *et al.* 2010). Catecholamines and GCs also switch one type of adaptive immunity to another. GCs inhibits IL-12 which triggers T helper (T_H) 1 production, shifting the adaptive response from T_H1 (involved in cellular immunity) to T_H2 (humoral immunity) type response (Elenkov & Chrousos 2002; Sternberg 2006). Catecholamines also suppress IL-12 and stimulate IL-10, together shift toward T_H2 type response (Elenkov & Chrousos 2002). The relationship between GC and immune system is often bi-directional: cytokines can also activate the HPA axis (Martin 2009). Normally, corticotropin releasing hormone (CRH) and arginine vasopressin together stimulate the release of ACTH from the anterior pituitary (Antoni 1993). ACTH then stimulates the adrenal cortex to secrete GCs. However, murine cytomegalovirus infection elevates GC in CRH-knockout mice (Silverman *et al.* 2004). The same study suppressed ACTH by treating mice with CRH-antibody. Mice with low ACTH exhibited elevation of GC via release of IL-6, demonstrating viral infection can directly modulate GC levels without an activation of the HPA axis.

As seen in the relationship among immune, nervous system, and HPA axis, virtually all mediators are intricately intertwined and together translate environment into gene expression and cellular signal (Lema & Kitano 2013). The relationship among them also changes with developmental environment, age, and experience (Martin 2009).

For instance, early exposure to certain pathogen can program the HPA axis as well as immune system. Postnatal exposure to endotoxin weakened negative feedback of the HPA axis, demonstrated by attenuated reduction of adrenocortical response to synthetic GC, dexamethasone (Shanks *et al.* 1995). This resulted in permanent alteration in the HPA axis as the endotoxin-exposed individual had greater adrenocortical responses to a psychological or restraint stressor and enhanced suppression of lymphocytes after a stressor (Shanks *et al.* 1995; Shanks *et al.* 2000). Future studies should address how immune programming of the HPA axis and HPA axis programming of the immune system may be altered by developmental environment to prioritize a certain components of the systems to maximize fitness in the local environment.

CONCLUSION

How well an individual can adjust to changing environments is critical to its fitness. Organismal responses to stressors are a result of interactions among genes, developmental plasticity, and phenotypic flexibility via epigenetic and cellular modifications. Thus understanding the permanent and temporary changes in physiological mediators and their regulators as well as connections among them can help us understand how physiological and behavioral stress responses relate to fitness. Shifting the focus of fitness estimation from offspring to parents, in light of developmental history, might also shed new light on the selection pressure maintaining individual variation in stress responses.

REFERENCES

Antoni, F.A. (1993) Vasopressinergic control of pituitary adrenocorticotropin secretion comes of age. *Frontiers in Neuroendocrinology*, **14**, 76–122.

Archer, T.K., Hager, G.L. and Omichinski, J.G. (1990) Sequence-specific DNA binding by glucocorticoid receptor "zinc finger peptides." *Proceedings of the National Academy of Sciences*, **87**, 7560–4.

Bagot, R.C., van Hasselt, F.N., Champagne, D.L. *et al.* (2009) Maternal care determines rapid effects of stress mediators on synaptic plasticity in adult rat hippocampal dentate gyrus. *Neurobiol Learn Mem*, **92**, 292–300.

Barker, D.J.P. (2004) The developmental origins of adult disease. *Journal of the American College of Nutrition*, **23**, 588S–595.

Bateson, P., Barker, D., Clutton-Brock, T. *et al.* (2004) Developmental plasticity and human health. *Nature*, **430**, 419(413).

Bokony, V., Lendvai, A.Z., Liker, A. *et al.* (2009) Stress response and the value of reproduction: Are birds prudent parents? *American Naturalist*, **173**, 589–98.

Bonier, F., Martin, P.R., Moore, I.T. and Wingfield, J.C. (2009a) Do baseline glucocorticoids predict fitness? *Trends in Ecology & Evolution*, **24**, 634–42.

Bonier, F., Moore, I.T., Martin, P.R. and Robertson, R.J. (2009b) The relationship between fitness and baseline glucocorticoids in a passerine bird. *General and Comparative Endocrinology*, **163**, 208–13.

Breuner, C.W. (2011) Stress and reproduction in birds, in *Hormones and Reproduction of Vertebrates: Birds* (eds D.O. Norris and K.H. Lopez), Academic Press, London, pp. 129–51.

Breuner, C.W., Lynn, S.E., Julian, G.E., Cornelius, J.M., Heidinger, B.J., Love, O.P., Sprague, R.S., Wada, H. & Whitman, B.A. (2006) Plasma-binding globulins and acute stress response. *Hormone and Metabolic Research*, 260–8.

Breuner, C.W. and Orchinik, M. (2001) Seasonal regulation of membrane and intracellular corticosteroid receptors in the house sparrow brain. *Journal of Neuroendocrinology*, **13**, 412–20.

Breuner, C.W., Patterson, S.H. and Hahn, T.P. (2008) In search of relationships between the acute adrenocortical response and fitness. *General and Comparative Endocrinology*, **157**, 288–95.

Brown, C.R., Brown, M.B., Raouf, S.A. *et al.* (2005) Effects of endogenous steroid hormone levels on annual survival in cliff swallows. *Ecology*, **86**, 1034–46.

Caldji, C., Tannenbaum, B., Sharma, S. *et al.* (1998) Maternal care during infancy regulates the development of neural systems mediating the expression of fearfulness in the rat. *Proceedings of the National Academy of Sciences*, **95**, 5335–40.

Claessens, S.E., Daskalakis, N.P., van der Veen, R. *et al.* (2011) Development of individual differences in stress responsiveness: an overview of factors mediating the outcome of early life experiences. *Psychopharmacology (Berl)*, **214**, 141–54.

Cohen, A.A., Martin, L.B., Wingfield, J.C. *et al.* (2012) Physiological regulatory networks: ecological roles and evolutionary constraints. *Trends Ecol Evol*, **27**, 428–35.

Cornelius, J.M., Boswell, T., Jenni-Eiermann, S. *et al.* (2013) Contributions of endocrinology to the migration life history of birds. *Gen Comp Endocrinol*, **190**, 47–60.

Creighton, J.C., Heflin, N.D. and Belk, M.C. (2009) Cost of reproduction, resource quality, and terminal investment in a burying beetle. *Am Nat*, **174**, 673–84.

Crespi, E.J. and Denver, R.J. (2005) Ancient origins of human developmental plasticity. *American Journal of Human Biology*, **17**, 44–54.

Crespi, E.J., Williams, T.D., Jessop, T.S. *et al.* (2013) Life history and the ecology of stress: how do glucocorticoid hormones influence life-history variation in animals? *Functional Ecology*, **27**, 93–106.

Crews, D. (2008) Epigenetics and its implications for behavioral neuroendocrinology. *Frontiers in Neuroendocrinology*, **29**, 344–57.

Crews, D., Gillette, R., Scarpino, S.V. *et al.* (2012) Epigenetic transgenerational inheritance of altered stress responses. *Proc Natl Acad Sci U S A*, **109**, 9143–8.

Dallman, M.F., Akana, S.F., Levin, N. *et al.* (1994) Corticosteroids and the control of function in the hypothalamo-pituitary-adrenal (HPA) axis. *Annals of the New York Academy of Sciences*, **746**, 22–31.

de Kloet, E.R., Oitzl, M. and Joëls, M. (1993) Functional implications of brain corticosteroid receptor diversity. *Cellular and Molecular Neurobiology*, **13**, 433–55.

Dimitrov, S., Benedict, C., Heutling, D. *et al.* (2009) Cortisol and epinephrine control opposing circadian rhythms in T cell subsets. *Blood*, **113**, 5134–43.

Ekins, R. (1990) Measurement of free hormones in blood. *Endocrine Reviews*, **11**, 5–46.

Elenkov, I.J. and Chrousos, G.P. (2002) Stress hormones, proinflammatory and antiinflammatory cytokines, and autoimmunity. *Annals of the New York Academy of Sciences*, **966**, 290–303.

Ellis, B.J., Jackson, J.J. and Boyce, W.T. (2006) The stress response systems: Universality and adaptive individual differences. *Developmental Review*, **26**, 175–212.

Gonzalez, A., Lovic, V., Ward, G.R. *et al.* (2001) Intergenerational effects of complete maternal deprivation and replacement stimulation on maternal behavior and emotionality in female rats. *Developmental Psychobiology*, **38**, 11–32.

Grafen, A. (1988) On the uses of data on lifetime reproductive success, in *Reproductive success* (ed T. Clutton-Brock), University of Chicago Press, Chicago, IL, pp. 454–71.

Grebe, K.M., Takeda, K., Hickman, H.D. *et al.* (2010) Cutting edge: Sympathetic nervous system increases proinflammatory cytokines and exacerbates influenza A virus pathogenesis. *J Immunol*, **184**, 540–4.

Guimont, F.S. and Wynne-Edwards, K.E. (2006) Individual variation in cortisol responses to acute "on-back" restraint in an outbred hamster. *Hormones and Behavior*, **50**, 252–60.

Hayes, T., Chan, R. and Licht, P. (1993) Interactions of temperature and steroids on larval growth, development, and metamorphosis in a toad (*Bufo boreas*). *Journal of Experimental Zoology*, **266**, 206–15.

Johnson, J.D., O'Connor, K.A., Deak, T. *et al.* (2002) Prior stressor exposure primes the HPA axis. *Psychoneuroendocrinology*, **27**, 353–65.

Jutapakdeegul, N., Casalotti, S.O., Govitrapong, P. and Kotchabhakdi, N. (2003) Postnatal touch stimulation acutely alters corticosterone levels and glucocorticoid receptor gene expression in the neonatal rat. *Developmental Neuroscience*, **25**, 26–33.

Kalinichev, M., Easterling, K.W., Plotsky, P.M. and Holtzman, S.G. (2002) Long-lasting changes in stress-induced corticosterone response and anxiety-like behaviors as a consequence of neonatal maternal separation in Long–Evans rats. *Pharmacology Biochemistry and Behavior*, **73**, 131–40.

Kitaysky, A.S., Piatt, J.F., Hatch, S.A. *et al.* (2010) Food availability and population processes: severity of nutritional stress during reproduction predicts survival of long-lived seabirds. *Functional Ecology*, **24**, 625–37.

Koolhaas, J.M., Bartolomucci, A., Buwalda, B. *et al.* (2011) Stress revisited: a critical evaluation of the stress concept. *Neurosci Biobehav Rev*, **35**, 1291–1301.

Landfield, P.W., Waymire, J.C. and Lynch, G. (1978) Hippocampal aging and adrenocorticoids: Quantitative correlations. *Science*, **202**, 1098–1102.

Landys, M.M., Ramenofsky, M. and Wingfield, J.C. (2006) Actions of glucocorticoids at a seasonal baseline as compared to stress-related levels in the regulation of periodic life processes. *Gen Comp Endocrinol*, **148**, 132–49.

Ledon-Rettig, C.C., Richards, C.L. and Martin, L.B. (2012) Epigenetics for behavioral ecologists. *Behavioral Ecology*, **24**, 311–24.

Lema, S.C. and Kitano, J. (2013) Hormones and phenotypic plasticity: Implications for the evolution of integrated adaptive phenotypes. *Current Zoology*, **59**, 506–25.

Lendvai, A.Z. and Chastel, O. (2008) Experimental mate-removal increases the stress response of female house sparrows: the effects of offspring value? *Horm Behav*, **53**, 395–401.

Lendvai, Á.Z., Giraudeau, M. and Chastel, O. (2007) Reproduction and modulation of the stress response: an experimental test in the house sparrow. *Proceedings of the Royal Society B: Biological Sciences*, **274**, 391–7.

Liu, D., Diorio, J., Day, J.C. *et al.* (2000) Maternal care, hippocampal synaptogenesis and cognitive development in rats. *Nat Neurosci*, **3**, 799–806.

Liu, D., Diorio, J., Tannenbaum, B. *et al.* (1997) Maternal care, hippocampal glucocorticoid receptors, and hypothalamic-pituitary-adrenal responses to stress. *Science*, **277**, 1659–62.

Love, O.P., Chin, E.H., Wynne-Edwards, K.E. and Williams, T.D. (2005) Stress hormones: A link between maternal condition and sex-biased reproductive investment. *American Naturalist*, **166**, 751–66.

Love, O.P. and Williams, T.D. (2008) The adaptive value of stress-induced phenotypes: Effects of maternally derived corticosterone on sex-biased investment, cost of reproduction, and maternal fitness. *American Naturalist*, **172**, E135–E149.

Lynn, S.E., Breuner, C.W. and Wingfield, J.C. (2003) Short-term fasting affects locomotor activity, corticosterone, and corticosterone binding globulin in a migratory songbird. *Hormones and Behavior*, **43**, 150–7.

Marshall, D.J. and Uller, T. (2007) When is a maternal effect adaptive? *Oikos*, **116**, 1957–63.

Martin, L.B. (2009) Stress and immunity in wild vertebrates: timing is everything. *Gen Comp Endocrinol*, **163**, 70–6.

Martin, L.B., Liebl, A.L., Trotter, J.H. *et al.* (2011) Integrator networks: illuminating the black box linking genotype and phenotype. *Integr Comp Biol*, **51**, 514–27.

McEwen, B.S. and Wingfield, J.C. (2003) The concept of allostasis in biology and biomedicine. *Hormones and Behavior*, **43**, 2–15.

Meaney, M.J. (2001) Maternal care, gene expression, and the transmission of individual differences in stress reactivity across generations. *Annual Review of Neuroscience*, **24**, 1161–92.

Meaney, M.J., Aitken, D.H., Berkel, C.V. *et al.* (1988) Effect of neonatal handling on age-related impairments associated with the hippocampus. *Science*, **239**, 766–8.

Monaghan, P. (2008) Early growth conditions, phenotypic development and environmental change. *Philos Trans R Soc Lond B Biol Sci*, **363**, 1635–45.

Moore, F.L. and Orchinik, M. (1994) Membrane receptors for corticosterone: A mechanism for rapid behavioral responses in an amphibian. *Hormones and Behavior*, **28**, 512–19.

Moore, I.T. and Jessop, T.S. (2003) Stress, reproduction, and adrenocortical modulation in amphibians and reptiles. *Hormones and Behavior*, **43**, 39–47.

Pianka, E.R. (1992) Reproductive tactics, in *Sex origin and evolution. International symposium on origin and evolution of sex* (ed R. Dallai), Mucchi Editore, Siena, Italy, pp. 189–209.

Plotsky, P.M. and Meaney, M.J. (1993) Early, postnatal experience alters hypothalamic corticotropin-releasing factor (CRF) mRNA, median eminence CRF content and stress-induced release in adult rats. *Molecular Brain Research*, **18**, 195–200.

Rensel, M.A. and Schoech, S.J. (2011) Repeatability of baseline and stress-induced corticosterone levels across early life stages in the Florida scrub-jay (Aphelocoma coerulescens). *Hormones and Behavior*, **59**, 497–502.

Robson, A.C., Leckie, C.M., Seckl, J.R. and Holmes, M.C. (1998) 11[beta]-Hydroxysteroid dehydrogenase type 2 in the postnatal and adult rat brain. *Molecular Brain Research*, **61**, 1–10.

Romero, L.M. and Butler, L.K. (2007) Endocrinology of stress. *International Journal of Comparative Psychology*, **20**, 89–95.

Romero, L.M., Dickens, M.J. and Cyr, N.E. (2009) The reactive scope model – A new model integrating homeostasis, allostasis, and stress. *Hormones and Behavior*, **55**, 375–89.

Rose, J.D., Moore, F.L. and Orchinik, M. (1993) Rapid neurophysiological effects of corticosterone on medullary neurons – relationship to stress-induced suppression of courtship clasping in an amphibian. *Neuroendocrinology*, **57**, 815–24.

Rosner, W. (1990) The functions of corticosteroid-binding globulin and sex hormone-binding globulin: Recent advances. *Endocrine Reviews*, **11**, 80–91.

Sachs, C., Hamberger, B. and Kaijser, L. (1985) Cardiovascular responses and plasma catecholamines in old age. *Clinical Physiology*, **5**, 553–65.

Sapolsky, R.M., Krey, L.C. and McEwen, B.S. (1983) Corticosterone receptors decline in a site-specific manner in the aged rat brain. *Brain Research*, **289**, 235–40.

Sapolsky, R.M., Krey, L.C. and McEwen, B.S. (1984) Glucocorticoid-sensitive hippocampal neurons are involved in terminating the adrenocortical stress response. *Proceedings of the National Academy of Sciences of the United States of America*, **81**, 6174–7.

Sapolsky, R.M., Romero, L.M. and Munck, A.U. (2000) How do glucocorticoids influence stress responses? Integrating permissive, suppressive, stimulatory, and preparative actions. *Endocrine Reviews*, **21**, 55–89.

Schmidt, K.L., Pradhan, D.S., Shah, A.H. *et al.* (2008) Neurosteroids, immunosteroids, and the Balkanization of endocrinology. *Gen Comp Endocrinol*, **157**, 266–74.

Schoech, S.J., Romero, L.M., Moore, I.T. *et al.* (2013) Constraints, concerns and considerations about the necessity of estimating free glucocorticoid concentrations for field endocrine studies. *Functional Ecology*, **27**, 1100–6.

Schrandt, M.N. and Lema, S.C. (2011) Habitat-associated intraspecific variation in behavior and stress responses in a demersal coral reef fish. *Marine Ecology Progress Series*, **443**, 153–66.

Schwartz, T.S. and Bronikowski, A.M. (2011) Molecular stress pathways and the evolution of life histories in reptiles, in *Molecular Mechanisms of Life History Evolution* (eds T. Flatt and A. Heyland), Oxford University Press, Oxford, pp. 193–209.

Schwarzkopf, L. and Andrews, R.M. (2012) Are moms manipulative or just selfish? Evaluating the "maternal manipulation hypothesis" and implications for life-history studies of reptiles. *Herpetologica*, **68**, 147–59.

Seckl, J.R. (2001) Glucocorticoid programming of the fetus; adult phenotypes and molecular mechanisms. *Molecular and Cellular Endocrinology*, **185**, 61–71.

Shanks, N., Larocque, S. and Meaney, M. (1995) Neonatal endotoxin exposure alters the development of the hypothalamic- pituitary-adrenal axis: early illness and later responsivity to stress. *The Journal of Neuroscience*, **15**, 376–84.

Shanks, N., Windle, R.J., Perks, P.A. *et al.* (2000) Early-life exposure to endotoxin alters hypothalamic-pituitary-adrenal function and predisposition to inflammation. *Proc Natl Acad Sci U S A*, **97**, 5645–50.

Sheriff, M.J. and Love, O.P. (2013) Determining the adaptive potential of maternal stress. *Ecol Lett*, **16**, 271–80.

Silverman, M.N., Miller, A.H., Biron, C.A. and Pearce, B.D. (2004) Characterization of an interleukin-6- and adrenocorticotropin-dependent, immune-to-adrenal pathway during viral infection. *Endocrinology*, **145**, 3580–9.

Spencer, K.A., Buchanan, K.L., Goldsmith, A.R. and Catchpole, C.K. (2003) Song as an honest signal of developmental stress in the zebra finch (*Taeniopygia guttata*). *Hormones and Behavior*, **44**, 132–9.

Stearns, S.C. (1992) *The Evolution of Life History*, Oxford University Press, Oxford.

Sterling, P. and Eyer, J. (1988) Allostasis: a new paradigm to explain arousal pathology, in *Handbook of Life Stress, Cognition and Health* (eds S. Fisher and J. Reason), John Wiley & Sons, Inc, New York.

Sternberg, E.M. (2006) Neural regulation of innate immunity: a coordinated nonspecific host response to pathogens. *Nat Rev Immunol*, **6**, 318–28.

Tasker, J.G., Di, S. and Malcher-Lopes, R. (2006) Minireview: rapid glucocorticoid signaling via membrane-associated receptors. *Endocrinology*, **147**, 5549–56.

Taves, M.D., Gomez-Sanchez, C.E. and Soma, K.K. (2011) Extra-adrenal glucocorticoids and mineralocorticoids: evidence for local synthesis, regulation, and function. *American Journal of Physiology – Endocrinology and Metabolism*, **301**, E11–E24.

Tinnikov, A. (1999) Responses of serum corticosterone and corticosteroid-binding globulin to acute and prolonged stress in the rat. *Endocrine*, **11**, 145–50.

Wada, H. (2008) Glucocorticoids: Mediators of vertebrate ontogenetic transitions. *General and Comparative Endocrinology*, **156**, 441–53.

Wada, H., Salvante, K.G., Stables, C. *et al.* (2008) Adrenocortical responses in zebra finches (*Taeniopygia guttata*): individual variation, repeatability, and relationship to phenotypic quality. *Hormones and Behavior*, **53**, 472–80.

Weaver, I.C., Cervoni, N., Champagne, F.A. *et al.* (2004) Epigenetic programming by maternal behavior. *Nat Neurosci*, **7**, 847–54.

Weaver, I.C., Champagne, F.A., Brown, S.E. *et al.* (2005) Reversal of maternal programming of stress responses in adult offspring through methyl supplementation: altering epigenetic marking later in life. *J Neurosci*, **25**, 11045–54.

Weber, J.M. (2009) The physiology of long-distance migration: extending the limits of endurance metabolism. *J Exp Biol*, **212**, 593–7.

Williams, G.C. (1966) Natural selection, the cost of reproduction and a refinement of Lack's principle. *American Naturalist*, **100**, 687–90.

Wingfield, J.C., Maney, D.L., Breuner, C.W. *et al.* (1998) Ecological bases of hormone-behavior interactions: The "emergency life history stage.". *American Zoologist*, **38**, 191–206.

Wingfield, J.C. and Sapolsky, R.M. (2003) Reproduction and resistance to stress: When and how. *Journal of Neuroendocrinology*, **15**, 711–24.

Zhou, D.H., Kusnecov, A.W., Shurin, M.R. *et al.* (1993) Exposure to physical and psychological stressors elevates plasma interleukin-6 – relationship to the activation of hypothalamic-pituitary-adrenal axis. *Endocrinology*, **133**, 2523–30.

13

A COMMON FRAMEWORK FOR THE REGULATION OF GROWTH AND SIZE: STEPPING AWAY FROM THE TREES TO SEE THE FOREST

Goggy Davidowitz[1] and Bryan R. Helm[2,3]

[1]*Department of Entomology, The University of Arizona, Tucson, AZ, USA*
[2]*Department of Ecology and Evolutionary Biology, The University of Arizona, Tucson, AZ, USA*
[3]*Biological Sciences, North Dakota State University, Fargo, ND, USA*

INTRODUCTION

Life histories, the demographic patterns of the life cycle that make up growth, maturity, reproduction, and survival, are the basis of our understanding of how organisms cope with their environments and how populations and species evolve (Stearns 1992; Roff 1992). Much of life history theory does not distinguish among taxa; indeed, it is common to discuss life history traits as if it were irrelevant whether the organism is a plant, an insect, or a mammal. In fact, biologists *expect* common patterns in life histories that extend beyond specific study systems. Even though life history theory is largely species independent, the details of the developmental, genetic, physiological, and endocrine mechanisms underlying life histories, are often specific to a particular study system.

The surge of interest in understanding the mechanistic detail underlying life histories is reflected in recent volumes by Gilbert and Epel (2009) and by Flatt and Heyland (2011), as well as this one. And there exists already a great deal of information on such mechanisms in a range of organisms. This wealth of mechanistic information does, however, have a significant drawback: it makes it difficult to identify common themes, and to make explicit evolutionary predictions that cut across taxa. In Stearns's (2011: 370) words, "The result is a body of knowledge in which greater and greater detail threatens to obscure simplifying features needed to construct a general theory of the organism." Another complication is that the potential combinations of mechanism-components increases exponentially with each

Integrative Organismal Biology, First Edition.
Edited by Lynn B. Martin, Cameron K. Ghalambor, and H. Arthur Woods.

component added, making it experimentally intractable to identify tradeoffs that constrain life histories. It would be advantageous to narrow this wealth of information to a few key components that will enable predictions of life history evolution based on real mechanisms.

This approach has been applied to the tobacco hornworm (*Manduca sexta*: Sphingidae) where we have identified a physiological model to explain the regulation of body size and development time. This model has four main components, which encapsulate a complex suite of physiological and endocrine events. Variation in these components are sufficient to explain over 95% of the variation in body size and development time, as well as plasticity of body size (D'Amico *et al.* 2001; Davidowitz & Nijhout 2004; Davidowitz *et al.* 2003, 2004; Nijhout *et al.* 2006, 2010). It also generates explicit evolutionary predictions about how the underlying physiology enables or constrains the evolution of the two life history traits (Davidowitz *et al.* 2005, 2012). Changes in these components explain all the variation in body size and development time following simultaneous directional selection (Davidowitz *et al.* 2005 and manuscript in prep.).

The goal of this chapter is to use the *Manduca* model to build a more general framework for the regulation of body size that is applicable across taxa. Overall, we aim to show that such a framework is possible and that it can generate mechanistic insight into the ecology, evolution and constraints on major life history traits in a broad range of taxa. That is, it allows us to step away from system-specific detail (the "trees") to understand general life history phenomena (the "forest").

A GENERAL FRAMEWORK FOR THE REGULATION OF GROWTH AND BODY SIZE

For the majority of organisms, the juvenile stage is the primary growth phase, while reproduction occurs in the adult phase. The shift from growth to reproductive maturity is a major life history transition, with reproductive success often highly correlated with body size (Roff 1992; Stearns 1992). It has long been understood that body size is the product of the rate and duration of growth (Haber *et al.* 1964; Atkinson, 1994; Davidowitz *et al.* 2004). Even so, much of the work on the regulation of body size has focused on how growth rate, as well as the environmental and genetic factors that affect it, affect body size. Growth rate alone, however, cannot fully explain body size because a slow-growing individual can become large given enough time, and a fast-growing individual will stay small if not (e.g., Blanckenhorn 1999). Thus, a solid understanding of the regulation of body size requires an understanding not only of what regulates growth rate, but also of the duration of the growth period. The regulation of the growth period is ultimately a question of the mechanism that stops growth (Davidowitz & Nijhout 2004; Davidowitz *et al.* 2004; Edgar 2006; Mirth & Riddiford 2007), and for any growing individual with determinate growth, at some point in ontogeny, the "decision" is made to stop growth.

The general framework for the regulation of growth and size has four components, which, in ontogenetic order (Figure 13.1), are:

1. *The decision point*, the point in development when a well-defined cascade of events leading to reproductive maturation is initiated.
2. *The terminal growth period (TGP)*, the time interval between the decision to initiate the life history transition and when growth actually stops.
3. *The cessation of growth*, the point in development when the juvenile phase ends, after which there is little or no additional growth.
4. *Growth rate*, is defined as the biomass gained divided by the duration of the growth period.

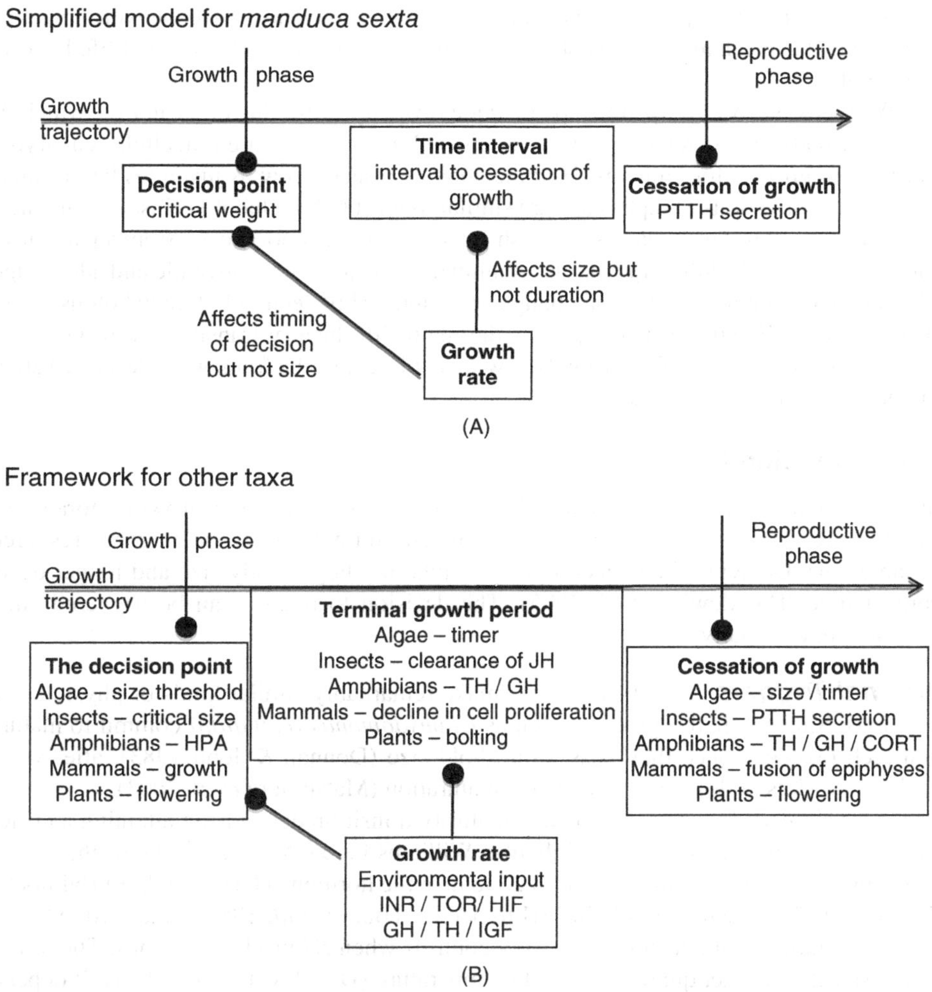

Figure 13.1. A simplified physiological framework for the regulation of body size and development time. The long horizontal arrow depicts the developmental trajectory of an organism. The four boxes depict the four main physiological components of the framework. Implicit in each of these is variation due to genetic and environmental effects.

For simplicity we focus on organisms with determinate growth that reproduce once. However, there is no conceptual difficulty in extending this framework to organisms with indeterminate growth, or to long-lived iteroparous organisms.

COMMONALITIES IN THE REGULATION OF BODY SIZE ACROSS TAXA

In this section, we identify the four components of the general framework in five taxa. Our discussion is by no means exhaustive, and in some taxa there is more mechanistic detail than in others. Identifying the four components can only be done in taxa for which there are already data on physiological, developmental and endocrine mechanisms of transitions among developmental stages. We suggest that correct identification of the components of the framework should be done by experts- those most familiar with the mechanistic detail of a given study system. While the present authors lack this expertise in systems other than

Manduca sexta, our purpose here is to argue this is feasible and an avenue worth pursuing to gain a better mechanistic understanding of the evolution of body size and life histories in general.

We focus on five groups that span a wide diversity of life histories and for which the control of body size is relatively well understood. Green algae are unicellular eukaryotes that need to grow before mitosis. Holometabolous insects grow as juveniles before metamorphosing into adults. Amphibians are similar to most holometabolous insects but unique among chordates because many species shift habitats at metamorphosis (from aquatic juveniles into terrestrial adults). In contrast, mammals grow in both the juvenile and adult stages without the drastic morphological change and habitat shift seen in holometabolous insects or amphibians. Finally, in many plants, the major life history transition from growth to reproduction, occurs with flowering. Below we explore how the framework described above can be applied to these five taxa.

1. The Decision Point

Juveniles acquire resources that are then divided among somatic growth, storage and maintenance. After maturity, reproduction competes for nutritional and energetic resources (Chapter 14). Delaying the decision point results in a larger body size and longer development time (Davidowitz *et al.* 2005). The decision to mature can be triggered either internally or externally.

Internal Triggers Insects and algae make remarkably similar kinds of physiological decisions. Single-celled organisms such as *Clamydomonas reinhardtii* commit to maturation (cell division) once they pass a threshold size (Donnan & John 1983), after which further growth is no longer necessary for maturation (Matsumura *et al.* 2003).

For a variety of insects, the growth-maturity transition depends on attaining a critical size late in the larval phase of life (Nijhout & Williams 1974). At the critical size, the corpora allata, the glands that synthesize and secrete juvenile hormone (JH), switch off (Nijhout & Williams, 1974) causing circulating JH to decline (Baker *et al.* 1987; Riddiford 1994). In *Manduca sexta*, no one factor definitively controls when JH production stops. The critical size is sensitive to diet quality but not to temperature (Davidowitz *et al.* 2004). It depends on the growth history of the larva (Nijhout *et al.* 2006) and perhaps also on tissue oxygen levels (Callier & Nijhout 2011). In *Drosophila melanogaster*, the critical size is influenced by insulin-signaling-dependent growth of the prothoracic ring gland, prothoracicotropic hormone (PTTH), and the developing imaginal discs (Mirth *et al.* 2005; Mirth & Riddiford 2007; Stieper *et al.* 2008; Shingleton 2011).

A threshold size also appears to initiate events that ultimately lead to maturity in mammals. A negative feedback loop, controlled by growth itself, causes the down-regulation of cell proliferation genes, halting growth (Lui *et al.* 2010; Lui & Baron 2011). It is not known what mechanism triggers this, but it appears to depend on growth of the individual rather than on age per se.

External Triggers In contrast to algae, insects, and mammals, in amphibians and plants, the decision to mature appears to be triggered externally. In amphibians, it depends on environmental cues transduced via the hypothalamo-pituitary-adrenal/thyroid (HPA) axis (Denver 2009; Chapter 12). Nutrition is the main determinant of the decision to initiate metamorphosis, but crowding, pond drying, predation, light, water movement and temperature can all contribute (Smith-Gill & Berven 1979; Newman 1989; Pfennig, 1992; Denver 1996, 2009; Rose 2005; Buchholz *et al.* 2011).

In plants, the transition from growth to reproduction occurs with flowering (Komeda 2004; Banta & Purugganan 2011). In *Arabidopsis thaliana*, the onset of flowering is controlled by four genetic pathways: (1) the autonomous pathway lifts flower suppression by the Flowering locus C (FLC), (2) the vernalization pathway responds to temperature, (3) the light-dependent pathway responds to photoperiod, and (4) the gibberellin pathway promotes growth and flowering (Komeda 2004; Banta & Purugganan 2011). *SOC1* (suppressor of overexpression of Constans1) and *FT* (flowering locus T) integrate these pathways to determine flowering (Mouradov *et al.* 2002).

2. The Terminal Growth Period (TGP)

In all the taxa we consider here, the time interval between the decision point and the actual cessation of growth affects final size and development time. An increase in the TGP results in a larger body size and a longer development time, as the organism has more time to feed and grow (Davidowitz *et al.* 2005).

In *Clamydomonas*, once the threshold size has been reached, the interval is determined by a timer that does not depend on additional growth (Donnan & John 1983; Matsumura *et al.* 2003). The timer is initiated in darkness, but is reset once the cells are re-illuminated.

The TGP in insects depends on two factors: the clearance of residual JH from the hemolymph and ecdysone-promoting PTTH signaling. In *Manduca* the TGP is called the Interval to Cessation of Growth (ICG) and is a special case of the more general TGP (Shingleton *et al.* 2007). In the last larval instar of *M. sexta*, JH inhibits the secretion of PTTH and is catabolized by juvenile hormone esterase (Browder *et al.* 2001). Once JH is cleared, PTTH signaling can occur, although its activation is restricted to a recurring daily photoperiodic gate (Truman 1972). The catabolism of JH and gating of PTTH create a lag between the decision to stop growth and when growth stops. During this lag, larvae usually continue to grow as in *M. sexta* (Davidowitz *et al.* 2004), or it may coincide with the decision point as in *Drosophila* (Mirth *et al.* 2005). The daily resetting of a time-based mechanism in *Clamydomonas* is reminiscent of the PTTH photoperiodic gate in *Manduca*.

Growth hormone (GH) and the insulin-signaling pathway regulate growth in all vertebrates, and are controlled by thyroid hormone (TH) (Buchholz *et al.* 2011). Unlike other vertebrates, however, amphibians regulate GH and TH independently. This disassociation may allow amphibians to free growth from the regulation of development (Rose 2005), thereby maximizing developmental plasticity of size and age at maturity in response to variation in the environment. In mammals, growth stops after a progressive decline in cell proliferation that begins early in life and is analogous to the *Manduca* ICG. It varies among species: in humans it can last 15 years, while in mice, a few weeks (Lui & Baron 2011).

The reproductive meristem in annual plants elongates (bolts) a few days before flowering (Banta & Purugganan 2011) and is thus analogous to the ICG in *Manduca*. In the sugar beet (*Beta vulgaris*), dominant alleles of the *B* gene that controls bolting are sensitive to photoperiod whereas the recessive allele is sensitive to vernalization (Mutasa-Gottgens *et al.* 2010).

3. The Cessation of Growth

In most of the taxa we consider here, growth cessation is generally controlled hormonally. In all holometabolous insects examined to date, growth stops when the first pulse of ecdysteroids is secreted by the prothoracic glands (Bollenbacher *et al.* 1979; Rountree & Bollenbacher 1986; Shingleton 2011). Triiodothyronine (T3), the active form of TH, regulates tissue development in amphibians. Corticosterone, released from the interrenal gland, interacts with TH to control the timing of tissue development (Denver 2009; Buchholz *et al.* 2011). TH's role in amphibian development is similar to the one played by ecdysteroids in insects (Shi 1996). In *Arabidopsis*, once a plant decides to flower, the shoot apical meristem that previously produced rosette leaves (the growth phase) is converted to a reproductive meristem via integration of photoperiod and gibberellins that control stem elongation (Feng *et al.* 2008; de Lucas *et al.* 2008).

Examples of nonhormone regulated growth cessation can be seen in mammals and *Chlamydomonas*. In mammals, growth stops not because of a systemic or hormone mediated mechanism, but because chondrocytes – cells in the long-bone growth plates – can divide only a finite number of times. After chondrocytes stop dividing, they differentiate into osteoblasts that leads to fusion of the long bone epiphyses, stopping growth (Nilsson & Baron 2004; Lui & Baron 2011). The threshold sizer mechanism used by *Chlamydomonas* to decide to switch to reproduction is involved in stopping growth (Matsumura *et al.* 2003). The control of division and hence the end of growth appear to be regulated by *p34cdc2,* a protein conserved across the plant kingdom (John *et al.* 1989). The point at which growth stops is the endpoint of the ICG and in the *Manduca* model is explicitly contained in it (Davidowitz & Nijhout 2004).

4. Growth Rate

A higher growth rate produces a larger body size and shorter development time, because the organism accumulates more resources and moves through development faster (Davidowitz *et al.* 2005). Growth rate in essentially all organisms is affected by many environmental factors.

Growth rate determines the time taken and mass accumulated before growth stops (Davidowitz & Nijhout 2004; Figure 13.1). Substantial advances have been made in our understanding of how the insulin/insulin-like growth factor (IGF) signaling (IIS) and Target of Rapamycin (TOR) pathway affects growth rate. For many taxa, growth and proliferation of cells depend on the interactions of gene/protein receptor cascades, especially those involved in nutrient-, oxygen-, and growth-factor sensing. Central components of these networks appear to be conserved across animals. Cellular carbohydrate uptake is coordinated by the insulin receptor, InR, and its ligands insulin or IGFs. When occupied, InR promotes cell growth and proliferation (LeRoith *et al.* 1991; Goberdhan & Wilson 2003; Edgar 2006). The TOR pathway regulates amino acid uptake, coordinates cellular energy fluxes, and influences cellular growth and differentiation (Fingar & Blenis 2004; Hay & Sonnenberg 2004; Tokunaga *et al.* 2004; Wang & Proud 2006; Howell & Manning 2011; Oldham 2011). The hypoxia-inducible factor (HIF) promotes cellular metabolism when oxygen conditions are favorable and suppresses it otherwise (Fähling, 2009; Harrison & Haddad 2011; Semenza 2012). Above the cellular-scale, growth is coordinated within organisms by a set of growth factors and hormones that promote and/or suppress cell growth and proliferation. In vertebrates, global growth rate and differentiation promoters include growth hormone (GH), thyroid hormones (TH, T3, T4), leptin, and IGFs, especially IGF-1 (Lupu *et al.* 2001; Rose 2005; Lui *et al.* 2010; Beckman 2011; Elias 2012). In insects, DILP, a homologue of IGF, has been found to have a similar function in the regulation of growth as is found in vertebrates (Goberdhan & Wilson 2003; Edgar 2006) and TOR uptake of amino acids is coordinated by the fat body (Colombani *et al.* 2003).

USING THE FRAMEWORK TO ADDRESS THE ECOLOGY AND EVOLUTION OF GROWTH AND SIZE

Can we use the framework, developed to capture phenomena in one model organism, to better understand the ecology and evolution of life history traits such as body size across diverse systems? We think that the answer is yes. It is encouraging that the mechanisms controlling key components can be identified in most groups. In protists, insects, and mammals, the decision to mature is measured by size, and once it is reached, the individual commits to maturing. The mechanism underlying cessation of growth can also be identified in most groups: long-bone chondrocyte proliferation in mammals, secretion of ecdysteroids in insects, and bolting and flowering in plants, for example.

Reducing mechanistic detail to a few key components shows differences in how organisms respond to sources of simultaneous environmental variation. Differential sensitivity of components of the same mechanism allow *M. sexta* to respond to multiple sources of environmental variation simultaneously: growth rate and the decision point respond to diet quality, whereas growth rate and the ICG respond to temperature (Davidowitz *et al.* 2004). In contrast, different plant pathways respond to different environmental stimuli. For example, the vernalization pathway responds to temperature, and the light-dependent pathway responds to photoperiod and light intensity. Both pathways converge on a single gene, Flowering Locus C (*FLC*), that regulates flowering time and the cessation of growth (Mouradov *et al.* 2002; Komeda 2004). Amphibians show yet a third mechanism: stress hormones of the HPA axis integrates numerous sources of environmental input (Denver 2009; Chapter 12).

There are also broad commonalities in how organisms respond to environmental variation. Perhaps the best examples are the IIS/TOR signaling system in response to nutrition which is conserved across all animals (Shingleton 2011) despite the 750-million-year divergence between insects and mammals (Hedges *et al.* 2006), and the hypothalamus-pituitary-adrenal axis that coordinates environmental effects on growth in all vertebrates (Denver 2009).

Mechanistic details illuminate both the constraints and opportunities for life history evolution (Braendle *et al.* 2011). It is very difficult, however, to identify patterns when the mechanistic details are so abundant, and when environmental effects are so pervasive at every step. The general framework identifies four components that encapsulate the details of one such life history transition, the decision to mature, which, in many organisms, ultimately determines body size. Most importantly, it generates explicit, mechanism-based, *a priori* predictions for life history evolution (Davidowitz *et al.* 2005, 2012). For example, natural selection generally acts to increase body size and decrease development time (Kingsolver & Pfennig 2004, Kingsolver & Huey 2008). The *Manduca* model predicts that the response to simultaneous selection on body size and development time will be enabled by an increase in growth rate and constrained by the decision point and the time interval (Davidowitz *et al.* 2005). A test of these predictions following ten generations of simultaneous selection shows that these predictions are upheld in *M. sexta* (Davidowitz, Roff, & Nijhout, manuscript in prep).

To integrate mechanism more strongly into our understanding of life history transitions, it may, at least initially, be profitable to encapsulate many of the mechanistic details into fewer, broader key components. This has worked well in the *Manduca* system and the framework we developed for this system seems at least superficially to have applicability to other taxa as well. It is clear that "hanging the details on the framework" can only be done correctly by experts in each particular system, and one framework may not fit all life history traits or taxa. Stearns (2011: 370) argues that "Those whose priority is description focus on detail; those whose priority is prediction focus on simplicity." Studying life history evolution through a mechanistically-informed reduced framework may bridge the gap between too much and too little detail. Such a framework allows us to use the trees to see the forest.

ACKNOWLEDGMENTS

This work was funded by NSF (USA) grant IOS-1053318 to GD. We thank Matthew Herron for bringing the Donnan and John (1983) paper to our attention, Jeffrey Baron and Ola Nilsson for discussion on growth in mammals and H. Arthur Woods, Lynn B. Martin, Cameron Ghalambor, Joel G. Kingsolver, Judith L. Bronstein and Christopher S. Rose for comments on previous versions.

REFERENCES

Atkinson, D. (1994) Temperature and organism size – a biological law for ectotherms. *Advances in Ecological Research*, **25**, 1–58.

Baker, F.C., Tsai, L.W., Reuter, C.C. and Schooley, D.A. (1987) Invivo fluctuation of JH, JH Acid, and ecdysteroid titer, and JH esterase-activity, during development of 5th stadium *Manduca sexta. Insect Biochemistry*, **17**, 989–96.

Banta, J.A. and Purugganan, M.D. (2011) The genetics and evolution of flowering time variation in plants: identifying genes that control a key life history transition, in *Mechanisms of Life History Evolution: The Genetics and Physiology of Life History Traits and Trade-offs* (eds T. Flatt and A. Heyland), Oxford University Press, Oxford, pp. 114–26.

Beckman, B. (2011) Perspective on concordant and discordant relations between insulin-like growth factor 1 (IGF1) and growth in fishes. *General and Comparative Endocrinology*, **170**, 233–52.

Blanckenhorn, W.U. (1999) Different growth responses to temperature and resource limitation in three fly species with similar life histories. *Evolutionary Ecology*, **13**, 395–409.

Bollenbacher, W.E., Agui, N., Granger, N.A. and Gilbert, L.I. (1979) Invitro activation of insect prothoracic glands by the prothoracicotropic hormone. *Proceedings of the National Academy of Sciences*, **76**, 5148–52.

Braendle, C., Heyland, A. and Flatt, T. (2011) Integrating mechanistic and evolutionary analysis of life history variation, in *Mechanisms of Life History Evolution: The Genetics and Physiology of Life History Traits and Trade-offs* (eds T. Flatt and A. Heyland), Oxford University Press, Oxford, pp. 3–10.

Browder, M.H., D'Amico, L.J. and Nijhout, H.F. (2001) The role of juvenile hormone esterase in the metamorphosis of *Manduca sexta. Journal of Insect Sciebce*, **1**, 11.

Buchholz, D., Moskalik, C., Kulkarni, S. *et al.* (2011) Hormone regulation and the evolution of frog metamorphic diversity, in *Mechanisms of Life History Evolution: The Genetics and Physiology of Life History Traits and Trade-offs* (eds T. Flatt and A. Heyland), Oxford University Press, Oxford, pp. 87–97.

Callier, V. and Nijhout, H.F. (2011) Control of body size by oxygen supply reveals size-dependent and size-independent mechanisms of molting and metamorphosis. *Proceedings of the National Academy of Sciences*, **108**, 14664–9.

Colombani, J., Raisin, S., Pantalacci, S. *et al.* (2003) A nutrient sensor mechanism controls Drosophila growth. *Cell*, **114**, 739–49.

D'Amico, L.J., Davidowitz, G. and Nijhout, H.F. (2001) The developmental and physiological basis of body size evolution in an insect. *Proceedings of the Royal Society of London, Series B*, **268**, 1589–93.

Davidowitz, G., D'Amico, L.J. and Nijhout, H.F. (2003) Critical weight in the development of insect body size. *Evolution & Development*, **5**, 188–97.

Davidowitz, G., D'Amico, L.J. and Nijhout, H.F. (2004) The effects of environmental variation on a mechanism that controls insect body size. *Evolutionary Ecology Research*, **6**, 49–62.

Davidowitz, G. and Nijhout, H.F. (2004) The physiological basis of reaction norms: the interaction between growth rate, the duration of growth and body size. *Integrative and Comparative Biology*, **44**, 443–9.

Davidowitz, G., Roff, D.A. and Nijhout, H.F. (2005) A physiological perspective on the response of body size and development time to simultaneous directional selection. *Integrative and Comparative Biology*, **45**, 525–31.

Davidowitz, G., Nijhout, H.F. and Roff, D.A. (2012) Predicting the response to simultaneous selection: genetic architecture and physiological constraints. *Evolution*, **66**, 2916–28.

de Lucas, M., Davière, J., Rodríguez-Falcón, M. *et al.* (2008) A molecular framework for light and gibberellin control of cell elongation. *Nature*, **451**, 480–4.

Denver, R.J. (1996) Neuroendocrine Control of Amphibian Metamorphosis, in *Metamorphosis: Postembryonic Reprogramming of Gene Expression in Amphibian and Insect Cells* (eds L.I. Gilbert, J.R. Tata and B.G. Atkinson), Academic Press, San Diego.

Denver, R.J. (2009) Stress hormones mediate environment-genotype interactions during amphibian development. *General and Comparative Endocrinology*, **164**, 20–31.

Donnan, L. and John, P.C.L. (1983) Cell-cycle control by timer and sizer in *Chlamydomonas*. *Nature*, **304**, 63, 0–633.

Edgar, B.A. (2006) How flies get their size: genetics meets physiology. *Nature Reviews Genetics*, **7**, 907–16.

Elias, C. (2012) Leptin action in pubertal development: recent advances and unanswered questions. *Trends in Endocrinology and Metabolism*, **23**, 9–15.

Fähling, M. (2009) Cellular oxygen sensing, signaling and how to survive translational arrest in hypoxia. *Acta Physiologica*, **195**, 205–30.

Feng, S., Martinez, C., Gusmaroli, G. *et al.* (2008) Coordinated regulation of *Arabidopsis thaliana* development by light and gibberellins. *Nature*, **451**, 475–9.

Fingar, D. and Blenis, J. (2004) Target of rapamycin (TOR): and integrator of nutrient and growth factor signals and coordinator of cell growth and cell cycle progression. *Oncogene*, **23**, 3151–71.

Flatt, T. and Heyland, A. (2011) *Mechanisms of life History Evolution: The Genetics and Physiology of Life History Traits and Trade-offs*, Oxford University Press, Oxford.

Gilbert, S.F. and Epel, D. (2009) *Ecological Developmental Biology: Integrating Epigenetics, Medicine, and Evolution*, Sinauer Associates Inc., Sunderland.

Goberdhan, D. and Wilson, C. (2003) The functions of insulin signaling: size isn't everything, even in Drosophila. *Differentiation*, **71**, 375–97.

Gomer, R.H. (2001) Not being the wrong size. *Nature Reviews Molecular Cell Biology*, **2**, 48–54.

Haber, A.H., Long, T.J. and Foard, D.E. (1964) Is final size determined by rate and duration of growth. *Nature*, **201**, 479–80.

Harrison, J. and Haddad, G. (2011) Effects of oxygen and growth and size: synthesis of molecular, organismal, and evolutionary studies with Drosophila melanogaster. *Annual Review of Physiology*, **73**, 95–113.

Hay, N. and Sonnenberg, N. (2004) Upstream and downstream of mTOR. *Genes and Development*, **18**, 1926–45.

Hedges, S.B., Dudley, J. and Kumar, S. (2006) TimeTree: a public knowledge-base of divergence times among organisms. *Bioinformatics*, **22**, 2971–2.

Howell, J. and Manning, B. (2011) mTOR couples cellular nutrient sensing to organismal metabolic homeostasis. *Trends in Endocrinology and Metabolism*, **22**, 94–102.

John, P.C.L., Sek, F.J. and Lee, M.G. (1989) A homolog of the cell cycle control protein p34cdc2 participates in the division cycle of *Chlamydomonas*, and a similar protein is detectable in higher plants and remote taxa. *The Plant Cell*, **1**, 1185–93.

Kingsolver, J.G. and Pfennig, D.W. (2004) Individual-level selection as a cause of Cope's rule phyletic size increase. *Evolution*, **58**, 1608–12.

Kingsolver, J.G. and Huey, R.B. (2008) Size, temperature, and fitness: three rules. *Evolutionary Ecology Research*, **10**, 251–68.

Komeda, Y. (2004) Genetic regulation of time to flower in Arabidopsis thaliana. *Annual Review of Plant Biology*, **55**, 521–35.

LeRoith, D., Adamo, M., Werner, J. and Roberts, C. (1991) Insulin-like growth actors and their receptors as growth regulators in normal physiology and pathologic states. *Trends in Endocrinology and Metabolism*, **2**, 134–9.

Lui, J.C. and Baron, J. (2011) Mechanisms limiting body growth in mammals. *Endocrine Reviews*, **32**, 422–40.

Lui, J.C., Forcinito, P., Chang, M. *et al.* (2010) Coordinated postnatal down-regulation of multiple growth-promoting genes: evidence for a genetic program limiting organ growth. *Faseb Journal*, **24**, 3083–92.

Lupu, F., Terwilliger, J., Lee, K. *et al.* (2001) Roles of growth hormone and insulin-like growth factor 1 in mouse postnatal growth. *Developmental Biology*, **229**, 141–62.

Matsumura, K., Yagi, T. and Yasuda, K. (2003) Role of timer and sizer in regulation of Chlamydomonas cell cycle. *Biochemical and Biophysical Research Communications*, **306**, 1042–9.

Mirth, C., Truman, J.W. and Riddiford, L.M. (2005) The role of the prothoracic gland in determining critical weight to metamorphosis in *Drosophila melanogaster*. *Current Biology*, **15**, 1796–1807.

Mirth, C. and Riddiford, L.M. (2007) Size assessment and growth control: how adult size is determined in insects. *Bioessays*, **29**, 344–55.

Mouradov, A., Cremer, F. and Coupland, G. (2002) Control of flowering time: Interacting pathways as a basis for diversity. *Plant Cell*, **14**, S111–S130.

Mutasa-Gottgens, E.S., Qi, A., Zhang, W., Schulze-Buxloh, G., Jennings A., Hohmann U., Muller A.E., & Hedden P. (2010) Bolting and flowering control in sugar beet: Relationships and effects of gibberellin, the bolting gene B and vernalization. *AoB Plants* 10.1093/aobpla/plq012.

Newman, R.A. (1989) Developmental plasticity of *Scaphiopus couchii* tadpoles in an unpredictable environment. *Ecology*, **70**, 1775–87.

Nijhout, H.F. and Williams, C.M. (1974) Control of molting and metamorphosis in tobacco hornworm, *Manduca sexta* (L.) – cessation of juvenile-hormone secretion as a trigger for pupation. *Journal of Experimental Biology*, **61**, 493–501.

Nijhout, H.F., Davidowitz, G. and Roff, D.A. (2006) A quantitative analysis of the mechanism that controls body size in *Manduca sexta*. *Journal of Biology*, **5**, 1–15.

Nijhout, H.F., Roff, D.A. and Davidowitz, G. (2010) Conflicting processes in the evolution of body size and development time. *Philosophical Transactions of the Royal Society B*, **365**, 567–75.

Nilsson, O. and Baron, J. (2004) Fundamental limits on longitudinal bone growth: growth plate senescence and epiphyseal fusion. *Trends in Endocrinology and Metabolism*, **15**, 370–4.

Oldham, S. (2011) Obesity and nutrient sensing pathway in flies and vertebrates: Functional conservation of genetic mecchanisms. *Trends in Endocrinology and Metabolism*, **22**, 45–52.

Pfennig, D.W. (1992) Proximate and functional causes of polyphenism in an anuran tadpole. *Functional Ecology*, **6**, 167–74.

Riddiford, L.M. (1994) Cellular and molecular actions of juvenile-hormone .1. general- considerations and premetamorphic actions. *Advances in Insect Physiology*, **24**, 213–74.

Roff, D.A. (1992) *The Evolution of Life Histories*, Chapman & Hall, New York.

Rose, C.S. (2005) Integrating ecology and developmental biology to explain the timing of frog metamorphosis. *Trends in Ecology and Evolution*, **20**, 129–35.

Rountree, D.B. and Bollenbacher, W.E. (1986) The release of the prothoracicotropic hormone in the tobacco hornworm, *Manduca sexta*, is controlled intrinsically by juvenile-hormone. *Journal of Experimental Biology*, **120**, 41–58.

Semenza, G. (2012) Hypoxia-inducible factors in physiology and medicine. *Cell*, **148**, 399–408.

Shi, Y.B. (1996) Thyroid hormone-regulated early and late genes during amphibian metamorphosis, in *Metamorphosis: Postembryonic Reprogramming of Gene Expression in Amphibian and Insect Cells* (eds L.I. Gilbert, J.R. Tata and B.G. Atkinson), Academic Press, San Diego, pp. 505–33.

Shingleton, A.W., Frankino, T., Flatt, T. *et al.* (2007) Size and shape: the developmental regulation of static allometry in insects. *Bioessays*, **29**, 536–48.

Shingleton, A.W. (2011) The regulation and evolution of growth and body size, in *Mechanisms of Life History Evolution: The Genetics and Physiology of Life History Traits and Trade-offs* (eds T. Flatt and A. Heyland), Oxford University Press, Oxford, pp. 43–55.

Smith-Gill, S.J. and Berven, K.A. (1979) Predicting amphibian metamorphosis. *American Naturalist*, **113**, 563–85.

Stearns, S.C. (1992) *The Evolution of Life Histories*, Oxford University Press, Oxford.

Stearns, S.C. (2011) Does impressive progress on understanding mechanisms advance life history theory?, in *Mechanisms of Life History Evolution: The Genetics and Physiology of Life History Traits and Trade-offs* (eds T. Flatt and A. Heyland), Oxford University Press, Oxford, pp. 365–73.

Stieper, B.C., Kupershtok, M., Driscoll, M.V. and Shingleton, A.W. (2008) Imaginal disc growth regulates the timing of metamorphosis in *Drosophila melanogaster*. *Developmental Biology*, **321**, 18–26.

Tokunaga, C., Yoshino, K. and Yonezawa, K. (2004) mTOR integrates amino acid- and energy sensing pathways. *Biochemical and Biophysical Research Communications*, **313**, 443–6.

Truman, J.W. (1972) Physiology of insect rhythms: I. circadian organization of the endocrine events underlying the moulting cycle of larval tobacco hornworms. *Journal of Experimental Biology*, **57**, 805–20.

Wang, X. and Proud, C. (2006) The mTOR pathway in the control of protein synthesis. *Physiology*, **21**, 362–9.

14

ADDING FUEL TO THE "FIRE OF LIFE": ENERGY BUDGETS ACROSS LEVELS OF VARIATION IN ECTOTHERMS AND ENDOTHERMS

Vincent Careau[1], Shaun S. Killen[2], and Neil B. Metcalfe[2]

[1]*Centre for Integrative Ecology, School of Life and Environmental Sciences, Deakin University, Victoria, Australia*
[2]*Institute of Biodiversity, Animal Health and Comparative Medicine, College of Medical, Veterinary and Life Sciences, University of Glasgow, Glasgow, UK*

INTRODUCTION

Energetics, within the realm of biology, is defined as *the study of the causes, mechanisms, and consequences of the acquisition, storage, and use of energy by biological organisms.* Energy metabolism – the "fire of life" – is the sum of the processes by which animals acquire energy, channel energy into useful functions, and dissipate energy from their bodies. Animals need energy to accomplish three main types of physiological work: biosynthesis (e.g., growth, lactation, gamete production), maintenance (e.g., circulation, respiration, tissue repair or turnover, nervous coordination), and generation of external work (e.g., locomotion). Kinetic energy (heat) is an inevitable by-product of this metabolism; the heat produced by an animal over an entire day, its *daily energy expenditure* (DEE; see Box 14.1 for definitions of bold terms), is a fundamental measure in biology because it provides quantitative information on how much food it needs, the total activity of all its physiological and behavioral mechanisms, and the energy it drains from its ecosystem (Speakman 2000).

Research initially concentrated on measurement of the *metabolic rate* (MR) of animals under standardized conditions to obtain a comparable measure of the minimal metabolic level required to maintain physiological homeostasis, such as *basal metabolic rate* (BMR), *resting metabolic rate* (RMR), *standard metabolic rate* (SMR), and *routine MR.* Physiologists have also quantified the MR of animals under challenging situations to estimate

Integrative Organismal Biology, First Edition.
Edited by Lynn B. Martin, Cameron K. Ghalambor, and H. Arthur Woods.

maximal metabolic rate (MMR). A major advent was the invention of the doubly-labelled water technique to measure DEE in free-living subjects (Lifson *et al.* 1955). The wide application of this technique over the last 35 years yielded hundreds of DEE measurements in free-ranging animals (Speakman & Król 2010).

As of today, identifying and understanding the factors causing inter-specific variation in BMR, RMR, SMR, MMR, and DEE remains a major research line in the field of energetics, yet the focus is shifting toward understanding differences among individuals (Burton *et al.* 2011; Careau & Garland 2012). The hypotheses put forward to explain variation in DEE, BMR, SMR, and MMR have invoked a wide variation of factors, both intrinsic and environmental, that range from the subcellular to the habitat level (recently reviewed in Konarzewski & Ksiazek 2013; White & Kearney 2013).

Box 14.1. Definitions and Abbreviations of Key Terms

Metabolic rate (MR): the amount of energy consumed by an animal in a given period, as measured by heat produced, O_2 consumed, or CO_2 produced. It represents the rate at which an animal converts chemical energy to heat and mechanical work. Energy is measured in joules (J), and accordingly, the fundamental unit in which MR is expressed is the Watt (W; = 1 $J{\cdot}s^{-1}$).

Basal metabolic rate (BMR): the lowest MR of an adult endotherm, post-absorptive, nonreproductive, and inactive while in its thermal neutral zone and inactive phase of its daily cycle.

Resting metabolic rate (RMR): the lowest MR of an endotherm while resting in its thermal neutral zone, when one or more of the conditions required for measuring BMR cannot be met (e.g., adult, postabsorptive, nonreproductive, resting phase).

Standard metabolic rate (SMR): the lowest MR of an ectotherm, post-absorptive, nonreproductive, and inactive while in its resting phase, measured at a specified ambient temperature.

Routine metabolic rate (routine MR): in some aquatic animals, the MR of post-absorptive, undisturbed animals that also includes the costs of spontaneous activity and the maintenance of posture and equilibrium.

Maximum metabolic rate (MMR): the highest MR that can be sustained by an animal over a very short period of time (i.e., 1 to 10 min), usually elicited by forced exercise (e.g., running or swimming). Also termed $VO_{2\text{-}max}$ or peak metabolic rate (PMR). In endotherms, MMR can also be elicited by cold exposure, then termed summit metabolism. In fishes, the typical tests used to quantify MMR, such as the U_{crit} test, will force the animal to swim at a relatively sustainable (nonburst) level. Alternatively, MMR can be quantified over short period during recovery after complete exhaustion, sometimes yielding measures that differ from those obtained with U_{crit} tests.

Daily energy expenditure (DEE): the total MR of an animal summed over 24 h, usually measured by metabolizable food intake and/or respirometry in captive animals, or heart rate monitoring and/or the doubly-labelled water technique in wild animals (then sometimes termed field metabolic rate (FMR)).

Absolute aerobic scope (AAS=MMR-BMR or SMR): the absolute amount of energy (in Watts) available to an individual to cover the costs of various short-term, O_2-consuming, physiological functions (e.g., digestion, locomotion, thermoregulation).

Factorial aerobic scope (FAS=MMR/BMR or SMR): the factor (in multiples of BMR or SMR) by which an individual can increase its metabolism above maintenance levels to cover the costs of various short-term, O_2-consuming, physiological functions (e.g., digestion, locomotion, thermoregulation).

Nonresting energy expenditure (NREE=DEE-BMR or SMR): the total energy spent over 24 h to cover the costs of various sustainable, O_2-consuming, physiological functions (e.g., reproduction, growth, locomotion, thermoregulation).

Sustained metabolic scope (SusMS=DEE/BMR or SMR): sustained levels of DEE expressed during periods long enough in duration that metabolism is powered by food intake rather than by transient depletion of energy reserves, expressed as a multiple of BMR or SMR.

INTRINSIC FACTORS THAT EXPLAIN VARIATION IN MR

Perhaps the most fundamental of intrinsic factors is body mass (M_b), since it accounts for over 90% of the inter-specific variation in BMR (White & Kearney 2013) and DEE (Speakman & Król 2010). However, the underlying cause of the strong effect of M_b on metabolism is still debated, as is the exact value of the allometric scaling exponent. There have been many attempts to find fundamental physiological or physical principles that can explain the relationship between M_b and metabolism (e.g., resource distribution theories such as the Metabolic Theory of Ecology (Brown *et al.* 2004), resource allocation theories such as Dynamic Energy Budgets (Nisbet *et al.* 2000), and physiological models based on constraints on cell size (Kozlowski *et al.* 2003) or heat dissipation (Speakman & Król 2010)). However, the current consensus is that the majority of these unifying principles are overly simplistic and that there is no single exponent that relates M_b to MR that can be applied across all animal groups or across the entire range of M_b (White & Kearney 2013). Instead it seems that different constraints take precedence at different M_bs, leading to complex exponents and nonlinear allometries (White & Kearney 2013). There is also evidence that environmental and ecological factors can modulate the effects of M_b on MR (Killen *et al.* 2010, Carey *et al.* 2013).

Even having controlled for the complex effect of M_b, there is still significant inter- and intra-specific variation in MR (Careau *et al.* 2008). Hulbert and Else (1999) highlighted the fact that the degree of phospholipid polyunsaturation of cell bilayer membranes affects the speed of key cellular processes and hence SMR or BMR (the "membrane pacemaker" hypothesis). Another important cellular mechanism is basal proton leakage across mitochondrial membranes: the greater the degree of leakage the less efficient are the mitochondria, resulting in a greater O_2 usage per unit of ATP produced. However, in light of the mixed empirical evidence, the general importance of both membrane pacemaker and mitochondrial functioning as unifying explanations for the variability in SMR and BMR is still debated (Polymeropoulos *et al.* 2012; Konarzewski and Ksiazek 2013).

Moving to a higher level of organization, it is commonly assumed that whole-animal MR should in some way be connected with body composition – and in particular with the relative size of organs presumed to have a high metabolic demand such as the intestine, liver, heart and kidneys. Broad support for this hypothesis comes from a range of studies relating variation in M_b-corrected BMR or SMR to relative organ size, but the details show inconsistency as organs identified as significant contributors to BMR or SMR differ widely among and even within studies (Russell & Chappell 2007). There are currently too few studies aimed at determining the reason for this variation, let alone the effect of organs on DEE (Meerlo *et al.* 1997).

A wide range of other individual factors can potentially generate variation in MR such as reproductive status (see Chapter 10, this volume), age (see Chapter 16, this volume), social rank, number of parasites, and "personality" (see Careau *et al.* 2008). As the list of potential factors continues to grow, it is becoming increasingly important to examine several factors simultaneously (including extrinsic factors, see next section) to estimate the *relative* importance of the contributors to BMR, SMR, MMR, and DEE, at both intra-specific (e.g., Fletcher *et al.* 2012; Careau *et al.* 2013b) and inter-specific levels (e.g., Anderson and Jetz 2005).

EXTRINSIC FACTORS THAT EXPLAIN VARIATION IN MR

Although some studies on ectotherms have found that species living at high latitude (cold environments) have higher SMR than species living at low latitudes (warm environments) when normalized to a common ambient temperature (T_a), the generality of this pattern remains very controversial (White *et al.* 2012). By contrast, in endotherms it is relatively well established that there is a positive interspecific correlation between latitude or T_a and BMR measured at thermoneutrality (Lovegrove 2000; Rezende *et al.* 2004). Interestingly, however, rodents that use torpor do not show this pattern, perhaps because they do not face the "problem" of maintaining high body temperature (T_b) in the cold (Careau 2013). Endotherms must also avoid over-heating if living at high T_a, which selects for a reduction in relative BMR (Speakman & Król 2010; White & Kearney 2013). This has been achieved in tropical birds through reductions in organ size: they have relatively smaller hearts, liver, and flight muscles than their counterparts in temperate regions, leading to a lower total body metabolism (Wiersma *et al.* 2012). The relationship between MR and species range is discussed more fully by Bozinovic and Naya (Chapter 17, this volume).

It has been suggested that a high and temporally stable net primary productivity (NPP) allows the evolution of high MMR and DEE (since fuel is abundant) – and that it results indirectly in a high BMR or SMR because this reflects the metabolic cost of maintaining this high-performance metabolic machinery (Mueller & Diamond 2001; see also increased-intake model below). BMR has been found to be correlated with the NPP of the native habitat at both an inter-specific (Mueller & Diamond 2001) and intra-specific level (Bozinovic *et al.* 2009). However, NPP often correlates closely with other environmental factors (e.g., rainfall, T_a) that might affect MR and no such relationship between NPP and BMR was found in an inter-specific analysis of birds that occupied habitats covering a much wider range of primary productivities (White *et al.* 2007). The suggestion that diet itself may influence BMR (the "food-habits" hypothesis) has been much debated (Bozinovic & Sabat 2010), but it has proved difficult to tease apart effects of food quality versus quantity (or indeed NPP), and to separate cause and effect (which comes first – the diet or the MR?). Predation risk is another extrinsic factor that potentially interacts with NPP and food quality (Lovegrove 2000). Predators can significantly constrain the foraging ability of prey, which in turn can result in physiological changes (Lovegrove 2000; Handelsman *et al.* 2013).

The effect of intrinsic and extrinsic factors may vary according to whether analyses consider intraspecific or interspecific variation. For example, the effect of T_a on DEE was positive in a wild rodent population (Fletcher *et al.* 2012), which is opposite to the negative relationship found in inter-specific studies (Humphries *et al.* 2005). Links with life-history traits show similar contrasts: while rodent species with higher-than-average MMR have higher-than-average BMR (Rezende *et al.* 2004), there is mixed support for this relationship at the individual level (Careau *et al.* 2012). In mammals, a positive relationship between mass-adjusted DEE and BMR is widely supported at the inter-specific level (Ricklefs *et al.* 1996), but such a relationship was rarely found at the among-individual level (Careau *et al.* 2013b).

ENERGY BUDGETS

An animal can only start investing energy toward growth, reproduction, or activity when energy intake is above maintenance requirements. This can be achieved by either increasing food intake or reducing maintenance requirements. The first possibility is known as the "increased-intake" (or acquisition, or additive) model and the second as the "allocation" (or compensation) model.

Increased-Intake and Allocation Models

To ingest and process more food, an individual may need to increase the size of its alimentary tract, which may come with an increase in maintenance costs because organs such as gut, intestines, and liver are very active metabolically (see previous section). The increased-intake model therefore predicts a positive association between any energy-demanding activity and BMR or SMR (the "engine"). Individuals may differ in their capacity and willingness to acquire resources from their environment, yielding positive correlations between different components of the energy budget. By contrast, in the allocation model BMR is seen as the "competitor" because it monopolizes energy that becomes unavailable for other energy-demanding activities. Compensation may be achieved by behavioral modifications, choice of environment, and/or by physiological mechanisms (see Chapter 10, this volume). Energy allocation constraints should impose a trade-off between energy-demanding activities and BMR or SMR, hence generating negative relationships.

The increased-intake and allocation models have been widely used to interpret the relationship obtained between BMR and fitness, life-history, and behavioral traits. However, if reasonable arguments can be made to expect a positive and/or a negative relationship between BMR and energy-demanding activities, one may wonder whether these models are really helpful in making sense of the mixed results obtained and move the field forward. Or perhaps these models are simply convenient ways to "escape" the possibility that BMR and SMR do "not have as great an influence on life history and fitness traits as the current theoretical mindset would have us believe" (Schimpf *et al.* 2012)? Many studies on BMR and SMR have been framed as if they were testing or supporting the increased-intake vs. allocation principles as alternative hypotheses. This is absolutely correct, but what if no relationship at all was obtained because both processes are cancelling each other out (e.g., if a high BMR is associated with both high energy intake and high maintenance costs)?

Context-Dependence Hypothesis and the Y-model

In an attempt to reconcile the diversity of results obtained, the "context-dependence" hypothesis predicts that high-BMR individuals will have higher fitness when environmental conditions are productive, whereas low-BMR individuals will have higher fitness when environmental conditions are unproductive (Burton *et al.* 2011). The context-dependence

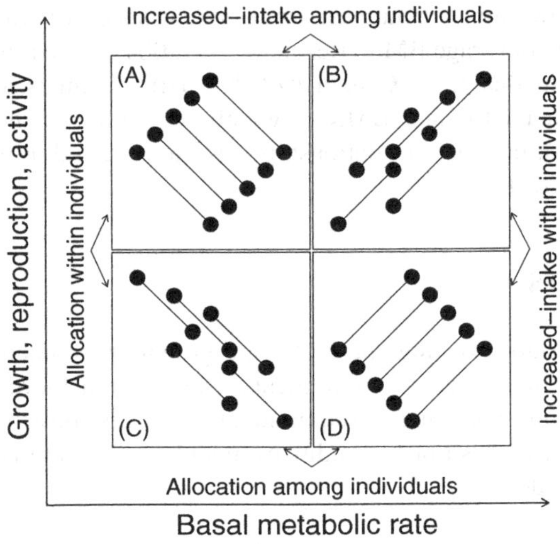

Figure 14.1. Schematic representations of among-individual correlations (r_{ind}) and within-individual correlations (r_e) between basal metabolic rate (BMR) and the energy invested in any energy-demanding activity (e.g., growth, reproduction, locomotion). Repeated pairs of measurements on each individual are represented by dots connected by a line. Panel A represents a scenario where the increased-intake principle applies among individuals (yielding a positive r_{ind}) and the allocation principle applies within individuals (yielding a negative r_e). The increased-intake principle may also apply within individuals (e.g., if increases in BMR allow a greater food intake, yielding a positive r_e; panel B). By contrast, the allocation principle may apply at both levels, yielding negative r_{ind} and r_e (panel C). Perhaps less intuitively, but still possible, would be the scenario in which high-BMR individuals have less energy available for other activities (negative r_{ind}), but when a given individual increases its intake it has more energy to invest into reproduction (positive r_e; panel D).

hypothesis can in fact be seen as a verbal representation of the Y-model envisioned by Van Noordwijk and de Jong (1986) applied to the case of BMR. Let's imagine that the increased-intake principle applies among individuals and the allocation principle applies within individuals (Figure 14.1A). Let's also imagine that as environmental productivity increases, variation in acquisition also *increases* to a point where there is relatively more variation in acquisition than allocation among individuals. This would generate a positive correlation between BMR and reproduction in productive environments, but a negative correlation in unproductive environments, as predicted by the context-dependence hypothesis (Burton *et al.* 2011).

However, we currently know little on the relative changes in acquisition vs. allocation across environmental gradients. What if the variation in resource acquisition *decreases* as environmental productivity increases? Assuming the pattern in Figure 14.1A still holds, then one should expect a negative correlation between BMR and reproduction in productive environments. To add even more complexity, we can also imagine other scenarios whereby plastic changes in BMR within an individual follow the increased-intake principle (Figure 14.1B) or that compensation occurs among individuals (Figures 14.1C and D). If the same principles apply at both levels of variation (Figures 14.1B and C), then the phenotypic correlation (r_P) will be reflective of the principle governing energy budgets. However, if different principles apply among vs. within individuals (Figures 14.1A, B, C and D), then the r_P will not tell much about the nature of the association between components of energy budgets and the processes that governs them as r_P depends on the relative amount of among- vs. within-individual variation (van Noordwijk & de Jong 1986). In the face of a nonsignificant r_P, for example, one could erroneously conclude that both increased-intake and allocation principles are not applicable, while both processes are occurring at the same time!

A r_P is shaped by correlations at the among-individual level (r_{ind}) and the within-individual level (r_e) (Dingemanse & Dochtermann 2013). Since repeatability of MR averages ~0.35 in fishes and wild mammals (White *et al.* 2013), r_P should be influenced 1.86x [i.e., =(1 – 0.35)/0.35] more strongly by r_e than r_{ind} in these taxa. By contrast, repeatability of MR in birds and reptiles appears to be relatively higher at 0.56 and 0.67 (White *et al.* 2013), respectively, and in these cases the influence of r_e on r_P should be 0.79x and 0.49x weaker than r_{ind}. All else being equal, differences in the sign and relative magnitude of r_{ind} and r_e can generate a range of values for r_P. Therefore, a better understanding of how the increased-intake and allocation models govern animals' energy budget can be reached by testing whether two components of the energy budget are correlated at the among- and within-individual levels (Dingemanse & Dochtermann 2013).

From Individuals to Species

The discrepancy between correlations obtained at the intra- and inter-specific levels of variation (see section 2) may be explained by the same logic as that described by van Noordwijk and de Jong (1986). One may be tempted to think that variation in allocation might be less important at the inter-specific level as this process is best thought of occurring within individuals. However, some species may need to allocate more energy to particular functions than others: in endotherms, those in colder environments may need to allocate more to heat production, iteroparous species presumably allocate less to reproductive tissue than semelparous species, those with poor quality diet (herbivores) may allocate more to digestion and so on (see also Chapter 18, this volume). In fact, some patterns of allocation may be more evident across species than within species, because there has been substantial time for selection to create larger relative differences in trait means.

A positive inter-specific correlation between maintenance costs and reproduction or activity may occur if species vary in total acquisition despite genetic tradeoffs that occur within species (Agrawal *et al.* 2010). To understand how processes occurring within species can determine patterns across species, we need information on the genetic variation in acquisition and allocation. Extending Van Noordwijk and de Jong's (1986) Y-model to genetic correlations, Houle (1991) showed theoretically that if there were more loci involved in resource acquisition than in allocation of that resource, the genetic correlation could be positive in spite of a fundamental allocation trade-off.

A positive genetic correlation between RMR and exploratory behavior in deer mice is concordant with a positive inter-specific correlation between the two traits among species of the Neotominae subfamily (Careau *et al.* 2011). As suggested by this and few other studies, genetic correlations may bias the direction of adaptive divergence across species along "genetic lines of least resistance" (Schluter 1996). These positive genetic and interspecific correlations between BMR and behavior do not imply that there is no energy allocation trade-off occurring, perhaps rather that Neotominae species have more energy to devote to both as they evolved toward higher energy budgets (Houle 1991, Mueller & Diamond 2001). Presumably, when selection or drift is strong enough the bias caused by genetic correlations can be broken, as for example the interspecific correlation between BMR and exploratory behavior is negative across a wider range of Muroid species (Careau *et al.* 2011).

METABOLIC SCOPES: A DIFFERENT KIND OF ENERGETIC BUDGETING

Acquisition and allocation processes deal with energy mainly as a quantity, but energy is absorbed or expended over a period of time (just as MR is measured in Watts; $J{\cdot}s^{-1}$); therefore, the most relevant way to conceptualize energy budgets is with the use of rate functions. Animals can only do so many things at once, and the ability to perform multiple O_2-consuming physiological tasks is limited by an animal's *absolute* or *factorial aerobic scopes* (AAS and FAS). Previous authors have argued that natural selection should act on the ability to maximize or optimize AAS (Fry 1975; Guderley & Pörtner 2010). In fact, the "energetic definition of fitness" is based on the principle of allocation (see section 3.1) and stipulates that natural selection should maximize the capacity to channel "residual energy" (i.e., energy in excess of maintenance, as represented by AAS) toward reproduction (Artacho & Nespolo 2009).

The boundaries of an animal's capacity to utilize energy aerobically are set by BMR or SMR (the metabolic "floor") and MMR (the "ceiling"). As these measures are made under contrasting circumstances, it is perhaps not surprising that the factors contributing to BMR (or SMR) and MMR can be somewhat different (i.e., the liver and kidney for BMR and musculature for MMR, Weibel *et al.* 2004, Wone *et al.* 2009). Analysing the difference between the two (AAS) or their ratio (FAS) may yield results otherwise undetectable from the individual analyses of the two components (see Boxes 14.2 and 14.3 for case studies).

Box 14.2. The Importance of FAS in Small Rodents

In a wild population of eastern chipmunks (*Tamias striatus*) FAS was not influenced by sex, age, or M_b, but significantly increased as the winter approached (Careau *et al.* 2012). Notably, FAS was strongly negatively correlated with the number of conspicuous botfly larvae that parasitize chipmunks, being the result of a differential effect of parasites on RMR (increase) and cold-induced MMR (decrease). Perhaps not coincidentally, botfly parasitism in this population negatively influences the likelihood of surviving winter (Careau *et al.* 2013a). Chipmunks survive winter by storing food in their burrow and using torpor, which simultaneously requires low maintenance costs (to conserve energy) and high thermogenic capacity (to warm up from a torpid state). Hence, parasitized chipmunks with low FAS may be doubly penalized as they have lower thermogenic capacity (cold-induced MMR) and higher maintenance costs (RMR). In fact, a comparative analysis of FAS in across rodent species reported a strong negative correlation between FAS and T_a, suggesting that torpor is an adaptation that entails a high FAS in cold climates (Careau 2013).

Box 14.3. The Importance of AAS in Young Marine Fish

Lumpfish (*Cyclopterus lumpus*) possess a ventral adhesive disc, which allows them to easily switch between actively swimming and remaining perfectly still even when exposed to wave action. Juveniles spend large amounts of time clinging to rocks and seaweed which probably reduces their predation risk, but also results in a 6–12% decrease in overall energetic costs. This decrease is modest, indicating that swimming at routine speeds while foraging is a relatively cheap activity. However, if individuals were to swim at speeds that would be most efficient, this activity alone would use approximately 46–70% of total AAS. This could restrict the ability to carry out other important aerobic processes while swimming, including growth, which is critical during this life-stage. Therefore, in addition to reducing overall energy expenditure, the behavioral flexibility to adjust locomotor strategies provides a means to save energy for other activities within a limited AAS (Killen *et al.* 2007a; Killen *et al.* 2007b).

The way that aerobic scopes are conceptualized and applied varies among researchers working on different taxonomic groups. In endotherms, for example, aerobic scope is generally presumed to be most important over short temporal scales, for trade-offs between functions such as thermoregulation and locomotion. For ectotherms, however, aerobic scope also encompasses more protracted aerobic processes such as growth and reproduction (Guderley and Pörtner 2010), whereas in endotherm research these functions are usually studied from the perspective of *sustained metabolic scope* (SusMS) or *nonresting energy expenditure* (NREE). Ectotherms have a much lower AAS as compared to endotherms of similar size, and so over prolonged periods, small differences in activity level or any other source of metabolic loading can accumulate to cause trade-offs with growth or reproduction within an individual's aerobic scope (Guderley & Pörtner 2010). Endotherms have a much greater AAS as compared to similar-sized ectotherms, such that the long-term effects of locomotory activity on growth or reproduction are relatively trivial.

An examination of the proxies used for estimating aerobic scope in various taxa could be an effective means of inspiring new ways of thinking about research examining links between metabolic and behavioral traits. For example, for endotherms there has been considerable interest in the levels of DEE that can be sustained for long periods (Peterson *et al.* 1990). For endotherms, SusMS represents the upper boundary on the sum total of activities in which endotherms can engage over a prolonged period (Speakman 2000). This is the precise reason why researchers studying ectotherms have become interested in the potential effects of AAS on behavior and ecological phenomena (Killen *et al.* 2007b; Killen *et al.* 2012). Many of the same behavioral constraints and ecological effects stemming from a limited AAS in ecotherms could apply to SusMS in endotherms and vice-versa.

ENERGETIC CONSTRAINTS ON INDIVIDUAL BEHAVIOR

A large number of recent studies have examined co-variation in metabolic and behavioral traits among individuals (reviewed in Careau & Garland 2012; Krams *et al.* 2013). A general finding stemming from this body of work is that there is no single cause-and-effect mechanism driving these relationships, but that the direction of the effects is dynamic and shifts in different contexts and environments (Killen *et al.* 2013). However, the remarkable energetic differences between endotherms and ectotherms are an unexploited basis for achieving a further understanding of such relationships, as we now illustrate.

Spontaneous Activity

The effects of T_a pervade the behavioral and physiological differences observed between endotherms and ectotherms. Ectotherms tend to decrease activity at cooler T_a, while endotherms may either attempt to decrease energetic costs by entering torpor (Geiser 2004) or reducing activity (Humphries *et al.* 2005) or they may increase metabolic heat production, either through shivering thermogenesis or via physical exercise. If the heat generated through activity can substitute for heat required for thermoregulation, then activity in cold environments may be energetically free for endotherms in some situations (Humphries & Careau 2011). Still, for endotherms, MR can increase as T_a becomes colder until animals approach MMR. This is in contrast to ectotherms, which tend to increase SMR as T_a becomes *warmer* until SMR is equal to MMR and AAS is equal to zero (Pörtner & Farrell 2008). A key difference is that the increase in MR with decreasing T_a in endotherms is due to increased thermoregulation and/or activity while the increase in MR in ectotherms with T_a is mainly due to increasing maintenance requirements, but the symmetry of these opposite responses is striking.

For both endotherms and ectotherms, the spontaneous activity of individuals will also affect the way that they experience environmental variation over short temporal scales. Exposure to different microhabitats might cause bold or active individuals to experience more variability in extrinsic factors (e.g., T_a, hypoxia, parasites). More active individuals may therefore require increased regulatory performance (Husak *et al.* 2009) to maintain activity and a high level of dynamic performance during acute exposure to a range of environmental conditions, or perhaps having a wider thermal performance breadth than individuals that are shier or less active. For endotherms, activity-thermoregulatory heat substitution (Humphries & Careau 2011) may allow individuals that are intrinsically more active to temporarily exploit colder environments that would be off-limits to more sedentary individuals.

Foraging

An interesting example of how energetic constraints affect foraging is the differential responses of endotherms and ectotherms to spatial variation in prey density (Helfman 1990; Killen *et al.* 2007b). In many species individuals switch between alternate foraging strategies that differ in the level of energetic investment (e.g., active searching vs. sit and wait ambush). Curiously, individual endotherms and ectotherms choose opposite strategies in relation to prey density. Endotherms switch to the more costly foraging strategy only at relatively high prey densities (Rudolph 1982) while ectotherms switch to the more expensive strategy when prey densities decrease (Killen *et al.* 2007b). The explanation is that while endotherms maximize energy intake to satisfy high BMR and activity costs, ectotherms need only meet some minimal rate of energy intake. A promising area for future research will be to examine how foraging strategies may vary among individuals of the same species in relation to metabolic traits. At certain prey densities some individuals may become more active and susceptible to predation as a function of intrinsic physiological traits such as metabolic demand or AAS.

An important constraint for ectotherms stemming from foraging is the proportion of AAS that must be directed toward digestion and assimilation of food and the resulting trade-offs with other physiological functions. In some reptiles and fishes, MR measured after feeding can be substantial and exceed that measured during peak exercise (Fu *et al.* 2008). Endotherms also require aerobic capacity to be re-routed while processing a meal, but the trade-offs with other physiological functions are much less severe (e.g., Nespolo *et al.* 2003). It is also important to note that in endotherms the heat generated through digestion can substitute for heat production required for thermoregulation (Lovvorn 2007). As a result, individual variability in the energetic costs and efficiency of digestion may be a much less important factor constraining the behavior of endotherms as compared to ectotherms, where it can be significant (Dupont-Prinet *et al.* 2009).

Habitat Selection

Although the capacity for performance in endotherms can be affected by abiotic factors such as T_a, mainly due to effects on muscular activity stemming from regional homeothermy, these effects are much more pronounced for ectotherms (Angilletta *et al.* 2010). It is therefore possible that, compared to ectotherms, habitat selection in individual endotherms might be less driven by abiotic environmental factors and instead primarily affected by factors such as food availability or predation. Individual endotherms with decreased thermosensitivity could have a wider range of microhabitat options, but will still be bound to habitats that can satisfy the demand of an elevated BMR. For individual endotherms, a major benefit of a reduced BMR could be continued activity and foraging in relatively cold habitats with a higher threshold for starvation tolerance before either torpor or thermal conformity become necessary.

Habitat selection for ectotherms will also be affected by food availability and safety, but for them abiotic characteristics of environment itself may have a much stronger effect on the performance capacity of individuals (Claireaux *et al.* 2000). An interesting possibility which has not been investigated is that among individual ectotherms, bolder or more active individuals might select conditions (e.g warmer T_a) that maximize AAS and facilitate an active lifestyle; shy individuals might prefer conditions (e.g., cooler T_a) that minimize energetic costs. For ectotherms more so than endotherms, the environment itself could facilitate the expression of behavioral phenotypes or even cause bold and shy individuals to have different priorities when selecting habitats.

CONCLUSIONS AND FUTURE PERSPECTIVES

As energy is quantifiable and comparable across many levels of biological organization (cells, organs, individuals, population, species), the field of energetics lends itself to integrative studies linking patterns and processes at multiple levels. Although it is well known that variation in MR is determined by several intrinsic and extrinsic factors, few studies have evaluated the relative importance of many factors simultaneously, let alone the causal relationships between them. As many studies on the increased-intake and allocation processes were conducted at the phenotypic level, we still do not have a clear picture of how these processes govern energy budgets of animal. Together, changes in the relative importance of these two processes may produce contrasting correlations across levels of variation (e.g., within individuals, among individuals, among species) and potentially underlie the context-dependence nature of the relationship between maintenance costs and fitness or behavior. Our attempt to compare and integrate differences in the energy budgets of ectotherms and endotherms yielded interesting possibilities as to how the relative magnitude of individual variability should differ in these taxa, with implications on ecologically relevant aspects such as spontaneous activity, foraging patterns, and habitat selection.

REFERENCES

Agrawal, A.A., Conner, J.K. and Rasmann, S. (2010) Tradeoffs and adaptive negative correlations in evolutionary ecology, in *Evolution after Darwin: the First 150 Years* (eds M. Bell, W. Eanes, D. Futuyma and J. Levinton), Sinauer Associates, pp. 243–68.

Anderson, K.J. and Jetz, W. (2005) The broad-scale ecology of energy expenditure of endotherms. *Ecology Letters*, **8**, 310–18.

Angilletta, M.J. Jr., Cooper, B.S., Schuler, M.S. and Boyles, J.G. (2010) The evolution of thermal physiology in endotherms. *Frontiers in Bioscience Elite Edition*, **2**, 861–81.

Artacho, P. and Nespolo, R.F. (2009) Natural selection reduces energy metabolism in the garden snail, *Helix aspersa* (*Cornu aspersum*). *Evolution*, **63**, 1044–50.

Bozinovic, F. and Sabat, P. (2010) On the intraspecific variability in basal metabolism and the food habits hypothesis in birds. *Current Zoology*, **56**, 759–66.

Bozinovic, F., Rojas, J.M., Broitman, B.R. and Vasquez, R.A. (2009) Basal metabolism is correlated with habitat productivity among populations of degus (*Octodon degus*). *Comparative Biochemistry and Physiology A*, **152**, 560–4.

Brown, J.H., Gillooly, J.F., Allen, A.P. *et al.* (2004) Toward a metabolic theory of ecology. *Ecology*, **85**, 1771–89.

Burton, T., Killen, S.S., Armstrong, J.D. and Metcalfe, N.B. (2011) What causes intraspecific variation in resting metabolic rate and what are its ecological consequences? *Proceedings of the Royal Society B*, **278**, 3465–73.

Careau, V. (2013) Basal metabolic rate, maximum thermogenic capacity and aerobic scope in rodents: interaction between environmental temperature and torpor use. *Biology Letters*, **9**, 20121104.

Careau, V. and Garland, T. Jr. (2012) Performance, personality, and energetics: correlation, causation, and mechanism. *Physiological and Biochemical Zoology*, **85**, 543–71.

Careau, V., Garant, D. and Humphries, M.M. (2012) Free-ranging eastern chipmunks (*Tamias striatus*) infected with bot fly (*Cuterebra emasculator*) larvae have higher resting but lower maximum metabolism. *Canadian Journal of Zoology*, **90**, 413–21.

Careau, V., Thomas, D., Humphries, M.M. and Réale, D. (2008) Energy metabolism and animal personality. *Oikos*, **117**, 641–53.

Careau, V., Bergeron, P., Garant, D. *et al.* (2013a) The energetic and survival costs of growth in free-ranging eastern chipmunks. *Oecologia*, **171**, 11–23.

Careau, V., Réale, D., Garant, D. *et al.* (2013b) Context-dependent correlation between resting metabolic rate and daily energy expenditure in wild chipmunks. *Journal of Experimental Biology*, **216**, 418–26.

Careau, V., Thomas, D., Pelletier, F. *et al.* (2011) Genetic correlation between resting metabolic rate and exploratory behaviour environment in deer mice (*Peromyscus maniculatus*). *Journal of Evolutionary Biology*, **24**, 2153–63.

Carey, N., Sigwart, J.D. and Richards, J.G. (2013) Economies of scaling: More evidence that allometry of metabolism is linked to activity, metabolic rate and habitat. *Journal of Experimental Marine Biology and Ecology*, **439**, 7–14.

Claireaux, G., Webber, D.M., Lagardère, J.P. and Kerr, S.R. (2000) Influence of water temperature and oxygenation on the aerobic metabolic scope of Atlantic cod (*Gadus morhua*). *Journal of Sea Research*, **44**, 257–65.

Dingemanse, N.J. and Dochtermann, N.A. (2013) Quantifying individual variation in behaviour: mixed-effect modelling approaches. *Journal of Animal Ecology*, **82**, 39–54.

Dupont-Prinet, A., Claireaux, G. and McKenzie, D.J. (2009) Effects of feeding and hypoxia on cardiac performance and gastrointestinal blood flow during critical speed swimming in the sea bass *Dicentrarchus labrax*. *Comparative Biochemistry and Physiology A*, **154**, 233–40.

Fletcher, Q.E., Speakman, J.R., Boutin, S. *et al.* (2012) Seasonal stage differences overwhelm environmental and individual factors as determinants of energy expenditure in free-ranging red squirrels. *Functional Ecology*, **26**, 677–87.

Fry, F.E.J. (1975) The aquatic respiration of fish, in *The Physiology of Fishes* (ed M.E. Brown), Academic Press, New York, pp. 1–64.

Fu, S.J., Cao, Z.D., Peng, J.L. and Wang, Y.X. (2008) Is peak postprandial oxygen consumption positively related to growth rate and resting oxygen consumption in a sedentary catfish *Silurus meridionalis*? *Journal of Fish Biology*, **73**, 692–701.

Geiser, F. (2004) Metabolic rate and body temperature reduction during hibernation and daily torpor. *Annual Review of Physiology*, **66**, 239–74.

Guderley, H. and Pörtner, H.O. (2010) Metabolic power budgeting and adaptive strategies in zoology: examples from scallops and fish. *Canadian Journal of Zoology*, **88**, 753–63.

Handelsman, C.A., Broder, E.D., Dalton, C.M. *et al.* (2013) Predator-induced phenotypic plasticity in metabolism and rate of growth: rapid adaptation to a novel environment. *Integrative and Comparative Biology*, **53** (6), 975–88.

Helfman, G.S. (1990) Mode selection and mode switching in foraging animals. *Advances in the Study of Behavior*, **19**, 249–98.

Houle, D. (1991) Genetic covariance of fitness correlates: what genetic correlations are made of and why it matters. *Evolution*, **45**, 630–48.

Hulbert, A.J. and Else, P.L. (1999) Membranes as possible pacemakers of metabolism. *Journal of Theoretical Biology*, **199**, 257–74.

Humphries, M. and Careau, V. (2011) Heat for nothing or activity for free? evidence and implications of activity-thermoregulatory heat substitution. *Integrative and Comparative Biology*, **51**, 419–31.

Humphries, M.M., Boutin, S., Thomas, D.W. *et al.* (2005) Expenditure freeze: the metabolic response of small mammals to cold environments. *Ecology Letters*, **8**, 1326–33.

Husak, J.F., Irschick, D.J., McCormick, S.D. and Moore, I.T. (2009) Hormonal regulation of whole-animal performance: Implications for selection. *Integrative and Comparative Biology*, **49**, 349–53.

Killen, S.S., Brown, J.A. and Gamperl, A.K. (2007a) The effect of prey density on foraging mode selection in juvenile lumpfish: balancing food intake with the metabolic cost of foraging. *Journal of Animal Ecology*, **76**, 814–25.

Killen, S.S., Atkinson, D. and Glazier, D.S. (2010) The intraspecific scaling of metabolic rate with body mass in fishes depends on lifestyle and temperature. *Ecology Letters*, **13**, 184–93.

Killen, S.S., Costa, I., Brown, J.A. and Gamperl, A.K. (2007b) Little left in the tank: metabolic scaling in marine teleosts and its implications for aerobic scope. *Proceedings of the Royal Society B*, **274**, 431–8.

Killen, S.S., Marras, S., Steffensen, J.F. and McKenzie, D.J. (2012) Aerobic capacity influences the spatial position of individuals within fish schools. *Proceedings of the Royal Society B*, **279**, 357–64.

Killen, S.S., Marras, S., Metcalfe, N.B. *et al.* (2013) Environmental stressors alter relationships between physiology and behaviour. *Trends in Ecology & Evolution*, **28** (11), 651–8.

Konarzewski, M. and Ksiazek, A. (2013) Determinants of intra-specific variation in basal metabolic rate. *Journal of Comparative Physiology B*, **183**, 27–41.

Kozlowski, J., Konarzewski, M. and Gawelczyk, A.T. (2003) Cell size as a link between noncoding DNA and metabolic rate scaling. *Proceedings of the National Academy of Sciences of the United States of America*, **100**, 14080–5.

Krams, I., Kivleniece, I., Kuusik, A. *et al.* (2013) Predation selects for low resting metabolic rate and consistent individual differences in anti-predator behavior in a beetle. *Acta Ethologica*, **163**, 163–72.

Lifson, N., Gordon, G.B. and McClintock, R. (1955) Measurement of total carbon dioxide production by means of $D_2{}^{18}O$. *Journal of Applied Physiology*, **7**, 704–10.

Lovegrove, B.G. (2000) The zoogeography of mammalian basal metabolic rate. *American Naturalist*, **156**, 201–19.

Lovvorn, J.R. (2007) Thermal substitution and aerobic efficiency: measuring and predicting effects of heat balance on endotherm diving energetics. *Philosophical Transactions of the Royal Society B*, **362**, 2079–94.

Meerlo, P., Bolle, L., Visser, G.H. *et al.* (1997) Basal metabolic rate in relation to body composition and daily energy expenditure in the field vole, *Microtus agrestis*. *Physiological Zoology*, **70**, 362–9.

Mueller, P. and Diamond, J. (2001) Metabolic rate and environmental productivity: well-provisioned animals evolved to run and idle fast. *Proceedings of the National Academy of Sciences of the United States of America*, **98**, 12550–4.

Nespolo, R.F., Bacigalupe, L.D. and Bozinovic, F. (2003) The influence of heat increment of feeding on basal metabolic rate in *Phyllotis darwini* (Muridae). *Comparative Biochemistry and Physiology A*, **134**, 141–7.

Nisbet, R.M., Muller, E.B., Lika, K. and Kooijman, S.A.L.M. (2000) From molecules to ecosystems through dynamic energy budget models. *Journal of Animal Ecology*, **69**, 913–26.

Peterson, C.C., Nagy, K.A. and Diamond, J. (1990) Sustained metabolic scope. *Proceedings of the National Academy of Sciences of the United States of America*, **87**, 2324–8.

Polymeropoulos, E.T., Heldmaier, G., Frappell, P.B. *et al.* (2012) Phylogenetic differences of mammalian basal metabolic rate are not explained by mitochondrial basal proton leak. *Proceedings of the Royal Society B*, **279**, 185–93.

Pörtner, H.O. and Farrell, A.P. (2008) Ecology: physiology and climate change. *Science*, **322**, 690–2.

Rezende, E.L., Bozinovic, F. and Garland, T. Jr. (2004) Climatic adaptation and the evolution of basal and maximum rates of metabolism in rodents. *Evolution*, **58**, 1361–74.

Ricklefs, R.E., Konarzewski, M. and Daan, S. (1996) The relationship between basal metabolic rate and daily energy expenditure in birds and mammals. *American Naturalist*, **147**, 1047–71.

Rudolph, S.G. (1982) Foraging strategies of american kestrels during breeding. *Ecology*, **63**, 1268–76.

Russell, G.A. and Chappell, M.A. (2007) Is BMR repeatable in deer mice? Organ mass correlates and the effects of cold acclimation and natal altitude. *Journal of Comparative Physiology B*, **177**, 75–87.

Schimpf, N.G., Matthews, P.G.D. and White, C.R. (2012) Standard metabolic rate is associated with gestation duration, but not clutch size, in speckled cockroaches *Nauphoeta cinerea*. *Biology Open*, **15**, 1185–91.

Schluter, D. (1996) Adaptive radiation along genetic lines of least resistance. *Evolution*, **50**, 1766–74.

Speakman, J.R. (2000) The cost of living: Field metabolic rates of small mammals. *Advances in Ecological Research*, **30**, 177–297.

Speakman, J.R. and Król, E. (2010) Maximal heat dissipation capacity and hyperthermia risk: neglected key factors in the ecology of endotherms. *Journal of Animal Ecology*, **79**, 726–46.

van Noordwijk, A.J. and de Jong, G. (1986) Acquisition and allocation of resources: their influence on variation in life history tactics. *American Naturalist*, **128**, 127–42.

Weibel, E.R., Bacigalupe, L.D., Schmitt, B. and Hoppeler, H. (2004) Allometric scaling of maximal metabolic rate in mammals: muscle aerobic capacity as determinant factor. *Respiratory Physiology & Neurobiology*, **140**, 115–32.

White, C.R. and Kearney, M.R. (2013) Determinants of inter-specific variation in basal metabolic rate. *Journal of Comparative Physiology B*, **183**, 1–26.

White, C.R., Alton, L.A. and Frappell, P.B. (2012) Metabolic cold adaptation in fishes occurs at the level of whole animal, mitochondria and enzyme. *Proceedings of the Royal Society B*, **279**, 1740–7.

White, C.R., Schimpf, N.G. and Cassey, P. (2013) The repeatability of metabolic rate declines with time. *Journal of Experimental Biology*, **216**, 1763–5.

White, C.R., Blackburn, T.M., Martin, G.R. and Butler, P.J. (2007) Basal metabolic rate of birds is associated with habitat temperature and precipitation, not primary productivity. *Proceedings of the Royal Society B*, **274**, 287–93.

Wiersma, P., Nowak, B. and Williams, J.B. (2012) Small organ size contributes to the slow pace of life in tropical birds. *Journal of Experimental Biology*, **215**, 1662–9.

Wone, B., Sears, M.W., Labocha, M.K. *et al.* (2009) Genetic variances and covariances of aerobic metabolic rates in laboratory mice. *Proceedings of the Royal Society B*, **276**, 3695–3704.

15

BIOLOGICAL TIMEKEEPING: INDIVIDUAL VARIATION, PERFORMANCE, AND FITNESS

Scott A. MacDougall-Shackleton[1], Heather E. Watts[2], and Thomas P. Hahn[3]

[1]*Department of Psychology, Advanced Facility for Avian Research, University of Western Ontario, London, ON, Canada*
[2]*Department of Biology, Loyola Marymount University, Los Angeles, CA, USA*
[3]*Department of Neurobiology, Physiology and Behavior, University of California Davis, Davis, CA, USA*

INTRODUCTION

Virtually all parts of our planet exhibit cyclic changes in environmental conditions, and these cycles have persisted throughout the evolution and diversification of life. In response, most biological processes exhibit rhythms, and these rhythms can be observed at molecular, cellular, whole-organism, and population scales. As well, these rhythms exist at multiple time scales, including short-term oscillations, tidal, daily, lunar, and annual cycles. Biological rhythms have been extensively studied for many decades at a range of levels, including examination of the molecular basis of circadian clocks, neural and endocrine control of circadian cycles, seasonality and annual rhythms. Comprehensive books on this topic include Dunlap *et al.* (2004), Koukkari and Sothern (2006), and Foster and Kreitzman (2005, 2009). In this chapter we briefly introduce fundamental concepts of the integrative physiology of circadian rhythms and seasonality with an emphasis on examples where individual variation in biological timekeeping relates to variation in performance or fitness.

ENDOGENOUS VERSUS ENVIRONMENTAL DRIVERS

Organisms could exhibit rhythmic changes in physiology and behavior just by responding to cyclic changes in the external world. However, in almost all cases examined, organisms

Integrative Organismal Biology, First Edition.
Edited by Lynn B. Martin, Cameron K. Ghalambor, and H. Arthur Woods.

also possess endogenous clocks that continue rhythmic activity in the absence of exogenous cues. For example, mangrove crickets (*Apteronemobius asahinai*) continue rhythmic patterns of activity with a period of approximately 12.4 h even when removed from their intertidal zone (Satoh *et al.* 2008). More commonly, organisms across taxonomic groups exhibit endogenous circadian rhythms that persist on a cycle of approximately 24 h when held in constant conditions, or endogenous circannual rhythms that persist on a cycle of about a year. When endogenous clocks are expressed in the absence of input from external cues, they are said to be *free-running*. *Entrainment* can fine-tune an endogenous cycle through changes in phase or period when the organism is exposed to an external cue known as a *zeitgeber*, or time-giver. For example, as one travels east or west across time zones one's endogenous daily rhythms of hormone concentrations, body temperature, and activity are gradually entrained to match the local external daily cycles.

The ubiquity of entrainable endogenous clocks begs the question as to their adaptive significance. Broadly, two kinds of adaptive value have been proposed for endogenous clocks: *intrinsic* and *extrinsic* value (Johnson 2005). The intrinsic value of endogenous clocks is thought to lie in the advantage of integrating internal physiological processes such as sleep–wake cycles and changes in body temperature, metabolism, and hormones that must be coordinated with one another for optimal function. The extrinsic value of endogenous clocks is thought to lie in permitting anticipation of changes in external conditions before the conditions actually change, or when the animal is prevented from detecting the change, such as for a rodent sleeping in a burrow insulated from the external light-dark cycle (photoperiod) without an internal timer. Similarly, processes such as migration or reproduction may require organisms to begin preparatory adjustments in physiology weeks to months ahead of time, and thus require a capacity to anticipate seasonal cycles. Although the intrinsic and extrinsic adaptive values of endogenous clocks would appear self-evident, there are relatively few studies that explicitly test these hypotheses (Johnson 2005). The vast majority of studies on endogenous clocks focus on molecular and physiological mechanisms, and, to a lesser extent, ecological outcomes.

RESPONSES TO ENTRAINMENT BY ENDOGENOUS CLOCKS VARIES OVER THEIR CYCLE

An important hallmark of endogenous clocks is that their capacity for entrainment varies across their cycle. This temporally varying sensitivity can be described with a phase response curve. We illustrate this concept with two examples: the phase response curve of circadian activity of rodents responding to light pulses, and photorefractoriness in seasonally breeding songbirds.

Wheel-running by nocturnal rodents exhibits an endogenous circadian rhythm with a free-running periodicity of slightly longer than 24 h. Brief pulses of light can entrain this endogenous rhythm (reviewed in Johnson *et al.* 2004). However the effect of a light pulse can vary from phase-advancing, to phase-delaying, to having no effect on the free-running cycle (Figure 15.1). Exposure to light during the animal's subjective day – the inactive phase of the endogenous circadian rhythm – has little effect on subsequent cycles, and is sometimes referred to as a dead zone. In contrast, exposure to pulses of light during the animal's subjective night entrains the free-running rhythm. Light pulses during the early part of the animal's subjective night induce a phase delay of subsequent cycles and light pulses during the latter part of the animal's subjective night induce a phase advance. Thus the effect of the zeitgeber (light) varies over the endogenous cycle. Put another way, one property of

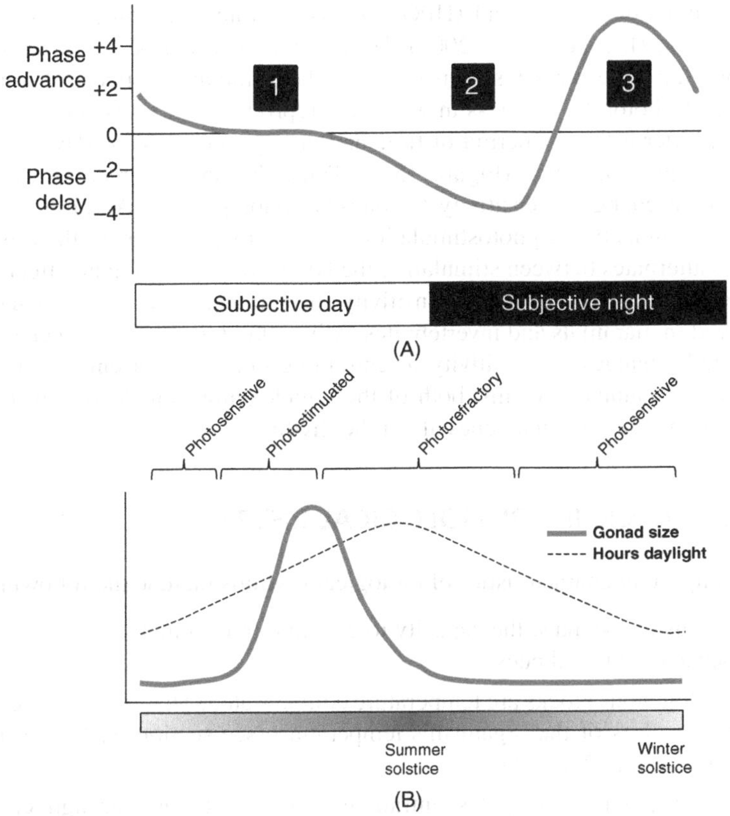

Figure 15.1. The entrainment response of endogenous rhythms varies over the course of their cycle. A: For endogenous circadian rhythms the effect of a zeitgeber varies over the subjective day-night cycle in a free-running animal, as illustrated in this generalized phase response curve. For a free-running nocturnal rodent, a light pulse experienced in the middle of the subjective day (1) when the animal is inactive does not entrain the rhythm. A light pulse late in the subjective day or early in the subjective night (2) will induce a phase delay in the free-running rhythm. A light pulse late in the subjective night (3) will induce a phase advance. B: Seasonally breeding songbirds vary in their response to photoperiod across the annual cycle. In late winter or early spring, birds have small gonads, but are photosensitive and will respond to long days with gonadal recrudescence. As day-length increases in spring, the long days induce a state of photostimulation. However, while days are still long birds cease being stimulated by long-day photoperiods and become photorefractory. Thus the stimulatory effect of long days varies over the course of annual cycle.

this endogenous circadian clock is a changing sensitivity to environmental cues. The clock cycles between insensitivity to light, being accelerated by light, and being delayed by light.

Changing sensitivity to environmental cues is a property of longer duration rhythms as well. For example, most seasonally breeding songbirds use photoperiod as the primary cue to time reproduction (Wingfield 1983; Dawson *et al.* 2001). In winter, on a short-day photoperiod, birds are in a physiological state of photosensitivity, in which they are reproductively responsive to the duration of the light phase of the day (photophase). If this photophase is sufficiently long, birds become photostimulated and the hypothalamus-pituitary-gonad (HPG) axis is activated, resulting in gonadal growth and reproduction (Dawson *et al.* 2001). However, after a period of weeks of exposure to these long days the birds lose sensitivity to the stimulatory effects of long days and enter a state of photorefractoriness in which the reproductive axis is shut down (Nicholls *et al.* 1988). After a further period of time and/or exposure to short days the birds once again become photosensitive (Figure 15.1). Thus, the annual breeding cycle of these birds involves a changing sensitivity to long-day photoperiods. As birds cycle through periods of photosensitivity, photostimulation, and photorefractoriness the effects of long photophases alternates between stimulating the HPG axis and having no effect. In addition to avian photoperiodism, changing sensitivity to entrainers over a circannual cycle has been observed in mammals and invertebrates (Miyazaki *et al.* 2005; Monecke *et al.* 2009). Thus, although changes in sensitivity to entraining cues are not entirely analogous for circadian and circannual rhythms, both of these endogenous clocks do involve a change over time in how environmental cues affect the rhythm.

GENERAL PROPERTIES OF BIOLOGICAL RHYTHMS

In general, important characteristics of biological rhythms include the following features:

1. Endogenous clocks have the capacity to generate free-running rhythms that persist in the absence of external cues.
2. These endogenous clocks are temperature compensated. That is, they have a consistent period regardless of the organism's temperature (see section on 'Extrinsic Adaptive Value', below, pp. 242–44).
3. External environmental cycles entrain endogenous timers through environmental drivers or cues (zeitgebers).
4. Sensitivity to these environmental cues varies over the cycle of the endogenous rhythm.

These characteristics are most extensively studied for circadian rhythms, but are also observed across a range of time scales and can be instantiated by different underlying physiological processes. The vast majority of research on biological rhythms has focused on circadian rhythms and many fundamental properties of circadian rhythms have been characterized (Johnson *et al.* 2004). Seasonality, photoperiodism and circannual rhythms have also been widely studied. In the following two sections we briefly introduce some of the physiological processes underlying circadian and circannual rhythmicity, then review studies that focus on individual variation, performance and fitness.

CIRCADIAN RHYTHMS

Molecular and Physiological Mechanisms

The day/night cycle is perhaps the most conspicuous environmental cycle on the planet. Organisms in all kingdoms of life exhibit endogenous circadian rhythms, and these rhythms have been studied at the molecular, cellular, physiological, and ecological levels. However, despite the ubiquity of circadian rhythms and the epic scientific endeavor to understand the molecular genetic and neuroendocrine bases of circadian timing there is still much we do not know.

Molecular Genetic Models The molecular genetic circadian model was initially developed in *Drosophila* and later explored in other taxa including fungi, cyanobacteria, and mammals. Put simply, these models posit that circadian rhythmicity is generated by a group of clock genes that participate in a transcription-translation delayed negative feedback loop (Figure 15.2; Dunlap 1999). A simplified model of the proposed molecular clock of *Drosophila* involves the transcription and translation of two clock genes, *per* and *tim*, whose protein products dimerize, translocate to the cell nucleus, and inhibit further expression of *per* and *tim*. Reduced expression of these genes, combined with metabolism of the PER and TIM proteins, leads to reduction in the amount of these proteins and thus reduced inhibition of *per* and *tim*, and an eventual increase in protein levels. Similar molecular clocks have now been identified in other taxa (Dunlap 1999). The fact that circadian clocks appear ubiquitous, but that there are nonhomologous clock genes in different taxa, suggests that endogenous circadian clocks may have evolved multiple times independently. Moreover, circadian clocks are an emergent property of groups of genes, rather than dependent on the effects of any single genes (Foster & Kreitzman 2005).

The discovery of clock genes, development of models of the molecular circadian clock, and clarification of how these transcription–translation feedback loops are modulated by light exposure, have been a triumph of modern biology. However, we currently lack a full understanding of how these mechanisms vary among species, or how they are linked

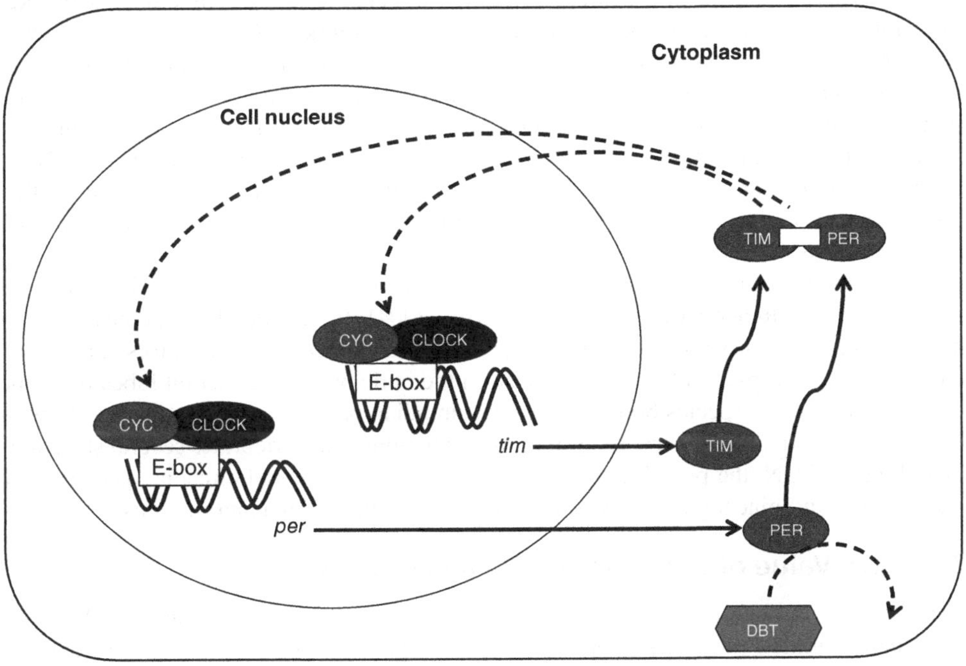

Figure 15.2. A simplified model of the molecular circadian clock in *Drosophila*. CLOCK and CYCLE (CYC) proteins bind to each other and promote expression of *per* and *tim* by binding to upstream enhancer-box (E-box) regions. Following translation PER and TIM protein dimerize and inhibit (dashed arrows) CYC and CLOCK, and thus their own production. PER is degraded (dashed arrow) by the kinase doubletime (DBT). PER and TIM thus exhibit cyclical increases and decreases in production with a periodicity of about 24 hours.

to whole-animal performance and fitness. Different properties of endogenous circadian timers (endogenous rhythm, entrainability, temperature compensation) may each have different molecular mechanisms that vary across taxa. For example, endogenous circadian rhythms of photosynthesis can be expressed in the algae *Acetabularia* even when the cell nucleus has been removed, casting doubt on the hypothesis that this rhythm is generated by a transcription-translation feedback loop (Edmunds 1988). Even for *Drosophila* there remain open questions (Lakin-Thomas 2000). For example, how the biochemical kinetics of transcription-translation-feedback loops produce a 24 h period as opposed to a cycle of another duration is not clear. In addition, clock genes are not simply clock genes: PER protein is synthesized in a variety of fly tissues and in some of these tissues it is neither found in the nucleus nor rhythmically produced (Hall 1995, 1996). Thus, despite the breakthroughs made in our understanding of molecular circadian processes, we still lack a full understanding of how these clocks generate an endogenous circadian rhythm and how they relate to whole animal performance.

Neural and Endocrine Organization Beyond molecular genetic models, the neural and endocrine physiology underlying circadian rhythms is also extensively studied, though our understanding of interspecies variation and links to whole organism performance and fitness are limited (see next section). In mammals a master circadian clock in the bilateral suprachiasmatic nuclei (SCN) of the hypothalamus drives all other circadian rhythms. There is unequivocal evidence that the SCN are the primary circadian pacemaker in mammals. Lesions of the SCN disrupt circadian rhythmicity (e.g., Stephan & Zucker 1972; Albers *et al.* 1984; Cohen & Albers 1991). More conclusively, transplantation of the SCN from a hamster with a mutant rhythm into a wild-type animal induces the mutant rhythm of circadian activity, and vice versa (Ralph *et al.* 1990). The SCN is entrained by light, which is detected by nonvisual photoreceptors in the retina that project to the SCN via the retinohypothalamic tract, and has a variety of primary and secondary projections that regulate physiological and behavioral circadian rhythms (Figure 15.3; reviewed in Meijer & Rietveld 1989).

In nonmammalian animals neural circadian pacemakers are less centralized, and are less extensively studied (reviewed in DeCoursey 2004). In gastropods, large basal retinal neurons appear to serve as both pacemakers and entrainment photoreceptors. In insects, the primary neural pacemakers are localized in the optic or cerebral neural lobes for some species, but in other species both the primary circadian pacemakers and entrainment photoreceptors lie outside the nervous system. In nonmammalian vertebrates several structures including the SCN, the pineal gland and the retina act as potential circadian pacemakers, and the eyes and pineal gland may transduce light to entrain circadian rhythms.

Adaptive Value of Endogenous Circadian Clocks

Regardless of the neural or other physiological substrates of circadian biological clocks, they share common functional properties including generation of an endogenous, entrainable rhythm that produces a variety of outputs. Below we review the link between such circadian systems and whole animal performance and fitness. Vaze and Sharma (2013) provide comprehensive coverage of this topic. The adaptive values of endogenous circadian rhythms are thought to lie in the intrinsic value of synchronizing and organizing internal processes and/or the extrinsic value of allowing organisms to entrain to and predict environmental conditions. To test these hypotheses, researchers have manipulated environmental conditions or the endogenous clock itself with the prediction that these manipulations should reduce performance.

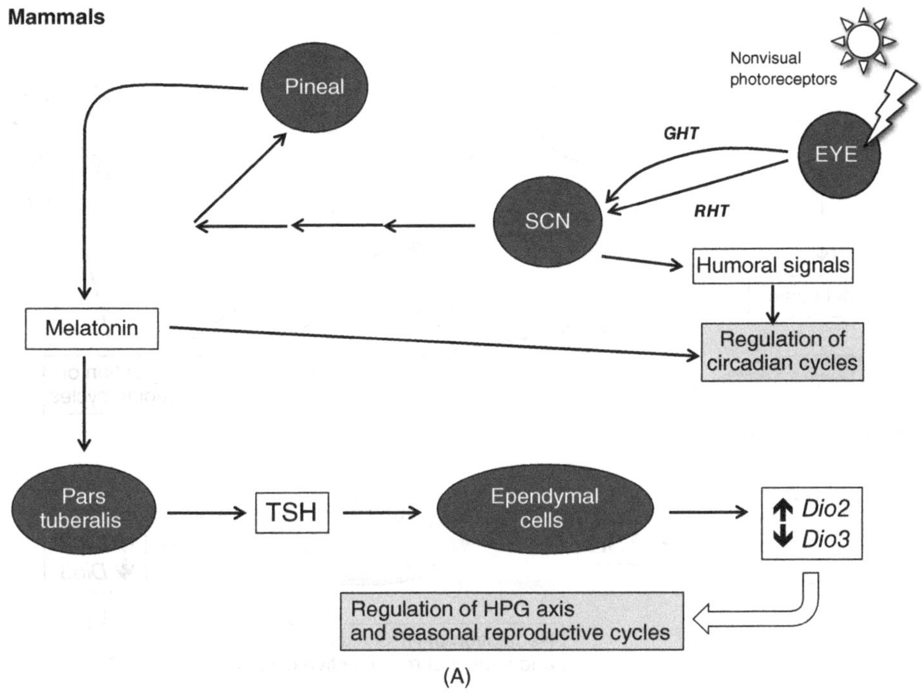

Figure 15.3. Schematic of photoperiodic regulation of seasonal reproduction in mammals and birds. A: In mammals, light is detected by nonvisual photoreceptors in the eye and synchronizes the circadian cycle of the SCN via the retinohypothalamic tract (RHT) and the more indirect geniculohypothalamic tract (GHT). SCN outputs include humoral signals that regulate circadian rhythms and a multisynaptic projection to the pineal gland. Melatonin is secreted in the dark phase. Long-days induce TSH which acts on ependymal cells to upregulate *Dio2* and downregulate *Dio3*, increasing levels of the bioactive thyroid hormone T3 which acts at the median eminence to regulate activity of the hypothamic-pituitary-gonad (HPG) axis, and thus seasonal reproduction. B: (see p. 242) In birds TSH modulation of Dio2 and Dio3 also regulate seasonal changes in the HPG axis. However, light acts by acting directly on deep brain photoreceptors, such as Opsin 5 in the paraventricular organ. Other photoreceptive structures such as the eye and pineal allow light to regulate circadian cycles. Note that in birds melatonin is involved in circadian rhythms but does not play a direct role in seasonal timing. Rather, seasonal photoperiod changes are detected directly by deep brain photoreceptors. In mammals photoperiod is detected by retinal photoreceptors and melatonin thus modulates both circadian and seasonal rhythms.

Intrinsic Adaptive Value Perhaps the most compelling evidence for the intrinsic adaptive value of endogenous circadian rhythms as synchronizers and organizers comes from the biomedical literature documenting the deleterious effects of disrupted circadian rhythms (Sack *et al.* 2007). Shift work, jet lag, and artificial lighting have all been suggested to have deleterious consequences for health. Rotating shift work, in particular, involves repeated and chronic mismatches between endogenous circadian clocks and external environmental drivers (Bøggild & Knutsson 1999). Thus, individuals are perpetually in a state of re-entrainment. Because different endogenous rhythms re-entrain at different rates, physiological rhythms become desynchronized from each other.

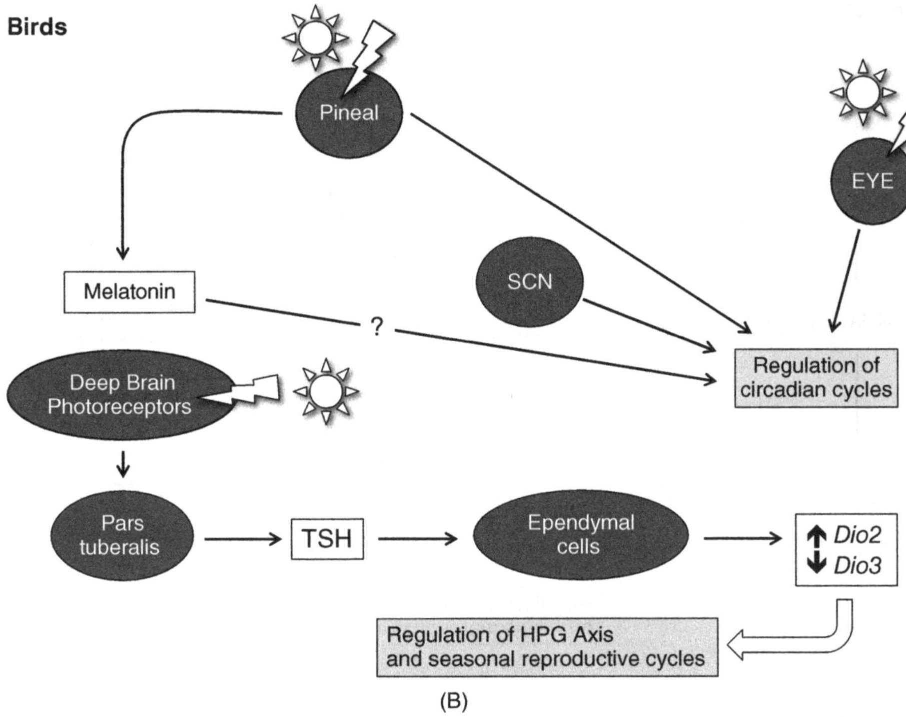

Figure 15.3. *(continued)*

Rotating shift work is associated with a variety of negative health consequences including increased risk of cardiovascular disease comparable to smoking a package of cigarettes per day (Bøggild & Knutsson 1999). Chronic repeated jet lag has also been associated with cognitive deficits and temporal lobe atrophy in humans (Cho 2001), and cognitive deficits and reduced neurogenesis in rodents (Gibson *et al.* 2010). Shift work is also associated with increased risk of metabolic disorders (Karlsson *et al.* 2001). Thus, a variety of increased health risks associated with disrupted circadian rhythms support the hypothesis that normal endogenous circadian function has intrinsic adaptive value.

Extrinsic Adaptive Value For an endogenous clock to be useful in anticipating the external environment it must keep good time. That is, the endogenous clock must have a free-running period that is consistent across different conditions of the organism. Most biochemical reactions are profoundly influenced by temperature, often doubling or tripling with increases in temperature of 10 kelvin (i.e., a Q_{10} of 2 or 3). Remarkably, free-running circadian rhythms typically exhibit a Q_{10} close to 1 (Sweeney & Hastings 1960). One limitation of our understanding of the molecular and physiological mechanisms of circadian rhythms (see above, 'Molecular and Physiological Mechanisms', pp. 238–39) is that how such mechanisms exhibit temperature compensation is a mystery. An endogenous clock that was not temperature compensated would provide little extrinsic adaptive value because it would not keep good time.

Other research that compares endogenous clocks that do or do not match the external environmental cycles has provided support for an extrinsic adaptive value of circadian rhythms. Malaria parasites match the circadian clocks of their hosts (Boyd 1929; Gautret *et al.* 1995), typically exhibiting synchronous cell cycles with durations that are a multiple of 24 hours (Mideo *et al.* 2013). A recent experiment examined the fitness consequences of synchrony between host and parasite rhythms in the rodent malaria parasite *Plasmodium chabaudi* (O'Donnell *et al.* 2011). When *P. chabaudi* rhythms were mismatched to the host circadian rhythm replication and transmission potential (production of gametocytes) were dramatically reduced. These results demonstrate a direct fitness benefit of circadian rhythms in malaria parasites, allowing them to match and exploit the rhythms of their host, supporting the idea of extrinsic adaptive value. Recent attention has also been paid to links between the molecular circadian clock and immune function, and most immune parameters exhibit circadian rhythmicity (Arjona *et al.* 2012). This opens the possibility of an evolutionary arms race in the circadian rhythmicity of pathogens and that of the immune systems of their hosts.

A similar experimental strategy to test the adaptive significance of circadian rhythms assessed competition between mutant strains of the cyanobacterium *Synechococcus elongatus* that have different endogenous circadian periods (Ouyang *et al.* 1998; Woelfle *et al.* 2004). These different strains had similar growth rates when cultured individually – whether in constant light or under various photoperiods. However, when different strains were mixed and cultured together under different photoperiods competition arose. The strains whose endogenous rhythm most closely matched the exogenous photoperiod tended to have the highest fitness (Ouyang *et al.* 1998). Moreover, when wild-type *Synechococcus* were placed in competition with an arrhythmic mutant strain the wild-type strain outcompeted the mutant under normal light-dark cycles, but the arrhythmic mutant outcompeted the wild-type under constant light (Woelfle *et al.* 2004). This suggests that the adaptive value of circadian rhythms in these cyanobacteria is of an extrinsic nature – strains whose rhythms most closely matched the external environmental cycles outcompeted strains whose rhythms did not. Similarly, across rodent and primate species there is increased longevity for taxa whose endogenous circadian rhythm most closely matches a 24 h cycle (Wyse *et al.* 2010).

Studies explicitly testing the adaptive value of circadian rhythms in vertebrates are logistically more complex, and are more rare. However, a series of studies on SCN-lesioned rodents living in semi-natural enclosures has shed some light on this (reviewed in DeCoursey 2004). DeCoursey *et al.* (1997) maintained groups of SCN-lesioned and intact antelope ground squirrels (*Ammospermophilus leucurus*) in a large outdoor enclosure under natural ambient conditions and monitored their behavior and survival via telemetry. The SCN-lesioned animals were more active at night, confirming that their circadian activity patterns were disrupted. Further, when a predator accidentally infiltrated the enclosure it killed far more SCN-lesioned animals than controls. Thus, failure to engage in normal circadian patterns of activity increased predation risk. A subsequent and larger-scale study of eastern chipmunks (*Tamias striatus*) yielded similar results (DeCoursey *et al.* 2000). SCN-lesioned chipmunks were more restless at night, and suffered higher predation by weasels during a year of high population density. Thus, circadian rhythms may have extrinsic adaptive value in that they allow animals to organize their behavior to a schedule adapted to reduce predation risk.

A variety of other studies in plants and animals indicate that organisms with disrupted circadian rhythms suffer reduced performance and fitness. Alterations in circadian rhythms are associated with reduced longevity in rodents and fruit flies (Hurd & Ralph 1998; Klarsfeld & Rouyer 1998). In *Arabidopsis* plants a mutation that disrupts the free-running rhythm (but not the ability to respond to light-dark cycles) also is associated with reduced fitness measures (Green *et al.* 2002). These mutants flowered later when on long-day light schedules, and had reduced viability under short day conditions. Disruption of endogenous circadian rhythms thus appears to reduce fitness in a variety of taxa.

Individual- and Population-Variation, and Measurements of Natural Selection Although the studies above demonstrate that a manipulated rhythm results in impaired fitness, they do not demonstrate selection of natural variation in the wild. There is some evidence of such selection for circadian rhythmicity (reviewed in DeCoursey 2004). First, circadian rhythms are reduced (though still present) in a millipede species that lives deep in caves with no circadian environmental cycle (Koilraj *et al.* 2000). This, and other studies demonstrating regressions of circadian rhythms in constant environments, suggests that there is a benefit to endogenous clocks that is not present for organisms that live in near-constant conditions. Second, geographic variation in circadian systems associated with geographic variation in circadian environmental fluctuations has been reported for *Drosophila, Arabidopsis,* and other taxa (Costa *et al.* 1992; Michael *et al.* 2003; Tauber *et al.* 2007; reviewed in Hut *et al.* 2013). This geographic variation suggests that endogenous circadian rhythms may respond to differential selection across different populations. In addition, urban and rural European blackbirds (*Turdus merula*) exhibit differing endogenous rhythms, suggesting adaptation to anthropogenic factors on small geographic scales (Dominoni *et al.* 2013).

In addition to these population differences, individual differences in circadian rhythm parameters such as the phase angle and period give rise to chronotypes (Kerkhof 1985; Kerkhof & Van Dongen 1996). In many cases such individual variation in circadian rhythm parameters is heritable, and a recent study in birds found that young produced from extra-pair sires had shorter circadian periods (Helm & Visser 2010). This suggests that males with shorter circadian periods, who perhaps rise earlier, are more attractive mates.

CIRCANNUAL CYCLES AND SEASONALITY

Overview and Introduction

Over the course of each year most parts of the planet experience a conspicuous change in abiotic and biotic environmental conditions, or seasons. Organisms organize processes such as reproduction, migration, immune function, and changing metabolism (e.g., periods of dormancy, or periods of accelerated metabolism) in response to these seasonal changes. Just as organisms exhibit endogenous circadian rhythms that entrain to the day-night cycle, so too there exist endogenous seasonal and circannual processes that entrain to the seasons. The exogenous environmental factors that directly affect fitness (survival and reproductive success) are referred to as *ultimate* factors, whereas the environmental cues that organisms actually use to adjust endogenous cycles through entrainment are referred to as *proximate* factors (Baker 1938). For example, in many temperate-zone seasonally breeding birds and rodents food is the ultimate factor that determines offspring survival, however, the seasonal change in photoperiod and other factors are the proximate cues that drive seasonal changes in reproductive physiology (Dawson *et al.* 2001). Although typically not described this way, proximate factors can be thought of as zeitgebers (Immelmann 1971). Rather than entrain a circadian rhythm, these proximate factors or environmental cues entrain longer period circannual processes.

Similar to circadian rhythms, there is clear evidence for endogenous circannual clocks. These endogenous circannual rhythms are logistically much more difficult to measure.

However, there are well-documented examples of endogenous circannual rhythms of hibernation, migratory behavior, and reproductive physiology, that are expressed when animals are held under constant conditions (Gwinner 1986). Similar to circadian rhythms, endogenous free-running circannual rhythms exhibit individual variation in their period, and free-running individuals become desynchronized from each other. Such desynchronization proves that circannual rhythms are endogenous, and not just a response to an uncontrolled environmental factor. In many cases endogenous circannual rhythms are expressed in only certain constant conditions. For example, circannual free-running rhythms of antler growth of sika deer (*Cervus nippon*) are exhibited in long- and short-day photoperiods but not on 12:12 light:dark (Goss 1969), but in other species circannual rhythms are expressed in a variety of conditions (Gwinner 1986).

In many ways, then, circannual rhythms and seasonality are analogous to circadian cycles. The external seasonal environment entrains endogenous processes through particular environmental cues (analogous to zeitgebers). The efficacy of these entraining cues varies over the endogenous cycle (see above, 'Molecular and Physiological Mechanisms', pp. 238–39; Figure 15.1) Although there are many analogies between circadian and circannual rhythms at a functional level, the underlying molecular and physiological mechanisms are very likely quite different. Below, we briefly review some of the known physiological mechanisms underlying seasonality, photoperiodism, and annual cycles. We then discuss examples where variation in these processes has been linked to performance and fitness.

Physiological Mechanisms of Annual Cycles

Unlike the case of the SCN in mammals for circadian rhythms, there are no identified central physiological oscillators with an endogenous circannual free-running rhythm. Rather, most research on the physiology of annual cycles focuses on the physiological transduction of the environmental factors that entrain annual rhythms and trigger transitions between different phases of the annual cycle. Indeed, SCN-lesioned ground squirrels continue to show evidence for endogenous circannual rhythms (Ruby *et al.* 1998), suggesting that at least in some mammals the SCN is not required for the internal circannual clock to function. Further, ablations of various other neural and endocrine structures have failed to indicate a single endogenous circannual pacemaker. Thus, the physiological underpinnings of endogenous circannual rhythmicity in vertebrates likely involve feedback loops across multiple sites within and without the central nervous system (Goldman *et al.* 2004). These physiological mechanisms remain a mystery (but see Wikelski *et al.* 2008; Hazlerigg & Lincoln 2011 for recent hypotheses).

Of the environmental cues that synchronize annual cycles, photoperiod is, by far the most important proximate factor in a diverse range of seasonal species of plants, invertebrates and vertebrate animals (Goldman *et al.* 2004). Indeed, photoperiods can completely over-ride endogenous rhythms. European starlings (*Sturnus vulgaris*) provide an example. Starlings will display an endogenous free-running circannual cycle of reproductive physiology and feather molt when held on a 12:12 L:D photoperiod (reviewed in Gwinner 1986). However, birds held on sinusoidal photoperiod cycles with periods of 8, 6, 4 or 3 months will track these cycles in reproductive physiology and molt (Gwinner 1986). Housing starlings under constant 11:13 L:D locks birds into a permanent state of photosensitivity, and housing photorefractory starlings on a constant photoperiod with a photophase greater than 12 h locks them permanently in a state of nonreproductive photorefractoriness (Dawson & Goldsmith 1997; Gwinner *et al.* 1985). Birds housed in these conditions will remain in the same reproductive condition for years without entering feather molt. Thus, the external cue (photoperiod) can over-ride the endogenous annual clock. Photoperiod provides a powerful environmental cue that can drive seasonal changes in physiology and behavior in a range of species, just as it is a powerful zeitgeber for circadian rhythms.

Most research on the physiology of seasonality has focused on the neuroendocrine mechanisms of photoperiodism and seasonal reproduction. In photoperiodic species the light cycle can drive profound changes in the HPG axis and thus seasonal changes in reproductive physiology and behavior. Gonadotropin-releasing hormone (GnRH) acts as a master regulator of this axis, and the pulsatile release of GnRH, as well as GnRH synthesis, can vary across different photoperiodic conditions (reviewed in Goldman *et al.* 2004; Stevenson *et al.* 2012). Ewes in nonbreeding condition exhibit less frequent pulses of GnRH, and subsequently lower gondadotropin levels, due to heightened negative feedback to estradiol (Karsch *et al.* 1993). Other species of mammals exhibit decreased GnRH release during nonbreeding periods that is independent of steroid feedback, as do several bird species. Seasonally breeding birds that become absolutely photorefractory exhibit almost complete shutdown of GnRH production when fully photorefractory (Dawson & Goldsmith 1997), but there is much less seasonal variation in GnRH in opportunistically breeding species (MacDougall-Shackleton *et al.* 2001, 2009). It thus appears that much of the influence of photoperiod on seasonal reproduction in vertebrate animals occurs via modulation of GnRH in the central nervous system.

How is photoperiod transduced to control GnRH and the HPG axis? In mammals, melatonin is a key regulator (Figure 15.3; Goldman *et al.* 2004). Photoperiod is detected by nonvisual photoreceptors in the retina that project to the SCN via the retinohypothalamic tract. One of the many outputs from the SCN is a multisynaptic projection to the pineal gland that influences melatonin secretion (Klein *et al.* 1983). As a result, melatonin is secreted primarily at night, and thus on a long-day photoperiod there is a short duration signal of melatonin. Conversely, on a short-day photoperiod melatonin secretion occurs throughout the longer night. This nocturnal melatonin secretion is conserved across species, but short-day and long-day breeding animals differ in their response to it (Goldman 2001). Infusion of melatonin in pinealectomized long-day-breeding rodents for 8 h per day or longer suppresses reproduction. However, the same manipulation in short-day breeding sheep stimulates reproduction. Thus melatonin does not inherently inhibit or stimulate reproduction, but acts as an endocrine signal of photoperiod, and species vary in how they respond to this signal.

Photoperiodic time measurement in birds is less well understood (Figure 15.3). Melatonin is not a critical determinant of reproductive function in birds (Gwinner *et al.* 1997), and photoperiod is transduced by deep-brain photoreceptors in the medial hypothalamus, not the eyes (Menaker *et al.* 1970; Foster *et al.* 1985). Recently expression of several genes in the rhodopsin superfamily has been identified in the avian hypothalamus, and these vary in spectral sensitivity (reviewed in Ikegami & Yoshimura 2012).

In mammals and birds, melatonin or deep brain photoreceptors, respectively, result in changes in expression of genes for enzymes that convert the prohormone thyroxine (T4) to the active form (T3) and vice versa (reviewed in Ikegami & Yoshimura 2012). Under long-day photoperiods upregulation of the gene *Dio2* occurs, whose protein product catalyzes T3 production from T4. On short days *Dio3* is upregulated, which has the opposite effect. T3 is thought to act at the median eminence, altering the ultrastructure of glia and facilitating GnRH release into the portal blood stream (Yamamura *et al.* 2004). It has long been known that photoperiodic responses in birds depended on thyroid hormones (Follett & Nicholls 1985) and these recent studies provide a mechanistic link between thyroid hormones and photoperiodic regulation of the HPG axis.

We thus have an emerging picture, for both birds and mammals, of how photoperiod is transduced by the central nervous system to regulate reproductive function. As yet, however, this work has focused on relatively few species. Further, we still have little information on the underlying physiology of endogenous circannual clocks. We also have very few links between these mechanisms, and whole organism performance and fitness (see next section). We have linked species variation in seasonal GnRH plasticity to species differences in photoperiodism and breeding schedules for some groups of birds (Hahn & MacDougall-Shackleton 2008; MacDougall-Shackleton *et al.* 2009). However, most studies of the physiology of photoperiodism in animals have been carried out in quail and rodents in a noncomparative context. Below, we review some research areas that have attempted to link photoperiodism and seasonality to fitness.

Adaptive Value of Seasonality

In many cases, photoperiod-response systems appear to provide a good match to an organism's environment. However, such matches may be a result of adaptation, a result of phenotypic plasticity, or an epiphenomenon of photoperiod response mechanisms (MacDougall-Shackleton & Hahn 2007, Hahn & MacDougall-Shackleton 2008). For example, spotted antbirds (*Hylophylax naevioides*) living near the equator can remarkably use very subtle changes in the annual photoperiod to adjust reproduction (Hau *et al.* 1998). However, it is unclear if this is an adaptive specialization or if such exquisite photoperiod sensitivity is a general property of birds, as nontropical species exhibit similar resolution (Dawson 1991, 2007). With that caveat in mind, there are lines of research that have linked variation in seasonality and photoperiodism to performance and fitness, and there is comparative evidence for the evolution of photoperiod response systems. However, we are unaware of any studies that directly assess the intrinsic or extrinsic adaptive value of endogenous circannual clocks, as has been done for circadian clocks.

Photoperiodic Polyphenism in Rodents In seasonally breeding rodents, animals are typically sensitive to changes in photoperiod, with short day-lengths inducing gonadal regression. However, considerable variation in photoperiodic responsiveness has been found both within and across populations (*Peromyscus leucopus*, Lynch *et al.* 1981; *Peromyscus maniculatus*, Dark *et al.* 1983; *Peromyscus leucopus*, Desjardins & Lopez 1983; *Microtus pennsylvanicus*, Kerbeshian *et al.* 1994) with some individuals (termed nonresponders) failing to regress their gonads on short days. The mechanisms underlying photoperiodic nonresponsiveness appear to vary across species, and include variation in the circadian clock, the GnRH system, and sensitivity to melatonin (reviewed in Prendergast *et al.* 2001; Heideman & Pittman 2009). Few studies have directly examined the fitness consequences of this variation in photoperiodic responsiveness. Nonresponsiveness could facilitate breeding in winter, if environmental conditions were suitable for successful rearing of young during a mild winter. Nelson (1987) has suggested that the occurrence of photoperiodic nonresponders may be an important driver of the dramatic fluctuations in rodent population densities. On the other hand, photoperiodic nonresponsiveness may impose considerable energetic costs. Unlike photoperiodic responders, nonresponders fail to reduce their body mass on short days (Moffatt *et al.* 1993; Ruf *et al.* 1997). Furthermore, nonresponders show reduced thermoregulatory behaviors, including reduced nest building and an absence of communal nesting (McShea 1990; Moffatt *et al.* 1993). Taken together, these findings suggest that photoperiodic nonresponsiveness is energetically costly. Thus, it is possible that year-to-year variation in the severity of winters may produce fluctuating selection. In mild years, nonresponders may produce more young than animals that undergo gonadal regression. But in colder years these nonresponders may have reduced survival. Indeed, in a warming climate nonresponders may have higher fitness. Confirming or rejecting these adaptive hypotheses awaits further study.

Photoperiod Response Systems in Birds Photoperiod response systems of a variety of avian taxa share fundamental features, including (1) activation of the reproductive neuroendocrine system when day length exceeds some minimum value, (2) persistence of physiological reproductive competence for some period of time determined by a combination of the response to photoperiod and to nonphotic cues, (3) termination of reproductive competence by the onset of a period of refractoriness to the stimulatory effects of photoperiod and other cues, and (4) dissipation of the refractory state by exposure to shortened day lengths and the re-activation of the reproductive neuroendocrine system (Hahn *et al.* 1997, 2008, 2009; Dawson *et al.* 2001). Although this basic photoperiodic system is strikingly flexible regarding the range of different reproductive cycles it can produce under different environmental conditions (termed "conditional plasticity," Hahn & MacDougall-Shackleton 2008; Hahn *et al.* 2009), some key features appear to be adaptive specializations (Hahn & MacDougall-Shackleton 2008). The most prominent species differences are those distinguishing relatively rigid seasonally breeding species from much more temporally flexible or opportunistic taxa. These differences include: (1) dependence on long day exposure for reproductive development in spring, (2) degree of sensitivity to nonphotic cues (e.g., temperature, rainfall, food), (3) degree to which the onset of photorefractoriness can be modified by nonphotic cues, and (4) the nature of photorefractoriness (Hahn *et al.* 2009). Although virtually all birds display some capacity for temporal flexibility of reproduction, the most flexible taxa, though often photoperiodic, display relaxed long-day requirements for reproductive development in winter/spring combined with a strong responsiveness to nonphotic cues such as food and social factors. These birds also display an ability to flexibly modulate the timing of reproductive termination and onset of feather molt in response to nonphotic cues, and tend to display a relative form of photorefractoriness (Hahn *et al.* 1997, 2009). Thus, relatively modest evolutionary changes to a fundamentally consistent environmental cue response system can lead to dramatic differences in degree of temporal flexibility of reproduction between species. Natural selection on the above parameters of the photoperiod response system could generate opportunistic breeders from seasonal breeders, and vice versa (Hahn *et al.* 2009).

Population- and Individual-Level Variation in Photoperiod Response Systems in Birds Photoperiodic response systems have been found to vary not only across species of birds, but also within species. Studies of both free-living and captive tits (family Paridae) have yielded important insights into population-level variation in patterns of seasonal breeding and photoperiodic response systems. Both great tits (*Parus major*) and blue tits (*Cyanistes caeruleus*) use photoperiodic cues to time reproduction such that the nestling period coincides with the seasonal peak in availability of caterpillars (Silverin *et al.* 1993; Lambrechts *et al.* 1996), and fitness is influenced by the accuracy of their timing with respect to caterpillar abundance (Dias & Blondel 1996; Thomas *et al.* 2001; Verboven *et al.* 2001). In great tits, the timing of breeding varies across Europe with breeding occurring later at higher latitudes (Sanz 1998). These population differences in reproductive timing correspond to differences in photoperiodic responsiveness, with birds at higher latitudes requiring longer photoperiods to stimulate gonadal development (Silverin *et al.* 1993). Among blue tits, population differences in reproductive timing have been found even among populations breeding at similar latitudes. These population differences are heritable (Blondel *et al.* 1990) and, as in the great tits, are due to population differences in photoperiodic response (Lambrechts *et al.* 1997). Thus, as in the case of geographic variation in circadian mechanisms in Drosophila and Arabidopsis (see above, 'Individual- and Population-Variation, and Measurements of Natural Selection', p. 244), population-level variation in photoperiod response systems in Paridae provides indirect evidence for adaptive specialization.

Studies of the photoperiodic response systems in birds have largely focused on the mean response of groups of individuals to a particular cue or set of cues, but individuals in such studies typically vary in their responsiveness. Several studies of cardueline finch species have measured gonadal response to an increase in photoperiod (Hahn 1995; Hahn *et al.* 2004; Hahn *et al.* 2005; Furlonger *et al.* 2012; Watts & Hahn 2012). These studies indicate considerable individual-level variation within treatment groups in response to increasing photoperiod. Specifically, across these studies, the mean coefficient of variation in gonadal response within treatment groups was about 44% (Watts *et al.* in revision). In another study, Pereyra *et al.* (2005) report considerable individual-level variation in the response of the GnRH system to long-day photoperiods among common redpolls (*Acanthis flammea*) and pine siskins (*Spinus pinus*). In both species, a subset of individuals maintained relatively high levels of GnRH immunoreactivity with correspondingly high circulating levels of LH and large testis sizes. Overall, the extent of such individual variation, as well as its underlying causes, remains poorly understood. Individual variation in response to photoperiod represent a reaction norm (Chapter 2, this volume) but understanding how these reaction norms evolve requires further study. Intriguingly, studies of both captive and free-living birds have found heritable variation in the timing of egg-laying within populations (van Noordwijk *et al.* 1981; Visser *et al.* 2011), which suggests that there is heritable variation in these reaction norms. This individual variation in photoperiod responsiveness provides a potential source of variation for natural selection.

Individual Variation in Human Photoperiodism As noted above, a hallmark of studies of annual cycles, seasonality, and photoperiodism, is the existence of extensive individual and population level variation. Indeed, it is the existence of individual variation in free-running circannual rhythms and desynchronization across individuals that proves that animals are not responding to uncontrolled exogenous cues (Gwinner 1986). Studies of human seasonality are particularly characterized by high levels of variation and inconsistency (Bronson 2004). For example, historic patterns of seasonal reproduction in humans are found in some populations but not others (Lam & Miron 1991). Bronson (2004) proposed that much of the inconsistency among studies of human photoperiodism and seasonality might stem from individual and population variation in sensitivity to photoperiod. Photoperiod response mechanisms appear conserved across mammals, including humans, including a functioning retinohypothalamic-pineal axis (Arendt 2000). However, as human populations migrated across the globe selection pressure for photoperiod responsiveness would have varied among populations and resulted in large intraspecific variation in photoperiod response cues (Bronson 2004). Consistent with this idea is the fact that photoperiod sensitivity and melatonin regulation varies in clinical populations including those suffering from seasonal affective disorder (Thompson *et al.* 1990). Thus, beyond the studies of rodents and birds described above, individual- and population-level variation in photoperiod responsiveness, as well as individual variation in underlying endogenous rhythms, may provide a rich source of variation on which natural selection could act to produce a range of annual cycles in physiology and behavior.

It is now clear that many species have endogenous circannual rhythms, however the physiological underpinnings of these rhythms are poorly understood. We have much more knowledge regarding the process of entrainment of annual cycles, particularly for photoperiodism and reproduction in birds and mammals. Further work is required to explore both the underlying physiology of circannual rhythms, and how selection has modified these mechanisms over the course of evolution.

CONCLUSION

Chronobiology is a large research field, and we have touched on only some highlights in this chapter. There is now a good understanding of the general principles of circadian and circannual rhythms, including the existence of endogenous timers, entrainability, and varying sensitivity to zeitgebers or proximate environmental cues. Biological rhythms provide a form of phenotypic plasticity and future work should explore the evolution of biological rhythms in the context of other phenotypic plasticity (Chapters 2 & 4, this volume). Biological rhythms allow organisms to alter their phenotype to match environmental cycles in the outside world (extrinsic adaptive value). As well, biological rhythms allow organisms to coordinate and synchronize internal processes (intrinsic adaptive value). Similarly, other phenotypically plastic physiological processes allow matching with the environment and coordination of internal physiological mechanisms, such as the stress response or an immune response (Chapter 12, this volume). Considering biological rhythms and these other response systems together may reveal important principles of whole organism regulatory mechanisms (Chapter 9, this volume).

ACKNOWLEDGMENTS

We thank the editors, Barbara Helm, and anonymous reviewers for improving this chapter. Our research has been funded by the NSF (USA) and NSERC (Canada).

REFERENCES

Albers, H.E., Lydic, R. and Moore-Ede, M.C. (1984) Role of the suprachiasmatic nuclei in the circadian timing system of the squirrel monkey. I. The generation of rhythmicity. *Brain Research*, **300**, 275–84.

Arendt, J. (2000) Is melatonin a photoperiodic signal in humans?, in *Melatonin after Four Decades* (ed J. Olese), Kluwer Academic/ Plenum, New York, pp. 417–24.

Arjona, A., Silver, A.C., Walker, W.E. and Fikrig, E. (2012) Immunity's fourth dimension: approaching the circadian-immune connection. *Trends Immunol*, **33**, 607–12.

Baker, J.R. (1938) The evolution of breeding seasons, in *Evolution: Essays on Aspects of Evolutionary Biology* (ed G.B. Debeer), Clarendon Press, Oxford, pp. 161–77.

Blondel, J., Perret, P. and Maistre, M. (1990) On the genetic basis of the laying-date in an island population of blue tits. *Journal of Evolutionary Biology*, **3**, 469–475.

Bøggild, H. and Knutsson, A. (1999) Shift work, risk factors and cardiovascular disease. *Scandinavian Journal of Work, Environment and Health*, **25**, 85–99.

Boyd, G.H. (1929) Induced variations in the asexual cycle of *Plasmodium cathemerium*. *American Journal of Hygiene*, **9**, 181–7.

Bronson, E.H. (2004) Are humans seasonally photoperiodic? *Journal of Biological Rhythms*, **19**, 180–92.

Cho, K. (2001) Chronic "jet lag" produces temporal lobe atrophy and spatial cognitive deficits. *Nature Neuroscience*, **4**, 567–8.

Cohen, R.A. and Albers, H.E. (1991) Disruption of human circadian and cognitive regulation following a discrete hypothalamic lesion: a case study. *Neurology*, **41**, 726–9.

Costa, R., Peixoto, A.A., Barbujani, G. and Kyriacou, C.P. (1992) A latitudinal cline in a *Drosophila* clock gene. *Proceedings of the Royal Society B-Biological Sciences*, **250**, 43–9.

Dark, J., Johnston, P.G., Healey, M. and Zucker, I. (1983) Latitude of origin influences photoperiodic control of reproduction in deer mice (*Peromyscus maniculatus*). *Biology of Reproduction*, **28**, 213–20.

Dawson, A. (1991) Photoperiodic control of testicular regression and molt in male house sparrows *Passer domesticus*. *Ibis*, **133**, 312–16.

Dawson, A. (2007) Seasonality in a temperate zone bird can be entrained by near equatorial photoperiods. *Proceedings of the Royal Society B-Biological Sciences*, **274**, 721–5.

Dawson, A. and Goldsmith, A.R. (1997) Changes in gonadotrophin-releasing hormone (GnRH-I) in the pre-optic area and median eminence of starlings (Sturnus vulgaris) during the recovery of photosensitivity and during photostimulation. *Journal of Reproduction and Fertility*, **111**, 1–6.

Dawson, A., King, V.M., Bentley, G.E. and Ball, G.F. (2001) Photoperiodic control of seasonality in birds. *Journal of Biological Rhythms*, **16**, 365–80.

DeCoursey, P.J. (2004) Cell physiology of circadian pacemaker systems in metazoan animals, in *Chronobiology: Biological Timekeeping* (eds J.C. Dunlap, J.J. Loros and P.J. Decoursey), Sinauer, Sunderland, MA, pp. 181–212.

DeCoursey, P.J., Krulas, J., Mele, G. and Holley, D. (1997) Circadian performance of suprachiasmatic nuclei (SCN)-lesioned antelope ground squirrels in a desert enclosure. *Physiology & Behavior*, **62**, 1099–1108.

DeCoursey, P.J., Walker, J.K. and Smith, S.A. (2000) A circadian pacemaker in free-living chipmunks: essential for survival? *Journal of Comparative Physiology A Sensory Neural and Behavioral Physiology*, **186**, 169–80.

Desjardins, C. and Lopez, M.J. (1983) Environmental cue evoke differential responses in pituitary-testicular function in deer mice. *Endocrinology*, **112**, 1398–1406.

Dias, P.C. and Blondel, J. (1996) Local specialization and maladaptation in the Mediterranean blue tit (*Parus caeruleus*). *Oecologia*, **107**, 79–86.

Dominoni, D.M., Helm, B., Lehmann, M. *et al.* (2013) Clocks for the city: circadian differences between forest and city songbirds. *Proceedings of the Royal Society B-Biological Sciences*, **280**, 1471–2954.

Dunlap, J.C. (1999) Molecular bases for circadian clocks. *Cell*, **96**, 271–90.

Dunlap, J.C., Loros, J.J. and DeCoursey, P.J. (eds) (2004) *Chronobiology: Biological Timekeeping*, Sinauer Associates, Sunderland, MA.

Edmunds, L.N. (1988) *Cellular and Molecular Bases of Biological Clocks*, Springer-Verlag, New York.

Follett, B.K. and Nicholls, T.J. (1985) Influences of thyroidectomy and thyroxine replacement on photoperiodically controlled reproduction in quail. *Journal of Endocrinology*, **107**, 211–21.

Foster, R.G. and Kreitzman, L. (2005) *Rhythms of Life*, Yale University Press, New Haven.

Foster, R.G. and Kreitzman, L. (2009) *Seasons of Life*, Yale University Press, New Haven.

Foster, R.G., Follett, B.K. and Lythgoe, J.N. (1985) Rhodopsin-like sensitivity of extra-retinal photoreceptors mediating the photoperiodic response in quail. *Nature*, **313**, 50–2.

Furlonger, A.A., Huynh, L., Luloff, T.W. *et al.* (2012) Effects of supplementation with preferred foods on the reproductive axis of American goldfiches (*Spinus tristis*). *Canadian Journal of Zoology*, **90**, 203–9.

Gautret, P., Deharo, E., Tahar, R. *et al.* (1995) The adjustment of the schizogonic cycle of *Plasmodium chabaudi chabaudi* in the blood to the circadian rhythm of the host. *Parasite*, **2**, 69–74.

Gibson, E.M., Wang, C., Tjho, S. *et al.* (2010) Experimental "jet lag" inhibits adult neurogenesis and produces long-term cognitive deficits in female hamsters. *PLoS ONE*, **5**, e15267.

Goldman, B.D. (2001) Mammalian photoperiodic systems: formal properties and neuroendocrine mechanisms of photoperiodic time measurement. *Journal of Biological Rhythms*, **16**, 283–301.

Goldman, B., Gwinner, E., Karsch, F.J. *et al.* (2004) Circannual rhythms and photoperiodism, in *Chronobiology: Biological Timekeeping* (eds J.C. Dunlap, J.J. Loros and P.J. DeCoursey), Sinauer, Sunderland, MA, pp. 107–44.

Goss, R.J. (1969) Photoperiodic control of antler cycles in deer I. Phase shift and frequency changes. *Journal of Experimental Zoology*, **170**, 311–24.

Green, R.M., Tingay, S., Wang, Z.-Y. and Tobin, E.M. (2002) Circadian rhythms confer a higher level of fitness to *Arabidopsis* plants. *Plant Physiology*, **29**, 301–2.

Gwinner, E. (1986) *Circannual Rhythms*, Springer-Verlag, Berlin.

Gwinner, E., Dittami, J., Gänshirt, G. *et al.* (1985) Endogenous and exogenous components in the control of the annual reproductive cycle of the European starling, in *Proceedings of the 18th International Ornithological Congress* (eds V.D. Ilyichev and V.M. Gavrilov), Nauka, Moscow, pp. 501–15.

Gwinner, E., Hau, M. and Heigl, S. (1997) Melatonin: generation and modulation of avian circadian rhythms. *Brain Research Bulletin*, **44**, 439–44.

Hahn, T.P. (1995) Integration of photoperiodic and food cues to time changes in reproductive physiology by an opportunistic breeder, the red crossbill, *Loxia curvirostra* (Aves: Carduelinae). *Journal of Experimental Zoology*, **272**, 213–226.

Hahn, T.P., Boswell, T., Wingfield, J.C. and Ball, G.F. (1997) Temporal flexibility in avian reproduction: Patterns and mechanisms. Current Ornithology 14: 39-80. *Current Ornithology*, **14**, 39–80.

Hahn, T.P., Pereyra, M.E., Sharbaugh, S.M. and Bentley, G.E. (2004) Physiological responses to photoperiod in three cardueline finch species. *General and Comparative Endocrinology*, **137**, 99–108.

Hahn, T.P., Pereyra, M.E., Katti, M. *et al.* (2005) Effects of food availability on the reproductive system, in *Functional Avian Endocrinology* (eds A. Dawson and P.J. Sharp), Narosa Publishing House, New Delhi, pp. 167–80.

Hahn, T.P. and MacDougall-Shackleton, S.A. (2008) Adaptive specialization, conditional plasticity and phylogenetic history in the reproductive cue response systems of birds. *Philosophical Transactions of the Royal Society B-Biological Sciences*, **363**, 267–86.

Hahn, T.P., Watts, H.E., Cornelius, J.M. *et al.* (2009) Evolution of environmental cue response systems: adaptive variation in photorefractoriness. *General and Comparative Endocrinology*, **163**, 193–200.

Hall, J.C. (1995) Tripping along the trail to the molecular mechanisms of biological clocks. *Trends in Neurosciences*, **18**, 230–40.

Hall, J.C. (1996) Are cycling gene products as internal zeitgebers no longer the zeitgeist of chronobiology? *Neuron*, **17**, 979–90.

Hau, M., Wikelski, M. and Wingfield, J.C. (1998) A neotropical forest bird can measure the slight changes in tropical photoperiod. *Proceedings of the Royal Society B-Biological Sciences*, **265**, 89–95.

Hazlerigg, D.G. and Lincoln, G.A. (2011) Hypothesis: cyclical histogenesis is the basis of circannual timing. *Journal of Biological Rhythms*, **26**, 471–85.

Heideman, P.D. and Pittman, J.T. (2009) Microevolution of neuroendocrine mechanisms regulating reproductive timing in *Peromyscus leucopus*. *Integrative and Comparative Biology*, **49**, 550–62.

Helm, B. and Visser, M.E. (2010) Heritable circadian period length in a wild bird population. *Proceedings of the Royal Society B-Biological Sciences*, **277**, 3335–42.

Hurd, M.W. and Ralph, M.R. (1998) The significance of circadian organization for longevity in the golden hamster. *Journal of Biological Rhythms*, **13**, 430–6.

Hut, R.A., Paolucci, S., Dor, R. *et al.* (2013) Latitudinal clines: an evolutionary view on biological rhythms. *Proceedings of the Royal Society B-Biological Sciences*, **280**, 1471–2954.

Ikegami, K. and Yoshimura, T. (2012) Circadian clocks and the measurement of daylength in seasonal reproduction. *Molecular and Cellular Endocrinology*, **349**, 76–81.

Immelmann, K. (1971) Ecological aspects of periodic reproduction, in *Avian Biology vol 1* (eds D.S. Farner and J.R. King), Academic Press, New York, pp. 341–89.

Johnson, C.H. (2005) Testing the adaptive value of circadian rhythms. *Methods in Enzymology*, **393**, 818–37.

Johnson, C.H., Elliott, J., Foster, R. *et al.* (2004) Fundamental properties of circadian rhythms, in *Chronobiology: Biological Timekeeping* (eds J.C. Dunlap, J.J. Loros and P.J. Decoursey), Sinauer, Sunderland, MA, pp. 67–106.

Karlsson, B., Knutsson, A. and Lindahl, B. (2001) Is there an association between shift work and having a metabolic syndrome? Results from a popoulation based study of 27 485 people. *Occupational and Environmental Medicine*, **58**, 747–52.

Karsch, F.J., Dahl, G.E., Evans, N.P. *et al.* (1993) Seasonal changes in gonadotropin-releasing hormone secretion in the ewe: alteration in response to the negative feedback action of estradiol. *Biology of Reproduction*, **49**, 1377–83.

Kerbeshian, M.C., Bronson, F.H. and Belliss, E.D. (1994) Variation in reproductive photoresponsiveness in a wild population of meadow voles. *Biology of Reproduction*, **50**, 745–50.

Kerkhof, G.A. (1985) Inter-individual differences in the human circadian system: A review. *Biological Psychology*, **20**, 83–112.

Kerkhof, G. and Van Dongen, H. (1996) Morning-type and evening-type individuals differ in the phase position of their endogenous circadian oscillator. *Neuroscience Letters*, **218**, 135–56.

Klarsfeld, A. and Rouyer, F. (1998) Effects of circadian mutations and LD periodicity on the life span of *Drosophila melanogaster*. *Journal of Biological Rhythms*, **13**, 471–8.

Klein, D.C., Smoot, R., Weller, J.L. *et al.* (1983) Lesions of the paraventricular nucleus area of the hypothalamus disrupt the suprachiasmatic→spinal cord circuit in the melatonin rhythm generating system. *Brain Research Bulletin*, **10**, 647–52.

Koilraj, A.J., Sharma, V.K., Marimuthu, G. and Chandrashekaran, M.K. (2000) Presence of circadian rhythms in the locomotor activity of a cave-dwelling millipede. *Glyphiulus cavernicolus sulu. Chronobiology International*, **17**, 757–65.

Koukkari, W.L. and Sothern, R.B. (2006) *Introducing Biological Rhythms*, Springer, New York.

Lakin-Thomas, P.L. (2000) Circadian rhythms: new functions for old clock genes? *Trends in Genetics*, **16**, 135–42.

Lam, D.A. and Miron, J.A. (1991) Seasonality of births in human populations. *Social Biology*, **38**, 51–78.

Lambrechts, M.M., Blondel, J. and Perret, P. (1996) Adaptive differences in the timing of egg laing between different populations of birds result from variation in photoresponsiveness. *Proceedings of the Royal Society B-Biological Sciences*, **263**, 19–22.

Lambrechts, M.M., Blondel, J., Maistre, M. and Perret, P. (1997) A single response mechanism is responsible for evolutionary adaptive variation in a bird's laying date. *Proceedings of the National Academy of Sciences USA*, **94**, 5153–5.

Lynch, G.R., Heath, H.W. and Johnston, C.M. (1981) Effect of geographical origin on the photoperiodic control of reproduction in the white-footed mouse, *Peromyscus leucopus*. *Biology of Reproduction*, **25**, 475–80.

MacDougall-Shackleton, S.A., Deviche, P.J., Crain, R.D. *et al.* (2001) Seasonal changes in brain GnRH immunoreactivity and song-control nuclei volumes in an opportunistically breeding songbird. *Brain Behavior and Evolution*, **58**, 38–48.

MacDougall-Shackleton, S.A. and Hahn, T.P. (2007) Adaptation and evolution of photoperiod response systems in birds. *Journal of Ornithology*, **148**, S219–S224.

MacDougall-Shackleton, S.A., Stevenson, T.J., Watts, H.E. *et al.* (2009) The evolution of photoperiod response systems and seasonal GnRH plasticity in birds. *Integrative and Comparative Biology*, **49**, 580–9.

McShea, W.J. (1990) Social tolerance and proximate mechanisms of dispersal among winter groups of meadow voles, *Microtus pennsylvanicus*. *Animal Behaviour*, **39**, 346–51.

Meijer, J.H. and Rietveld, W.J. (1989) Neurophysiology of the suprachiasmatic circadian pacemaker in rodents. *Physiological Reviews*, **69**, 671–707.

Menaker, M., Roberts, R., Elliott, J. and Underwood, H. (1970) Extraretinal light perception in the sparrow. III. The eyes do not participate in photoperiodic photoreception. *Proceedings of the National Academy of Sciences USA*, **67**, 320–5.

Michael, T.P., Salomé, P.A., Yu, H.J. *et al.* (2003) Enhanced fitness conferred by naturally occurring variation in the circadian clock. *Science*, **302**, 1049–53.

Mideo, N., Reece, S.E., Smith, A.L. and Metcalf, C.J. (2013) The Cinderella syndrome: why do malaria-infected cells burst at midnight? *Trends in Parasitology*, **29**, 10–16.

Miyazaki, Y., Nisimura, T. and Numata, H. (2005) A phase response curve for circannual rhythm in the varied carpet beetle Anthrenus verbasci. *J Comp Physiol A Neuroethol Sens Neural Behav Physiol*, **191**, 883–7.

Moffatt, C.A., Devries, C. and Nelson, R.J. (1993) Winter adaptations of male deer mice (*Peromyscus maniculatus*) and prairie voles (*Microtus ochrogaster*) that vary in reproductive responsiveness to photoperiod. *Journal of Biological Rhythms*, **8**, 221–32.

Monecke, S., Saboureau, M., Malan, A. *et al.* (2009) Circannual phase response curves to short and long photoperiod in the European hamster. *Journal of Biological Rhythms*, **24**, 413–26.

Nelson, R.J. (1987) Photoperiod-nonresponsive morphs: A possible variable in microtine population-density fluctuations. *American Naturalist*, **130**, 350–69.

Nicholls, T.J., Goldsmith, A.R. and Dawson, A. (1988) Photorefractoriness in birds and comparison with mammals. *Physiological Reviews*, **68**, 133–76.

O'Donnell, A.J., Schneider, P., McWatters, H.G. and Reece, S.E. (2011) Fitness costs of disrupting circadian rhythms in malaria parasites. *Proc Biol Sci*, **278**, 2429–36.

Ouyang, Y., Anderson, C.R., Kondo, T. *et al.* (1998) Resonating circadian clocks enhance fitness in cyanobacteria. *Proceedings of the National Academy of Sciences USA*, **95**, 8660–4.

Pereyra, M.E., Sharbaugh, S.M. and Hahn, T.P. (2005) Interspecific variation in photo-induced GnRH plasticity among nomadic cardueline finches. *Brain Behavior and Evolution*, **66**, 35–49.

Prendergast, B.J., Kriegsfeld, L.J. and Nelson, R.J. (2001) Photoperiodic polyphenisms in rodents: Neuroendocrine mechanisms, costs, and functions. *Quarterly Review of Biology*, **76**, 293–325.

Ralph, M.R., Foster, R.G., Davis, F.C. and Menaker, M. (1990) Transplanted suprachiasmatic nucleus determines circadian period. *Science*, **247**, 975–8.

Ruby, N.F., Dark, J., Heller, H.C. and Zucker, I. (1998) Suprachiasmatic nucleus: role in circannual body mass and hibernation rhythms of ground squirrels. *Brain Research*, **782**, 63–72.

Ruf, T., Korytko, A.I., Stieglitz, A. *et al.* (1997) Phenotypic variation in seasonal adjustments of testis size, body weight, and food intake in deer mice: role of pineal function and ambient temperature. *Journal of Comparative Physiology B Biochemical Systemic and Environmental Physiology*, **167**, 185–92.

Sack, R.L., Auckley, D., Auger, R.R. *et al.* (2007) Circadian rhythm sleep disorders: Part I, basic principles, shift work and jet lag disorders. *Sleep*, **30**, 1460–83.

Sanz, J.J. (1998) Effects of geographic location and habitat on breeding parameters of great tits. *Auk*, **115**, 1034–51.

Satoh, A., Yoshioka, E. and Numata, H. (2008) Circatidal actiivity rhythm in the mangrove cricket *Apteronemobius asahinai*. *Biology Letters*, **4**, 233–6.

Silverin, B., Massa, R. and Stokkan, K.A. (1993) Photoperiodic adaptation to breeding at different latitudes in great tits. *General and Comparative Endocrinology*, **90**, 14–22.

Stephan, J.K. and Zucker, I. (1972) Circadian rhythms in drinking behavior and locomotor activities are eliminated by hypothalamic lesions. *Proceedings of the National Academy of Sciences USA*, **69**, 1583–6.

Stevenson, T.J., Hahn, T.P., MacDougall-Shackleton, S.A. and Ball, G.F. (2012) Gonadotropin-releasing hormone plasticity: a comparative perspective. *Frontiers in Neuroendocrinology*, **33**, 287–300.

Sweeney, B.M. and Hastings, J.W. (1960) Effects of temperature upon diurnal rhythms. *Cold Spring Harbor Symposia on Quantitative Biology*, **25**, 87–104.

Tauber, E., Zordan, M., Sandrelli, F. *et al.* (2007) Natural selection favors a newly derived timeless allele in *Drosophila melanogaster*. *Science*, **316**, 1895–8.

Thomas, D.W., Blondel, J., Perret, P. *et al.* (2001) Energetic and fitness costs of mismatching resource supply and demand in seasonally breeding birds. *Science*, **291**, 2598–2600.

Thompson, C., Stinson, D. and Smith, A. (1990) Seasonal affective disorder and season-dependent abnormalities of melatonin suppression of light. *Lancet*, **336**, 703–6.

van Noordwijk, A.J., van Balen, J.H. and Scharloo, W. (1981) Genetic variation in the timing of reproduction in the great tit. *Oecologia*, **49**, 158–66.

Vaze, K.M. and Sharma, V.K. (2013) On the adaptive significance of circadian clocks for their owners. *Chronobiology International*, **30**, 413–33.

Verboven, N., Tinbergen, J.M. and Verhulst, S. (2001) Food, reproductive success and multiple breeding in the great tit *Parus major*. *Ardea*, **89**, 387–406.

Visser, M.E., Schaper, S.V., Holleman, L.J.M. *et al.* (2011) Genetic variation in cue sensitivity involved in avian timing of reproduction. *Functional Ecology*, **25**, 868–77.

Watts, H.E. and Hahn, T.P. (2012) Non-photoperiodic regulation of reproductive physiology in the flexibly breeding pine siskin (*Spinus pinus*). *General and Comparative Endocrinology*, **178**, 259–64.

Watts, H.E., MacDougall-Shackleton, S.A. & Hahn, T.P. (in revision) Individual variation in avian photoperiod responses: effects of breeding schedule, photoperiod, and experience. *Biology Letters*.

Wikelski, M., Martin, L.B., Scheuerlein, A. *et al.* (2008) Avian circannual clocks: adaptive significance and possible involvement of energy turnover in their proximate control. *Philosophical Transactions of the Royal Society of London B Biological Sciences*, **363**, 411–23.

Wingfield, J.C. (1983) Environmental and endocrine control of reproduction: An ecological approach, in *Avian Endocrinology: Environmental and Ecological Perspectives* (eds S.I. Mikami, K. Homma and K. Wada), Japan Scientific Society and Springer-Verlag, Tokyo and Berlin, pp. 265–88.

Woelfle, M.A., Ouyang, Y., Phanvijhitsiri, K. and Johnson, C.H. (2004) The adaptive value of circadian clocks: an experimental assessment in cyanobacteria. *Current Biology*, **14**, 1481–6.

Wyse, C.A., Coogan, A.N., Selman, C. *et al.* (2010) Association between mammalian lifespan and circadian free-running period: the circadian resonance hypothesis revisited. *Biological Letters*, **6**, 696–8.

Yamamura, T., Hirunagi, K., Ebihara, S. and Yoshimura, T. (2004) Seasonal morphological changes in the neuro-glial interaction between gonadotropin-releasing hormone nerve terminals and glial endfeet in Japanese quail. *Endocrinology*, **145**, 4264–7.

16

SENESCENCE: INTEGRATING BIOLOGY FROM CRADLE TO THE GRAVE

Mark F. Haussmann and Lisa A. Treidel

Department of Biology, Bucknell University, Lewisburg, PA, USA

INTRODUCTION

Life is characterized by change. Change may be developmental and permanent, like an undifferentiated cell becoming a muscle cell with the ability to contract, or change may be physiological and temporary, such as the seasonal growth and regression of song nuclei in the brains of songbirds. Both of these types of changes are beneficial and confer fitness advantages to the organism that possess them. However, a third group of changes, the progressive accumulation of deleterious changes that occur near the end of life, referred to as aging or senescence, is different. Senescence manifests as an age-specific increase in mortality or decrease in fertility and it is a unique change in the lifespan of an organism because unlike the other types of changes it is decidedly disadvantageous and maladaptive. The inherent complexity of senescence makes it one of the least understood disciplines in biology though many speculations have been put forth. An attempt by Medvedev (1990) to reconcile the many different ideas about the causes of the senescence led to more than 300 theories of aging. This problem was defined earlier by Alexander Comfort, one of senescence's founding fathers. He explained that in almost any biological field it is possible to show a steady progression from a large number of speculative theories, to one or two highly probable, main hypotheses. However, in the case of senescence this couldn't easily be done (Comfort 1979). While we have made substantial progress in the past forty years, the field of senescence is still one with many uncertainties and many lively debates.

Our goal here is not to try to review all of the aging literature, but instead to use an interdisciplinary approach to report on the current state of the field and suggest new

Integrative Organismal Biology, First Edition.
Edited by Lynn B. Martin, Cameron K. Ghalambor, and H. Arthur Woods.

promising areas of inquiry that are likely to lead to novel insights in the study of senescence. Senescence is clearly an integrative field in biology. Questions directed at how, at what rate, and why organisms age and die link disciplines such as evolutionary ecology and physiology to those of cell biology, demography, and gerontology. An understanding of senescence demands both insight into its evolutionary basis and the identification of its molecular and physiological underpinnings. We review both of these areas in the context of ultimate and proximate theories of aging in the first two sections of this chapter. In the last section, we focus on the integrative nature of the discipline. Specifically we address two broad but key areas for future research: (1) *senescence in the wild*: investigating senescence in natural conditions and nontraditional organisms to better understand the ways in which senescence evolved, and (2) *senescence and the developmental environment*: exploring of how environmental variation can influence mechanisms and patterns of senescence.

ULTIMATE THEORIES OF AGING

The process of senescence poses an evolutionary puzzle. Because senescence results in an increase in age-specific mortality and a decrease in fertility one would expect that natural selection would eliminate traits that promote senescence from populations. Yet, senescence is a widespread phenomenon in plants and animals. The answer to this puzzle is provided by evolutionary theories of senescence, which attribute the functional breakdown of organisms to the age-specific effects of extrinsic mortality – mortality that is imposed by the environment such as predation, starvation, or accident. Based upon these extrinsic hazards wild animals may not live long enough to grow old and thus, natural selection has limited opportunities to remove these traits from populations (Figure 16.1a, Kirkwood & Austad 2000).

Medawar (1952) was the first to formalize an evolutionary theory of senescence by explaining that the force of natural selection on an individual weakens with increasing age. In other words, natural selection has little power to act on germ-line mutations with late-acting deleterious effects because by the time that they impose their detrimental effects on organismal function, the individuals that bear these mutations have already passed their genes on to the next generation. The result is that this "selection shadow" will allow late-acting deleterious mutations to accumulate over evolutionary time. This process is known as the "Mutation Accumulation" theory (Figure 16.1b, Kirkwood & Austad 2000). Implicit in this hypothesis is that the effects of such a mutation accumulation process would only become evident if extrinsic mortality lessened sufficiently for organisms to survive long enough to actually express these mutations as symptoms of aging. Experimental studies in fruit flies (*Drosophila melanogaster*), have found some empirical support for the occurrence of Mutation Accumulation (Hughes & Reynolds 2005; Hughes *et al.* 2002).

Williams (1957) extended the idea of Mutation Accumulation in a second theory by proposing the existence of antagonistically pleiotropic genes. Similar to Medawar's idea, these genes have deleterious effects in old age when selection is weak. However, antagonistically pleiotropic genes have beneficial effects on fitness early in life when selection is strong (Figure 16.1c). With the "Antagonistic Pleiotropy" theory, Williams clearly showed that because a trait's contribution to fitness is a combination of not only the size of the effect, but also the probability of surviving to be affected by it, a small beneficial effect in early life can outweigh a catastrophic deleterious event in late life. In other words, under Antagonistic Pleiotropy aging is a maladaptive byproduct of selection for survival and reproduction during youth. Recent work in wild swans (Charmantier *et al.* 2006), red deer (Nussey *et al.* 2008), and Asian elephants (Robinson *et al.* 2012) find increasing support for Antagonistic Pleiotropy in wild populations.

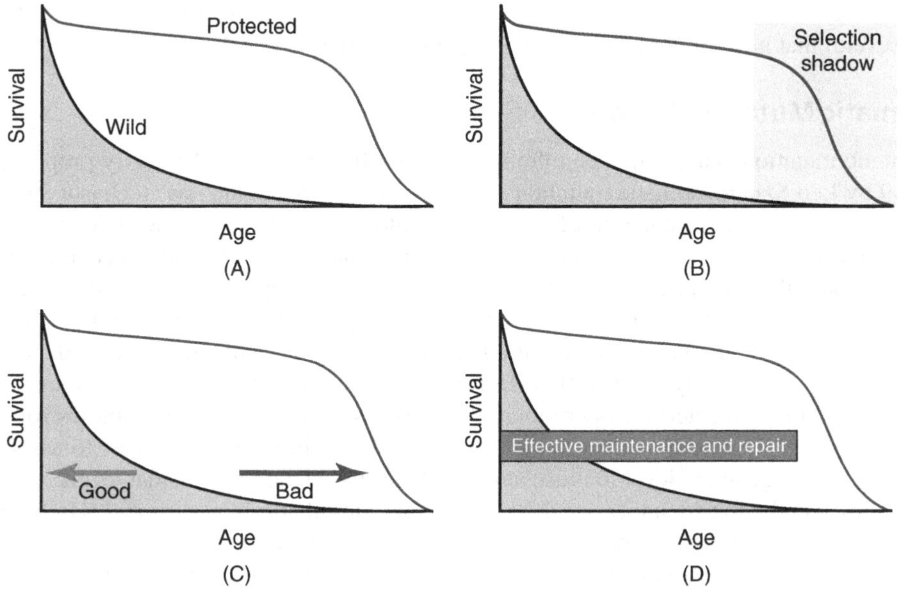

Figure 16.1. Survival curves illustrating the evolutionary theories of aging. *Source*: Thomas B.L. Kirkwood, Steven N. Austad (2000) Why do we age? *Nature*, **408**(6809). Reprinted by permission from Macmillan Publishers Ltd: *Nature*. Copyright 2000. (See plate section for color version.)

A final hypothesis similar to the Antagonistic Pleiotopy theory, is the "Disposable Soma" theory by Kirkwood (1977), which argues that senescence results from a trade-off between the optimal allocation of metabolic resources between somatic maintenance and reproduction (Figure 16.1d). Specifically, decline in function results from unrepaired damage to molecules and cells as a result of the normal process of living and this damage accumulates as we age (Westendorp & Kirkwood 1998). Kirkwood posits that the length of time that the soma should be maintained is a function of a specific environment over evolutionary time. Therefore, individuals living in safe environments with little worry of extrinsic mortality have a longer expectation of life and selection should increase allocation of resources to maintenance of the soma. In contrast, those in high-risk environments with high extrinsic mortality should invest in reproduction and, in the absence of extrinsic mortality, would be more vulnerable to senescence. The Disposable Soma theory is often viewed as a physiological explanation of senescence consistent with Antagonistic Pleiotropy theory and thus it is often considered a special case of Antagonistic Pleiotropy (Wilson *et al.* 2008). While current empirical support is much stronger for Antagonistic Pleiotropy and Disposable Soma theories as compared to Mutation Accumulation theory (Nussey *et al.* 2013), it is important to note that these three evolutionary theories of senescence are not mutually exclusive and their relative contributions to aging are likely to differ among species (Kirkwood & Austad 2000; Wilson *et al.* 2008; Nussey *et al.* 2013).

PROXIMATE THEORIES OF AGING

Whereas ultimate theories provide insight into "why we age" by explaining how selection pressures generate and maintain variability in senescence, in depth understanding requires insight into cellular and molecular mechanisms that regulate these evolutionary tradeoffs.

There are likely hundreds of proximate mechanisms that contribute to aging and we focus on several that have offered substantial insight into "how we age."

Somatic Mutation Theory

Somatic mutations have long been thought to contribute to aging. Originally proposed in 1959 by Leo Szilard, somatic mutation theory postulates that aging is a stochastic process resulting from the accumulation of mutations or alterations of the genetic code in somatic cells. With its focus on genetic mutations, somatic mutation theory readily accommodates the ultimate theories of aging. Szilard initially suggested that only mutations rendering whole chromosomes inactive are of functional importance for aging (Szilard 1959). However, the theory has been modified to include subtler mitochondrial and nuclear modifications (Kennedy *et al.* 2012). Buildup of genomic mutations is in turn expected to result in the characteristic operational declines associated with aging. Studies focused on the role of somatic mutation and genomic instability in aging thus strive to associate changes at the genomic level to those at the cellular, tissue, and organismal levels.

Historically, the soundness of the somatic mutation theory was disputed (Holliday & Kirkwood 1981). However, recent advancements in biotechnology have allowed for more direct assessments and confirmation that somatic mutations do in fact accumulate with aging. Through the application of *in vivo* reporter systems in humans (Martin *et al.* 1996) and transgenic mice (Hill *et al.* 2004; Dollé *et al.* 2000; Ono *et al.* 2000; Martin *et al.* 2001), the increase in mutation frequencies with age has been repeatedly demonstrated. In addition, in accordance with the somatic mutation theory a study by Odagiri *et al.* (1998) demonstrated a clear link between mutation frequencies and lifespan. They observed enhanced and reduced somatic mutation frequencies in strains of short-lived senescence prone and longer-lived senescence resistant mice lines respectively (Odagiri *et al.* 1998).

From an evolutionary perspective, the genetic and environmental factors mediating the generation and accumulation of somatic mutations are proximate factors that warrant attention. Ultimately, variation in the rates of mutation accumulation between tissues, individuals, and species is a combined function of improper replication, DNA damage induced by both endogenous (e.g., reactive oxygen species) and exogenous (e.g., UV light, radiation, chemicals) agents, and investment in genetic maintenance and repair (Morley & Turner 1999). DNA repair mechanisms are critically important in terms of mutation generation and represent a double-edged sword; replication of unrepaired mutated DNA and improper repair of mutated DNA are both major pathways to generating new mutations. Furthermore, life history evolution theory predicts that investment of resources into genomic maintenance comes at a cost to reproduction and growth. Nonetheless, greater investment in DNA repair is associated with extended life spans (Hart & Setlow 1974; Promislow 1994) and reduced aging-associated mutation accumulation (Ono *et al.* 2002).

Overall, while somatic mutations alone cannot account for all aspects of aging (Kirkwood & Proctor 2003) there is substantial evidence for their association with the aging process. Universal application of the somatic mutation theory is limited since studies have only been conducted in model systems and are either correlative or cross-sectional. To date no direct causative link between somatic mutations and aging has been established. Accordingly, future work on the somatic mutation theory of aging should focus on directly elucidating causality by longitudinally investigating the functional impact of somatic mutations via the potential application of new biotechnology such as high-throughput sequencing (Gundry & Vijg 2012).

Oxidative Stress

One process that remains a leading candidate mechanism for aging is the progressive loss of physiological function over time caused by cumulative oxidative damage (Harman 1956; Beckman & Ames 1998; Finkel & Holbrook 2000). To produce energy efficiently aerobic metabolism requires the use of oxygen as an electron acceptor. An associated consequence with this process is the production of reactive oxygen species (ROS). Whereas limited amounts of ROS serve important functions as regulators in signal processing, at higher concentrations ROS damage biomolecules (Finkel & Holbrook 2000; Barja 2004). The immune system also produces ROS, where they play a role in cellular apoptosis and defense against toxins and invading microorganisms (Finkel & Holbrook 2000). Mitochondrial ROS production can be reduced by altering membrane proton gradients (Brand 2000; Balaban *et al.* 2005), but this results in reduced efficiency of ATP production and increased overall energy requirement (Speakman *et al.* 2004). Because the production of ROS is such a routine consequence of cellular metabolism and reducing levels of ROS is costly, oxidative damage may account for the ubiquity of aging (but see Pérez *et al.* 2009).

Organisms have evolved mechanisms, such as antioxidant defenses and DNA repair, to reduce ROS propagation and oxidative damage (Surai 2002; Monaghan *et al.* 2009). Antioxidants can be grouped into two broad classes. In the first line of defense are enzymatic scavengers that act on specific ROS and include superoxide dismutase (SOD), catalase, and glutathione peroxidase (Finkel & Holbrook 2000). SOD converts certain ROS to hydrogen peroxide, whereas catalase and glutathione peroxidase convert hydrogen peroxide to water. The next level of defense involves the chain-breaking antioxidant compounds, which neutralize the propagation of oxidative damage initiated by ROS. These include both enzymatic (thioredoxin systems, ubiquinones, and glutathione) and dietary compounds (Vitamin C, Vitamin E, and carotenoids) (Halliwell & Gutteridge 2007).

The antioxidant defense systems cannot fully neutralize the ROS produced, inevitably resulting in oxidative damage to important biological molecules, notably DNA, proteins and lipids. Damage to DNA appears to have the most serious consequences to organismal function (Falnes *et al.* 2007) and is an important endogenous contributor to the mutations described by the Somatic Mutation theory. Due to both its proximity to the site of ROS generation and low repair rates, mitochondrial DNA is particularly vulnerable to ROS attack (Finkel & Holbrook 2000; Falnes *et al.* 2007). Damage to proteins, while less compromising than damage to DNA, results in conformational changes that can impair their function, and some amino acids are more susceptible to ROS attack than others (Droge 2002). The loss in protein function has been shown to compromise cellular viability (Mirzaei & Regnier 2008). Finally, because cellular membranes are largely composed of lipids, damage to these molecules has major consequences for membrane structure and function (Hulbert *et al.* 2007). Again, the composition of the lipids determines the extent of ROS attack as polyunsaturated fatty acids are much more vulnerable to oxidative assault than monounsaturated or saturated fatty acid acids. Thus, variation in the proportion of these different classes of lipids in membranes can influence the rate of oxidative damage (Hulbert *et al.* 2007).

In conclusion, the act of living pits aerobic organisms against ROS in a molecular battle that is at the center of life-history tradeoffs. Any energy invested toward damage mitigation or repair can only come at a cost to investment elsewhere. This has placed oxidative stress as an important mediator of life-history tradeoffs with wide reaching consequences for individual performance and aging rates (Monaghan *et al.* 2009; McGraw *et al.* 2010).

Telomeres

Telomeres are regions of non-coding but highly structured DNA at the end of linear eukaryotic chromosomes consisting of a tandem repeated DNA sequence (5′-TTAGGG-3′)$_n$ (Aubert & Lansdorp 2008). These evolutionarily conserved chromosome caps (Meyne *et al.* 1989; Bodnar 2009) play an important role in the protection of chromosome integrity (Blackburn 1991). Telomeres not only shorten with each cell replication event due to the end-replication problem (Capper *et al.* 2007) but compared with other regions of DNA, telomeres are particularly vulnerable to ROS attack (Rubio *et al.* 2004; Houben *et al.* 2008; von Zglinicki 2002). Telomeres that shorten to a critical length induce a permanent arrest in the cell cycle through a process called cellular senescence (Hornsby 2003; Capper *et al.* 2007).

Different taxa employ different mechanisms to retain or restore telomeres. Some mammalian cell lines are able to maintain telomere length through recombination (Aubert & Lansdrop 2008), while certain insects, such as *Drosophila melanogaster*, utilize transposable elements (Cenci 2009). However, the most common form of telomere restoration is through the enzyme telomerase. Telomerase activity in most cell lines is not sufficient to prevent telomere loss (Engelhardt *et al.* 1997; Lansdorp 2005) and thus telomeres in tissues with rapidly-dividing cells (like blood cells) shorten with age. Interestingly, oxidative stress also dramatically decreases telomerase activity (Borrás *et al.* 2004; Kurz *et al.* 2004) and therefore oxidative stress not only hastens telomere shortening by direct damage to telomeres, but also by inhibiting telomere restoration. Whereas telomerase activity appears to be essential for telomere maintenance it is repressed in most normal adult somatic tissues, most likely as a mechanism to prevent tumor growth (Taylor & Delany 2000; Parwaresch & Krupp 2002).

Substantial evidence is accumulating that telomeres are important to the aging phenotype (Campisi 2003; Patil *et al.* 2005). For example, senescent cells *in vivo* secrete degradative enzymes and inflammatory cytokines that disrupt nearby cells, contributing to aging and the threat of cancer (Wu *et al.* 2003; Campisi 2005; Capper *et al.* 2007). Work in mice has demonstrated that short telomeres result in multiple organismal defects caused by defective tissue regeneration (Blasco *et al.* 1997) and telomere dysfunction in these mice contributes to the nonreciprocal translocations that are common in adult carcinomas (Capper *et al.* 2007). Short telomeres are a risk factor in cardiovascular disease (Samani *et al.* 2001), liver cirrhosis (Mason *et al.* 2005), pulmonary fibrosis (Armanios *et al.* 2007), diabetes (Valdes *et al.* 2005), stroke (Martin-Ruiz *et al.* 2006), and Alzheimer's disease (Honig *et al.* 2006).

The big question in the context of life-history tradeoffs is how changes to telomeres at the cellular level influence organismal survival and lifespan. A number of studies in free-living birds have demonstrated that individuals with the shortest telomeres or the highest telomere loss rate have the poorest survival prospects (Haussmann *et al.* 2005; Salomons *et al.* 2009; Pauliny *et al.* 2006; Bize *et al.* 2009). Telomere length also predicts remaining lifespan in humans (Cawthon *et al.* 2003; Bakaysa *et al.* 2007; Kimura *et al.* 2007; but see Aviv 2009). While telomerase activity is not related to species lifespan, as shown by a phylogenetically-controlled comparison of 15 rodent species (Seluanov *et al.* 2007), the rate at which telomere erosion occurs appears to be (Haussmann *et al.* 2003). Based on a comparison of 5 avian species, the rate of telomere shortening is inversely related to maximum lifespan; shorter-lived species show greater telomere loss per year than longer-lived species (Haussmann *et al.* 2003). A survey of the mammalian literature also demonstrates a relationship between the rate of telomere shortening and lifespan (Haussmann *et al.* 2003). However, further comparative work is needed to determine whether this pattern holds in other taxa.

Hormonal Pathways

The endocrine system coordinates gene expression, which in turn regulates the physiological processes responsible for metabolism, growth, reproduction, and maintenance in response to varying environments. Therefore, disruptions within the endocrine system have pervasive and devastating effects. Not surprisingly, aging associated reductions in many hormonal axes (Chahal & Drake 2007) are recognized as candidate causative agents responsible for the progressive decline in physiology with age, a hallmark of senescence in all organisms.

To date, the insulin/insulin-like growth factor 1 (IGF-1) signaling pathway is the hormonal pathway for which we have the most evidence for a mechanistic role in aging. This conserved pathway plays a key part in promoting both growth and reproduction, while negatively impacting longevity in a wide range of vertebrate and invertebrate species (Dantzer & Swanson 2012). Single mutations in the IGF-1 signaling pathway are directly responsible for lifespan extensions in a number of species (Bartke 2005; Mukhopadhyay & Tissenbaum 2007; Buffenstein & Pinto 2009; Brown-Borg 2009; Toivonen & Partridge 2009; but see Kenyon 2010). For example, in *Caenorhabditis elegans* inhibition of IGF-1 activity more than doubles an individual's lifespan (Arantes-Oliveira *et al.* 2003). Mechanistically, IGF-1 acts via the IGFR receptor in mammals (or the DAF-2 receptor in *C. elegans*) to inhibit the activity and nuclear translocation of the forkhead box transcription factor, FOXO (or DAF-16 in *C. elegans*). Active FOXO in the nucleus can indirectly promote longevity through the expression of genes encoding antioxidants, chaperones, and metabolic proteins and concurrently prevent aging by inhibiting expression of pro-aging genes that promote growth and reproduction. Loss-of-function mutations introduced anywhere downstream in the IGF-1 signaling pathway ultimately act to prevent the inhibition of FOXO, allowing individuals to reap the benefits of its anti-aging activities (Russell & Kahn 2007; Kenyon 2010). Independent of changes within in the insulin/IGF-1 signaling pathway itself FOXO activity can also be directly up-regulated by a number of stress or developmental signals, such as sirtuins and heat shock proteins, all of which when over expressed result in delayed aging (Kenyon 2010). Together this suggests that FOXO is actually the key effector within the IGF-1 hormonal pathway responsible for its effects on longevity and aging. Importantly, at the organismal level, increased FOXO activity has numerous pleiotropic positive anti-aging effects including enhanced antioxidant protection (Brown-Borg *et al.* 2002; Holzenberger *et al.* 2003; Romanick *et al.* 2004), delayed degradation of immune function and collagen cross-linking (Flurkey *et al.* 2001), extended bone maintenance (Bronkowski *et al.* 2006), and cellular stress resistance (Henis-Korenblit *et al.* 2010). Having established a functional role of IGF-1 in aging and senescence in several laboratory model organisms it will be crucial for future work on IGF-1 to be performed in naturally occurring animal populations. Such work could both confirm the expected ubiquitous role of IGF-1 in aging and provide insight into its role in the evolution of senescence (Dantzer & Swanson 2012).

It is unlikely that IGF-1 is the sole hormonal pathway involved in regulating the rate of senescence. For instance, since the 1980s glucocorticoids have played a suspected role in regulating the timing of aging and death of semelparous species, such as pacific salmon and marsupial mice. Unique to individuals of semelparous species is the synchronous and predictable rapid physiological decline experienced following reproduction that ultimately results in the death of all mature individuals. This unusual synchrony of aging is expected to result from the also synchronous glucocorticoid hypersecretion that results from the combination of stressful competition for mate access and preprogrammed changes in the endocrine system triggered by reaching reproductive maturity (Robertson *et al.* 1961; Diamond 1982). While, modest lifespan expansions of semelparous individuals can be achieved by preventing the mating of a reproductive mature individual (Diamond 1982), the rapid endocrine disruption and its associated physiological deterioration cannot be avoided without castration prior to adulthood (Robertson 1961; Diamond 1982). This finding provides evidence for a necessary combined role of both predetermined changes in the endocrine functioning and external stressors in promoting aging and death of semelparous species. Critically, the case study of aging of semelparous species highlights how in the wild, hormonal influences stand at the foreground of the interaction between environmental and genetic changes, which in turn will have direct effects in determining longevity. Additionally, age related elevations of glucocorticoids, such as those seen in semelparous animals, are also common in iteroparous organisms (Chahal & Drake 2007) suggesting consistent cross-species deleterious contributions of glucocorticoids in aging (Sapolsky *et al.* 2000).

A third class of hormones that has been proximately related to the regulation of longevity in animals are thyroid hormones (e.g. thyroxine). Work on small, long-lived mammals, such as the naked mole rat (*Heterocephalus glaber*), has shown that reductions in thyroxine occurring in response to environmental stressors including temperature changes and food availability are related not only to low metabolic rates but also to lifespan extension. Thyroxine appears to promote both mitochondrial inefficiency and ROS generation. Further, an interspecies comparison of four small mammals has shown that circulating thyroxine levels are negatively related to maximum lifespan (Buffenstein & Pinto 2009). More comparative work is needed to see if the relationship between thyroxine and lifespan is found across a wider diversity of taxa.

Taken together, hormonal pathways play an integral role in mediating the rate of senescence in organisms. Thus far the study of endocrine signaling in aging has provided insight into additional mechanistic contributors to aging as well as notable conservation of these mechanisms over evolutionary history. Perhaps most excitingly, the activity of hormonal pathways is extremely plastic resulting in coordinated and dynamic shifts between life-history stages (Wingfield *et al.* 1998). This flexibility allows both individuals over the course of their lifespans and species over the course of evolutionary time to optimize life history strategies in varied environments and maximize fitness.

Network Theories of Aging

Our above discussion of proximate contributors to aging only barely scratches the surface of the candidate mechanisms involved in regulating longevity. To date, the widespread application of a reductive experimental approach in studies of aging has generated mainly correlative evidence for a large number of contributors to aging and proximate theories of aging. It is highly unlikely that there is either a single causative agent responsible for aging or that proposed proximate theories of senescence are mutually exclusive. Thus, future work in proximate theories of aging must take a broader integrative approach in order to better understand the additive and synergistic roles of identified mechanisms of aging. The importance of such an approach is most appropriately demonstrated by looking at advancements made by the development of network theories of aging.

Rooted in the idea that proximate mechanisms of aging interact cumulatively and synergistically, network theories of aging seek to synthesize and model these interactions. Kowald and Kirkwood (1993, 1994, 1996) developed the earliest network theories by focusing on relations between intracellular mechanisms of aging and maintenance. Specifically, their models incorporated the pro-aging functions of the accumulation of defective mitochondria, aberrant proteins, and ROS with the mitigating activities of antioxidants and proteolysis of aberrant proteins that occur simultaneously in cells (Kirkwood 1997). Today, more realistic increasingly complex multiscale network models that incorporate intra- and intercellular molecular mechanisms of aging are being developed (Kirkwood *et al.* 2003; West & Bergman 2009; Kriete *et al.* 2011). Further, proposed physiological regulatory networks (PRN) provide a framework for explaining aging at an organismal level by incorporating physiological processes and environmental interactions in network models (Cohen *et al.* 2012; Chapter 9, this volume).

Network models of aging generated from our current understanding of senescence provide a number of insights, which will shape future empirical investigations of the proximate causes of aging (Kirkwood 2011). For instance, the incorporation of the observed progressive activity of mechanisms of aging into models generates clear predictions about the expected rates of physiological decline. To illustrate this idea, take the successive role of mtDNA mutations in aging which represents only a piece of a whole network theory (Kirkwood *et al.* 2003). Based on this isolated branch of one model, we can predict a baseline pattern of exponential accumulation of mtDNA mutations and the point at which mutant mitochondria begin to accumulate. This in turn will result in functional deficits in energy production in longitudinal studies. When branches of models are imbedded within larger networks at the organismal level, such as a physiological regulatory network (Cohen *et al.* 2012; Chapter 9, this volume), we now have predictions to test the extent to which suspected environmental and physiological mediators of mtDNA accumulation will alter our baseline predictions in natural populations. Furthermore, through the use of the computational techniques, such as sensitivity analysis that analyze molecular and physiological models of aging, we can precisely quantify and make testable hypotheses about the contributions of individual components within developed networks to aging (Kriete *et al.* 2011).

Overall, to advance the field of aging it will be crucial to continually integrate insight gained from network models into experimental designs and questions about senescence. In serving simply as hypotheses of pathways to senescence, empirical tests of network theories will in turn be critical to refine and resolve the accuracy of generated models. As work in network theories of aging progresses it will be important to remember that since there is a large amount of variation in observed lifespans and life history strategies it is also likely that there are a large number of corresponding networks of aging acting within and between species. Thus, there are theoretically an infinite number of accurate network theories of aging to be generated. With this in mind, it should not necessarily be our goal to identify all possible acting networks. Instead we should aim to generate models from which we can gain a thorough understanding of the roles of the major determinates of aging and the general patterns by which they act to result in senescence.

FUTURE DIRECTIONS

Senescence in the Wild

For almost as long as senescence has been studied the idea has persisted that wild animals do not senesce. As mentioned above this notion lay at the root of Medawar's Mutation

Accumulation theory. However, recent theoretical and empirical work from a wide diversity of animals has shown that this idea is wrong (for an excellent review, see Nussey *et al.* 2013; Figure 16.2). Additionally, through comparative studies of natural populations, the expected genetic basis of aging upon which evolution may be acting (Wilson *et al.* 2007) and the direct relationship between longevity and investment in cellular maintenance (Robert & Bronikowski 2010) have also been confirmed. Together the recent surge of work investigating senescence in the wild has already begun to provide added support for the underlying assumptions of evolutionary theories of aging; the rates of senescence, which result from conserved molecular pathways, are optimized to an individual's environment.

In addition to supporting current theories of the evolution of aging, studies of aging in the wild have also shed light onto unique and conserved roles of the proximate mechanisms of aging exploited by animals in nature. For example, storm-petrels with exceptionally long lifespans were shown to employ the novel age-related physiological mechanism of a depressed telomere attrition rate to enhance their longevity (Haussmann *et al.* 2003). Additionally, studies of social insects have provided important clues connecting senescence and life history evolution (Jemielity *et al.* 2007). Such findings illustrate how work in other taxonomic groups has both opened up doors to surprising discoveries of unsuspected aging strategies and provided support for the evolutionary conservation of mechanisms of aging. Work focused on understanding the causes of intra-population variation in aging patterns have also identified critical environmental and life history characteristics responsible for regulating species-specific rates of senescence (Bouwhuis *et al.* 2010; Massot *et al.* 2011; Lemaître *et al.* 2013). For example, Bouwhuis *et al.* (2010) showed that both immigration and early life reproduction independently promoted accelerated senescence of great tits. More studies focused on individual variation of aging will no doubt further contribute to our understanding to the causes of aging in the wild.

In general, while aging in the wild represents a rapidly expanding area in the field of senescence, much work remains. Two critical considerations for the experimental design of future work on aging in the wild must be made. First, while laboratory studies can be powerful, they have limits in terms of the applicability of their results (Partridge & Gems 2007). By contrast, natural studies of aging sacrifice control and pose many practical challenges, but better emphasize the impact of an individual's environmental context on aging (Reznick *et al.* 2004; Hayward *et al.* 2009; Reichert *et al.* 2010). Thus, the coupling of findings from both captive and wild animals can serve as an effective means to gain an understanding of senescence. Second, we must optimize and standardize methodologies utilized in the field. Doing so will facilitate future meta-analyses, allow for eventual cross-study comparisons, and better accommodate the synthesis of experimental discoveries.

Senescence and the Developmental Environment: Compensatory Growth, Maternal Effects and Hormesis

Early aging researchers Raymond Pearl and John Miner (1935) realized that environmental variability would have large effects on the pattern and pace of senescence. Given the Disposable Soma Theory, natural selection has optimized the allocation of resources between somatic maintenance and reproduction. However, if environmental circumstances change within an organism's lifetime, previous allocation patterns may no longer be optimal and the result may alter the onset and rate at which senescence progresses. We have known for some time that environmental conditions during growth and development have broad and robust effects on the aging phenotype (Metcalfe & Monaghan 2001). Many of these effects are thought to occur through antagonistic pleiotropies. For example, if an individual develops

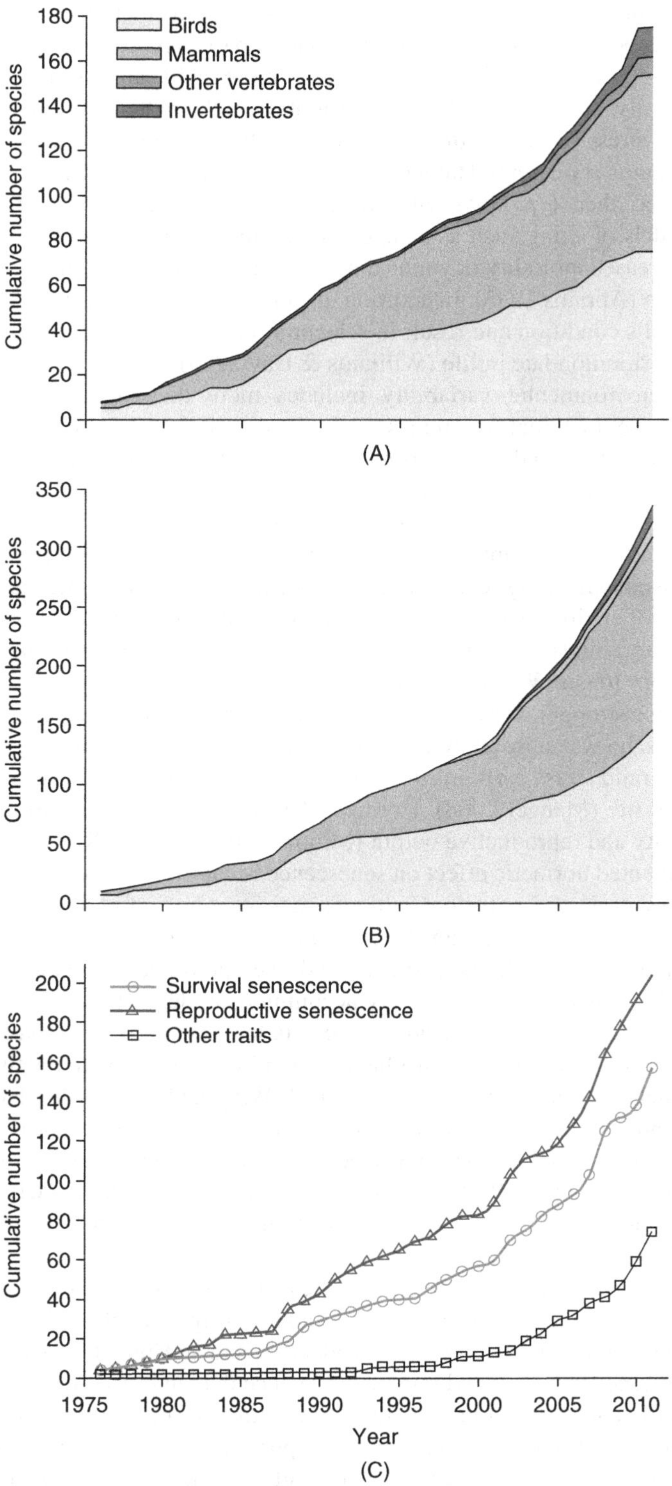

Figure 16.2. Cumulative evidence for aging in the wild. *Source*: Nussey *et al.* (2013) Senescence in natural populations of animals: widespread evidence and it's implications for bio-gerontology *Aging Research Reviews*, **12**(218). Reprinted with permission from Elsevier Limited.

in an environment where resources are scarce, but conditions then improve, that organism will normally respond by developing faster to catch-up to its peers. While these individuals may benefit by reproducing earlier they often show more rapid patterns of degeneration later in life. Specifically, environmental induced variation in the patterns of growth in early life effect oxidative stress (Blount *et al.* 2003; Alonso-Alvarez *et al.* 2006, 2007) and telomere dynamics (Jennings *et al.* 1999; Hall *et al.* 2004; Monaghan & Haussmann 2006). Environmentally induced phenotypic plasticity on aging has also recently been incorporated into traditional models of aging such as antagonistic pleiotropy. For example, high population density and increased mortality in young animals can result in selection that actually favors reduced lifespan (Abrams 1993). In addition, the effect of environmental stress can depend on an individual's condition and result in selection for slower senescence in early life, but more rapid deterioration late in life (Williams & Day 2003).

Relevant environmental variability includes more than just access to metabolic resources because maternal hormones like glucocorticoids (Haussmann *et al.* 2012; Entringer *et al.* 2011) and testosterone (Noguera *et al.* 2012) also appear to affect proximate mechanisms of aging such as oxidative stress and telomere dynamics. Because maternal effects depend on the mother's environment, the offspring's pattern of aging can be influenced by environmental factors experienced by the previous generation. This exciting new avenue in aging research has just begun and we need more studies looking at a wider variety of environments, hormones, and proximate mechanisms.

The German philosopher Friedrich Nietzsche opened his book *Twilight of the Idols* with an idea from Johann Wolfgang von Goethe, "From life's school of war: what does not kill me makes me stronger." From the perspective of senescence in the developmental environment, Nietzsche was absolutely correct. Hormesis is said to occur when an individual is exposed to a mild stress early in life, which leads to an increased resistance to stronger stresses later in life (Mangel 2008). Hormesis leads to an increase in fitness components such as longevity and reproductive output (Minois & Rattan 2003; Rattan 2004). Probably the best-documented hormetic effect on senescence is caloric restriction, which shows that any number of caloric and nutritional stresses activate a host of endogenous mechanisms that collectively extend the lifespan (Cysper *et al.* 2006). Other factors, such as temperature, radiation, a variety of drugs, and even hydrogen peroxide, also can induce hormesis (Mangel 2008). Ideas of hormesis readily accommodate ideas of stress physiology mediated by the hypothalamic-pituitary-adrenal axis (Chapters 8 & 12, this volume). Mild or acute stresses result in the release of glucocorticoid hormones, which activate protective mechanisms that can enhance immediate survival (Wingfield *et al.* 1998). However, just as hormesis depends on the first insult to be mild, overexposure to glucocorticoids, generally termed chronic stress, appears to hasten senescence. These effects may be realized via elevated metabolism and the ROS it generates (Haussmann & Marchetto 2010). Exposure to high levels of glucocorticoids also accelerates telomere shortening and cellular aging (Epel *et al.* 2004; Haussmann *et al.* 2012).

In summary, it's becoming increasingly clear that both genes and environment shape senescence. While genes capture the evolutionary response of optimal allocation of resources to the competing demand of reproduction and self-maintenance, environmental variability fine-tunes the onset and rate of senescence that an organism experiences.

The field of senescence continues to offer exciting avenues for new research as well as many challenges. It's time to put away our pursuit of a single unifying "fountain of youth" and embrace the diversity of traits and pathways that result in the marvelous complexity of the problem of senescence. Over the past decade, as senescence research has shifted to focus on natural populations, the marriage of field and laboratory methodologies and philosophies has continued to grow. Because senescence demands both evolutionary insight and mechanistic understanding it will continue to involve scientists from diverse backgrounds. Together, these advances will help us to continue to unravel the mysteries of why and how we grow old.

REFERENCES

Abrams, P.A. (1993) Does increased mortality favor the evolution of more rapid senescence? *Evolution*, **47**, 877–87.

Alonso-alvarez, C., Bertrand, S., Devevey, G. *et al.* (2006) An Experimental Manipulation of Life-History Trajectories and Resistance to Oxidative Stress. *Evolution*, **60**, 1913–24.

Alonso-Alvarez, C., Bertrand, S., Faivre, B. and Sorci, G. (2007) Increased susceptibility to oxidative damage as a cost of accelerated somatic growth in zebra finches. *Functional Ecology*, **21**, 873–9.

Arantes-Oliveira, N., Berman, J.R. and Kenyon, C. (2003) Healthy animals with extreme longevity. *Science*, **302**, 611.

Armanios, M., Chen, J., Cogan, J. *et al.* (2007) Telomerase mutations in families with idiopathic pulmonary fibrosis – NEJM. *New England Journal of Medicine*, **356**, 1317–26.

Aubert, G. and Lansdorp, P.M. (2008) Telomeres and Aging. *Physiol Rev*, **88**, 557–79.

Aviv, A., Chen, W., Gardner, J.P. *et al.* (2009) Leukocyte telomere dynamics: longitudinal findings among young adults in the Bogalusa Heart Study. *American Journal of Epidemiology*, **169**, 323–9.

Bakaysa, S.L., Mucci, L.A., Slagboom, P.E. *et al.* (2007) Telomere length predicts survival independent of genetic influences. *Aging cell*, **6**, 769–74.

Balaban, R.S., Nemoto, S. and Finkel, T. (2005) Mitochondria, oxidants, and aging. *Cell*, **120**, 483–95.

Barja, G. (2004) Free radicals and aging. *Trends in Neurosciences*, **27**, 595–600.

Bartke, A. (2005) Minireview: role of the growth hormone/insulin-like growth factor system in mammalian aging. *Endocrinology*, **146**, 3718–23.

Beckman, K.B. and Ames, B.N. (1998) The free radical theory of aging matures. *Physiological reviews*, **78**, 547–81.

Bize, P., Criscuolo, F., Metcalfe, N.B. *et al.* (2009) Telomere dynamics rather than age predict life expectancy in the wild. *Proceedings. Biological sciences / The Royal Society*, **276**, 1679–83.

Blackburn, E.H. (1991) Structure and function of telomeres. *Nature*, **350**, 569–73.

Blasco, M.A., Lee, H.-W., Hande, M.P. *et al.* (1997) Telomere shortening and tumor formation by mouse cells lacking telomerase RNA. *Cell*, **91**, 25–34.

Blount, J.D., Metcalfe, N.B., Arnold, K.E. *et al.* (2003) Neonatal nutrition, adult antioxidant defences and sexual attractiveness in the zebra finch. *Proceedings. Biological sciences / The Royal Society*, **270**, 1691–6.

Bodnar, A.G. (2009) Marine invertebrates as models for aging research. *Experimental Gerontology*, **44**, 477–84.

Bronkowski, M.S., Pamenter, R.W., Rocha, J.S. *et al.* (2006) Long-lived growth hormone receptor knockout mice show a delay in age-related changes of body composition and bone characteristics. *The Journals of Gerontology Series A: Biological Sciences and Medical Sciences*, **61**, 562–7.

Borrás, C., Esteve, J.M., Viña, J.R. *et al.* (2004) Glutathione regulates telomerase activity in 3T3 fibroblasts. *The Journal of Biological Chemistry*, **279**, 34332–5.

Bouwhuis, S., Charmantier, A., Verhulst, S. and Sheldon, B.C. (2010) Individual variation in rates of senescence: natal origin effects and disposable soma in a wild bird population. *The Journal of Animal Ecology*, **79**, 1251–61.

Brand, M. (2000) Uncoupling to survive? The role of mitochondrial inefficiency in ageing. *Experimental Gerontology*, **35**, 811–20.

Brown-Borg, H.M. (2009) Hormonal control of aging in rodents: the somatotropic axis. *Molecular and Cellular Endocrinology*, **299**, 64–71.

Brown-borg, H.M., Rakoczy, S.G., Romanick, M.A. and Kennedy, M.A. (2002) Effects of growth hormone and insulin-like growth factor-1 on hepatocyte antioxidant enzymes. *Experimental Biology and Medicine*, **227**, 94–104.

Buffenstein, R. and Pinto, M. (2009) Endocrine function in naturally long-living small mammals. *Molecular and Cellular Endocrinology*, **299**, 101–11.

Campisi, J. (2003) Cellular senescence and apoptosis: how cellular responses might influence aging phenotypes. *Experimental Gerontology*, **38**, 5–11.

Campisi, J. (2005) Suppressing cancer: the importance of being senescent. *Science (New York, N.Y.)*, **309**, 886–87.

Capper, R., Britt-Compton, B., Tankimanova, M. *et al.* (2007) The nature of telomere fusion and a definition of the critical telomere length in human cells. *Genes & development*, **21**, 2495–2508.

Cawthon, R.M., Smith, K.R., O'Brien, E. *et al.* (2003) Association between telomere length in blood and mortality in people aged 60 years or older. *The Lancet*, **361**, 393–5.

Cenci, G. (2009) Drosophila cell cycle under arrest: Uncapped telomeres plead guilty. *Cell Cycle*, **8**, 990–5.

Chahal, H.S. and Drake, W.M. (2007) The endocrine system and ageing. *Journal of Pathology*, **211**, 173–80.

Charmantier, A., Perrins, C., McCleery, R.H. and Sheldon, B.C. (2006) Quantitative genetics of age at reproduction in wild swans: support for antagonistic pleiotropy models of senescence. *Proceedings of the National Academy of Sciences of the United States of America*, **103**, 6587–92.

Cohen, A.A., Martin, L.B., Wingfield, J.C. *et al.* (2012) Physiological regulatory networks: ecological roles and evolutionary constraints. *Trends in Ecology & Evolution*, **27**, 428–35.

Comfort, A. (1979) *The Biology of Senescence, 3rd edition* edn, Churchill Livingstone, Edinburgh and London.

Cypser, J.R., Tedesco, P. and Johnson, T.E. (2006) Hormesis and aging in Caenorhabditis elegans. *Experimental Gerontology*, **41**, 935–9.

Dantzer, B. and Swanson, E.M. (2012) Mediation of vertebrate life histories via insulin-like growth factor-1. *Biological reviews of the Cambridge Philosophical Society*, **87**, 414–29.

Diamond, J.M. (1982) Big-bang reproduction and ageing in male marsupial mice. *Nature*, **298**, 115–16.

Dollé, M.E.T., Snyder, W.K., Gossen, J.A. *et al.* (2000) Distinct spectra of somatic mutations accumulated with age in mouse heart and small intestine. *Proceedings of the National Academy of Sciences of the United States of America*, **97**, 8403–8.

Droge, W. (2002) Free radicals in the physiological control of cell function. *Physiol Rev*, **82**, 47–95.

Engelhardt, M., Kumar, R., Albanell, J. *et al.* (1997) Telomerase regulation, cell cycle, and telomere stability in primitive hematopoietic cells. *Blood*, **90**, 182–93.

Entringer, S., Epel, E.S., Kumsta, R. *et al.* (2011) Stress exposure in intrauterine life is associated with shorter telomere length in young adulthood. *Proceedings of the National Academy of Sciences of the United States of America*, **108**, E513–518.

Epel, E.S., Blackburn, E.H., Lin, J. *et al.* (2004) Accelerated telomere shortening in response to life stress. *Proceedings of the National Academy of Sciences of the United States of America*, **101**, 17312–15.

Falnes, P.Ø., Klungland, A. and Alseth, I. (2007) Repair of methyl lesions in DNA and RNA by oxidative demethylation. *Neuroscience*, **145**, 1222–32.

Finkel, T. and Holbrook, N.J. (2000) Oxidants, oxidative stress and the biology of ageing. *Nature*, **408**, 239–47.

Flurkey, K., Papaconstantinou, J., Miller, R.A. and Harrison, D.E. (2001) Lifespan extension and delayed immune and collagen aging in mutant mice with defects in growth hormone production. *Proceedings of the National Academy of Sciences of the United States of America*, **98**, 6736–41.

Gundry, M. and Vijg, J. (2012) Direct mutation analysis by high-throughput sequencing: from germline to low-abundant, somatic variants. *Mutation research*, **729**, 1–15.

Hall, M.E., Nasir, L., Daunt, F. *et al.* (2004) Telomere loss in relation to age and early environment in long-lived birds. *Proceedings. Biological sciences / The Royal Society*, **271**, 1571–6.

Halliwell, B. and Gutteridge, J. (2007) *Free Radicals in Biology and Medicine*, Oxford University Press, Oxford.

Harman, D. (1956) Aging: A theory based on free-radical and radiation-chemistry. *Journals of Gerontology*, **11**, 298–300.

Hart, R.W. and Setlow, R.B. (1974) Correlation between deoxyribonucleic acid excision-repair and life-span in a number of mammalian species. *Proceedings of the National Academy of Sciences of the United States of America*, **71**, 2169–73.

Haussmann, M.F., Longenecker, A.S., Marchetto, N.M. *et al.* (2012) Embryonic exposure to corticosterone modifies the juvenile stress response, oxidative stress and telomere length. *Proceedings Biological Sciences / The Royal Society*, **279**, 1447–56.

Haussmann, M.F. and Marchetto, N.M. (2010) Telomeres: Linking stress and survival, ecology, and evolution. *Current Zoology*, **56**, 714–27.

Haussmann, M.F., Winkler, D.W., O'Reilly, K.M. *et al.* (2003) Telomeres shorten more slowly in long-lived birds and mammals than in short-lived ones. *Proceedings. Biological sciences / The Royal Society*, **270**, 1387–92.

Haussmann, M.F., Winkler, D.W. and Vleck, C.M. (2005) Longer telomeres associated with higher survival in birds. *Biology letters*, **1**, 212–14.

Hayward, A.D., Wilson, A.J., Pilkington, J.G. *et al.* (2009) Ageing in a variable habitat: environmental stress affects senescence in parasite resistance in St Kilda Soay sheep. *Proceedings. Biological Sciences / The Royal Society*, **276**, 3477–85.

Henis-Korenblit, S., Zhang, P., Hansen, M. *et al.* (2010) Insulin/IGF-1 signaling mutants reprogram ER stress response regulators to promote longevity. *Proceedings of the National Academy of Sciences of the United States of America*, **107**, 9730–5.

Hill, K.A., Buettner, V.L., Halangoda, A. *et al.* (2004) Spontaneous mutation in Big Blue mice from fetus to old age: tissue-specific time courses of mutation frequency but similar mutation types. *Environmental and Molecular Mutagenesis*, **43**, 110–20.

Holliday, R. and Kirkwood, T.B.L. (1981) Predictions of the somatic mutation and mortalization theories of cellular ageing are contrary to experimental observations. *Journal of Theoretical Biology*, **93**, 627–42.

Holzenberger, M., Dupont, J., Ducos, B. *et al.* (2003) IGF-1 receptor regulates lifespan and resistance to oxidative stress in mice. *Nature*, **421**, 182–7.

Honig, L.S., Schupf, N., Lee, J.H. *et al.* (2006) Shorter telomeres are associated with mortality in those with APOE epsilon4 and dementia. *Annals of Neurology*, **60**, 181–7.

Hornsby, P.J. (2003) Replicative senescence of human and mouse cells in culture: significance for aging research. *Mechanisms of Ageing and Development*, **124**, 853–5.

Houben, J.M.J., Moonen, H.J.J., Van Schooten, F.J. and Hageman, G.J. (2008) Telomere length assessment: Biomarker of chronic oxidative stress? *Free Radical Biology and Medicine*, **44**, 235–46.

Hughes, K.A., Alipaz, J.A., Drnevich, J.M. and Reynolds, R.M. (2002) A test of evolutionary theories of aging. *Proceedings of the National Academy of Sciences of the United States of America*, **99**, 14286–91.

Hughes, K.A. and Reynolds, R.M. (2005) Evolutionary and mechanistic theories of aging. *Annual Review of Entomology*, **50**, 421–45.

Hulbert, A.J., Pamplona, R., Buffenstein, R. and Buttemer, W.A. (2007) Life and death: metabolic rate, membrane composition, and life span of animals. *Physiological Reviews*, **87**, 1175–1213.

Jemielity, S., Kimura, M., Parker, K.M. *et al.* (2007) Short telomeres in short-lived males: what are the molecular and evolutionary causes? *Aging cell*, **6**, 225–33.

Jennings, B.J., Ozanne, S.E., Dorling, M.W. and Hales, C.N. (1999) Early growth determines longevity in male rats and may be related to telomere shortening in the kidney. *FEBS Letters*, **448**, 4–8.

Kennedy, S.R., Loeb, L.a. and Herr, A.J. (2012) Somatic mutations in aging, cancer and neurodegeneration. *Mechanisms of Ageing and Development*, **133**, 118–26.

Kenyon, C.J. (2010) The genetics of ageing. *Nature*, **464**, 504–12.

Kimura, M., Barbieri, M., Gardner, J.P. *et al.* (2007) Leukocytes of exceptionally old persons display ultra-short telomeres. *American Journal of Physiology. Regulatory, Integrative and Comparative Physiology*, **293**, R2210–17.

Kirkwood, T.B.L. (1977) Evolution of ageing. *Nature*, **270**, 301–4.

Kirkwood, T.B.L. (2011) Systems biology of ageing and longevity. *Philosophical Transactions of the Royal Society of London. Series B, Biological sciences*, **366**, 64–70.

Kirkwood, T.B.L. and Austad, S.N. (2000) Why do we age? *Nature*, **408**, 233–8.

Kirkwood, T.B.L., Boys, R.J., Gillespie, C.S. *et al.* (2003) Towards an e-biology of ageing: integrating theory and data. *Nature Reviews. Molecular Cell Biology*, **4**, 243–9.

Kirkwood, T.B.L. and Kowald, A. (1997) Network theory of aging. *Experimental Gerontology*, **32**, 395–9.

Kirkwood, T.B.L. and Proctor, C.J. (2003) Somatic mutations and ageing in silico. *Mechanisms of Ageing and Development*, **124**, 85–92.

Kowald, A. and Kirkwood, T.B.L. (1993) Mitochondrial mutations, cellular instability and ageing: modelling the population dynamics of mitochondria. *Mutation Research/DNAging*, **295**, 93–103.

Kowald, A. and Kirkwood, T.B.L. (1994) Towards a network theory of aging: A model combining the free radical theory and protein error theory. *Journal of theoretical biology*, **168**, 75–94.

Kowald, A. and Kirkwood, T.B.L. (1996) A network theory of ageing: the interactions of defective mitochondria, aberrant proteins, free radicals and scavengers in the ageing process. *Mutation Research/DNAging*, **316**, 209–236.

Kriete, A., Lechner, M., Clearfield, D. and Bohmann, D. (2011) Computational systems biology of aging. *Wiley Interdisciplinary Reviews. Systems Biology and Medicine*, **3**, 414–28.

Kurz, D.J., Decary, S., Hong, Y. *et al.* (2004) Chronic oxidative stress compromises telomere integrity and accelerates the onset of senescence in human endothelial cells. *Journal of Cell Science*, **117**, 2417–26.

Lansdorp, P.M. (2005) Major cutbacks at chromosome ends. *Trends in Biochemical Sciences*, **30**, 388–95.

Lemaître, J.-F., Gaillard, J.-M., Lackey, L.B. *et al.* (2013) Comparing free-ranging and captive populations reveals intra-specific variation in aging rates in large herbivores. *Experimental Gerontology*, **48**, 162–7.

Mangel, M. (2008) Environment, damage and senescence: modelling the life-history consequences of variable stress and caloric intake. *Functional Ecology*, **22**, 422–30.

Martin, S.L., Hopkins, C.L., Naumer, a. *et al.* (2001) Mutation frequency and type during ageing in mouse seminiferous tubules. *Mechanisms of ageing and development*, **122**, 1321–31.

Martin, G.M., Ogburn, C.E., Colgin, L.M. *et al.* (1996) Somatic mutations are frequent and increase with age in human kidney epithelial cells. *Human Molecular Genetics*, **5**, 215–21.

Martin-Ruiz, C., Dickinson, H.O., Keys, B. *et al.* (2006) Telomere length predicts poststroke mortality, dementia, and cognitive decline. *Annals of Neurology*, **60**, 174–80.

Mason, P., Wilson, D. and Bessler, M. (2005) Dyskeratosis congenita – a disease of dysfunctional telomere maintenance. *Current Molecular Medicine*, **5**, 159–70.

Massot, M., Clobert, J., Montes-Poloni, L. *et al.* (2011) An integrative study of ageing in a wild population of common lizards. *Functional Ecology*, **25**, 848–58.

McGraw, K.J., Cohen, A.A., Costantini, D. and Hõrak, P. (2010) The ecological significance of antioxidants and oxidative stress: a marriage between mechanistic and functional perspectives. *Functional Ecology*, **24**, 947–9.

Medwar, P.B. (1952) *An Unsolved Problem of Biology*, Lewis, London.

Medvedev, Z.A. (1990) An attempt at a rational classification of theories of ageing. *Biological Reviews*, **65**, 375–98.

Metcalfe, N.B. and Monaghan, P. (2001) Compensation for a bad start: grow now, pay later? *Trends in Ecology & Evolution*, **16**, 254–60.

Meyne, J., Ratliff, R.L. and Moyzis, R.K. (1989) Conservation of the human telomere sequence (TTAGGG)n among vertebrates. *Proceedings of the National Academy of Sciences*, **86**, 7049–53.

Minois, N. and Rattan, S.I.S. (2003) Hormesis in aging and longevity, in *Modulating Aging and Longevity* (ed S.I.S. Rattan), Kluwer Academic Publishers, Dordrecht, pp. 127–37.

Mirzaei, H. and Regnier, F. (2008) Protein:protein aggregation induced by protein oxidation. *Journal of Chromatography B*, **873**, 8–14.

Monaghan, P. and Haussmann, M.F. (2006) Do telomere dynamics link lifestyle and lifespan? *Trends in Ecology & Evolution*, **21**, 47–53.

Monaghan, P., Metcalfe, N.B. and Torres, R. (2009) Oxidative stress as a mediator of life history trade-offs: mechanisms, measurements and interpretation. *Ecology Letters*, **12**, 75–92.

Morley, A.A. and Turner, D.R. (1999) The contribution of exogenous and endogenous mutagens to in vivo mutations. *Mutation research*, **428**, 11–15.

Mukhopadhyay, A. and Tissenbaum, H.a. (2007) Reproduction and longevity: Secrets Revealed by C. elegans. *Trends in Cell Biology*, **17**, 65–71.

Noguera, J.C., Kim, S.-Y. and Velando, A. (2012) Pre-fledgling oxidative damage predicts recruitment in a long-lived bird. *Biology Letters*, **8**, 61–3.

Nussey, D.H., Froy, H., Lemaitre, J.-F. *et al.* (2013) Senescence in natural populations of animals: widespread evidence and its implications for bio-gerontology. *Ageing Research Reviews*, **12**, 214–25.

Nussey, D.H., Wilson, A.J., Morris, A. *et al.* (2008) Testing for genetic trade-offs between early- and late-life reproduction in a wild red deer population. *Proceedings. Biological sciences / The Royal Society*, **275**, 745–50.

Odagiri, Y., Uchida, H., Hosokawa, M. *et al.* (1998) Accelerated accumulation of somatic mutations in the senescence-accelerated mouse. *Nature Genetics*, **19**, 116–17.

Ono, T., Ikehata, H., Nakamura, S. *et al.* (2000) Age-associated increase of spontaneous mutant frequency and molecular nature of mutation in newborn and old lacZ-transgenic mouse. *Mutation Research*, **447**, 165–77.

Ono, T., Uehara, Y., Saito, Y. and Ikehata, H. (2002) Mutation theory of aging, assessed in transgenic mice and knockout mice. *Mechanisms of Ageing and Development*, **123**, 1543–52.

Partridge, L. and Gems, D. (2007) Benchmarks for ageing studies. *Nature*, **450**, 165–7.

Parwaresch, R. and Krupp, G. (2002) Molecular basis of aging. *Aging: Morphological, Biochemical, Molecular and Social Aspects*, **27**, 295–300.

Patil, C., Mian, S. and Campisi, J. (2005) The thorny path linking cellular senescence to organismal aging. *Mechanisms of Ageing and Development*, **126**, 1040–5.

Pauliny, A., Wagner, R.H., Augustin, J. *et al.* (2006) Age-independent telomere length predicts fitness in two bird species. *Molecular Ecology*, **15**, 1681–7.

Pearl, R. and Miner, J.R. (1935) Experimental studies on the duration of life XIV. The comparative mortality of certain lower organisms. *Quarterly Review of Biology*, **10**, 60–79.

Pérez, V.I., Buffenstein, R., Masamsetti, V. *et al.* (2009) Protein stability and resistance to oxidative stress are determinants of longevity in the longest-living rodent, the naked mole-rat. *Proceedings of the National Academy of Sciences of the United States of America*, **106**, 3059–64.

Promislow, D.E.L. (1994) DNA repair and the evolution of longevity: a critical analysis. *Journal of Theoretical Biology*, **170**, 291–300.

Rattan, S.I.S. (2004) Aging, anti-aging, and hormesis. *Mechanisms of Ageing and Development*, **125**, 285–9.

Reichert, B.E., Martin, J., Kendall, W.L. *et al.* (2010) Interactive effects of senescence and natural disturbance on the annual survival probabilities of snail kites. *Oikos*, **119**, 972–9.

Reznick, D.N., Bryant, M.J., Roff, D. *et al.* (2004) Effect of extrinsic mortality on the evolution of senescence in guppies. *Nature*, **431**, 1095–9.

Robert, K.A. and Bronikowski, A.M. (2010) Evolution of senescence in nature: physiological evolution in populations of garter snake with divergent life histories. *The American Naturalist*, **175**, 147–59.

Robertson, O.H. (1961) Prolongation of the life span of konkanee salmon (Oncorhynchus nerka kennerlyi) by castration before beginning of gonad development. *Proceedings of the National Academy of Sciences*, **47**, 609–21.

Robertson, O.H., Wexler, B.C. and Miller, B.F. (1961) Degenerative changes in the cardiovascular system of the spawning Pacific salmon (Oncorhynchus tshawytscha). *Circulation Research*, **9**, 826–34.

Robinson, M.R., Mar, K.U. and Lummaa, V. (2012) Senescence and age-specific trade-offs between reproduction and survival in female Asian elephants. *Ecology Letters*, **15**, 260–6.

Romanick, M.A., Rakoczy, S.G. and Brown-Borg, H.M. (2004) Long-lived Ames dwarf mouse exhibits increased antioxidant defense in skeletal muscle. *Mechanisms of Ageing and Development*, **125**, 269–81.

Rubio, M.A., Davalos, A.R. and Campisi, J. (2004) Telomere length mediates the effects of telomerase on the cellular response to genotoxic stress. *Experimental Cell Research*, **298**, 17–27.

Russell, S.J. and Kahn, C.R. (2007) Endocrine regulation of ageing. *Nature Reviews. Molecular Cell Biology*, **8**, 681–91.

Salomons, H.M., Mulder, G.A., Van de Zande, L. *et al.* (2009) Telomere shortening and survival in free-living corvids. *Proceedings. Biological sciences / The Royal Society*, **276**, 3157–65.

Samani, N.J., Boultby, R., Butler, R. *et al.* (2001) Telomere shortening in atherosclerosis. *The Lancet*, **358**, 472–3.

Sapolsky, R.M., Romero, L.M. and Munck, A.U. (2000) How do glucocorticoids influence stress responses? Integrating permissive, suppressive, stimulatory, and preparative action. *Endocrine Reviews*, **21**, 55–89.

Seluanov, A., Chen, Z., Hine, C. *et al.* (2007) Telomerase activity coevolves with body mass not lifespan. *Aging Cell*, **6**, 45–52.

Speakman, J.R., Talbot, D.A., Selman, C. *et al.* (2004) Uncoupled and surviving: individual mice with high metabolism have greater mitochondrial uncoupling and live longer. *Aging Cell*, **3**, 87–95.

Surai, P. (2002) *Natural Antioxidants in Avian Nutrition and Reproduction*, Nottingham University Press, Nottingham, UK.

Szilard, L. (1959) On the nature of the aging process. *Proceedings of the National Academy of Sciences*, **45**, 30–45.

Taylor, H.A. and Delany, M.E. (2000) Ontogeny of telomerase in chicken: Impact of downregulation on pre- and postnatal telomere length in vivo. *Development, Growth and Differentiation*, **42**, 613–21.

Toivonen, J.M. and Partridge, L. (2009) Endocrine regulation of aging and reproduction in Drosophila. *Molecular and Cellular Endocrinology*, **299**, 39–50.

Valdes, A.M., Andrew, T., Gardner, J.P. *et al.* (2005) Obesity, cigarette smoking, and telomere length in women. *The Lancet*, **366**, 662–4.

Von Zglinicki, T. (2002) Oxidative stress shortens telomeres. *Trends in Biochemical Sciences*, **27**, 339–44.

West, G.B. and Bergman, A. (2009) Toward a systems biology framework for understanding aging and health span. *The Journals of Gerontology. Series A, Biological Sciences and Medical Sciences*, **64**, 205–8.

Westendorp, R.G.J. and Kirkwood, T.B.L. (1998) Human longevity at the cost of reproductive success. *Nature*, **396**, 743–6.

Williams, G.C. (1957) Pleiotropy, natural selection and the evolution of senescence. *Evolution*, **11**, 398–411.

Williams, P.D. and Day, T. (2003) Antagonistic pleiotropy, mortality source interactions, and the evolutionary theory of senescence. *Evolution*, **57**, 1478–88.

Wilson, A.J., Charmantier, A. and Hadfield, J.D. (2008) Evolutionary genetics of ageing in the wild: Empirical patterns and future perspectives. *Functional Ecology*, **22**, 431–42.

Wilson, A.J., Nussey, D.H., Pemberton, J.M. *et al.* (2007) Evidence for a genetic basis of aging in two wild vertebrate populations. *Current Biology*, **17**, 2136–42.

Wingfield, J.C., Maney, D.L., Breuner, C.W. *et al.* (1998) Ecological bases of hormone – behavior interactions: The "Emergency Life History Stage." *Integrative and Comparative Biology*, **38**, 191–206.

Wu, X.F., Amos, C.I., Zhu, Y. *et al.* (2003) Telomere dysfunction: A potential cancer predisposition factor. *Journal of the National Cancer Institute*, **95**, 1211–18.

17

LINKING PHYSIOLOGY, CLIMATE, AND SPECIES DISTRIBUTIONAL RANGES

Francisco Bozinovic[1] and Daniel E. Naya[2]

[1]*Departamento de Ecología and Center of Applied Ecology and Sustainability (CAPES), Facultad de Ciencias Biológicas, Universidad Católica de Chile, Santiago 6513677, Chile*
[2]*Departamento de Ecología y Evolución, Facultad de Ciencias, Universidad de la República, Montevideo, Uruguay*

INTRODUCTION

> "As an ecological problem, the question of why a species has a restricted distribution, given its present physiological and other features, can be answered, even if with difficulty. But range limits pose an evolutionary problem that has not been solved"
>
> – *(Futuyma 1998: 535).*

The extent to which an organism's physiology determines its geographic range remains an unresolved problem in integrative biology. In a recent review, Bozinovic *et al.* (2011) concluded that the cause of geographic range limits is both an ecological problem focused on finding a relationship between physiological traits and geographic distributions, and an evolutionary problem focused on why populations do not evolve to expand their distributions. For example, the analysis of how physiological traits vary along a geographical range is considered a powerful approach for evaluating the selective pressures driving the evolution of physiological diversity (e.g., Spicer & Gaston 1999; Chown *et al.* 2004), but can also be viewed as a constraint determining the limits on geographic ranges (Bozinovic & Rosenmann 1989; McNab 2002). The variation in physiological traits across a gradient can also reflect evolved or plastic responses to the environment, making it difficult to know

Integrative Organismal Biology, First Edition.
Edited by Lynn B. Martin, Cameron K. Ghalambor, and H. Arthur Woods.

what the causal relationship are; do evolved physiological tolerances limit the environments an organism can occupy, or do organisms occupy and acclimatize to the environments in which they are found (see also Gaston 2003)? Finally, physiological tolerances to the abiotic environment may also influence the outcome of biotic interactions such as competition (see Chapter 18, this volume), and in turn shape species distributions and patterns of biodiversity along environmental gradients. Thus, identifying the specific factors that determine the geographic range of populations and species remains a difficult challenge. Nevertheless, we know the limits to distributional ranges must ultimately be the consequence of complex interactions between species-specific physiological, phenological and ecological traits, as well as dispersal ability and ecological interactions between species (Bozinovic *et al.* 2011). We also know that many of the underlying traits may be phylogenetically conserved amongst related species, which could also affect the ability to colonize new environments (Bozinovic *et al.* 2011). The relatively new discipline named macrophysiology seeks to incorporate these complicating factors that span different disciplines (e.g., physiology, ecology, evolution) to explain how levels of environmental variability encountered over large spatial scales shape physiological traits, and how physiology can provide mechanistic answers to macroecological patterns (Gaston *et al.* 2009). This interdisciplinary convergence compares physiological features between individuals possessing different geographical distributions and provides a conceptual framework for understanding how physiographic patterns are related to the mechanistic responses underling these large-scale patterns (Gaston *et al.* 2009). We take such a macrophysiology approach here and review what is known about how climatic variation shapes physiological traits and in turn how this can help explain distributional range limits.

Over the last decade a number of climate-based hypotheses have been put forward to explain variation in the distributional ranges of species and populations. These hypotheses typically use spatial and temporal changes in temperature to predict geographic ranges. For example, the annual range in temperature (the climate variability hypothesis), the lowest minimum or maximum temperature (the climate extreme hypothesis) or the average temperature over some time period (the optimal climate hypothesis) can all be used to build predictive models of geographic range size (Pither 2003). These hypotheses all assume that variation in physiological traits across latitude and altitude reflect on organisms tolerance (Spicer & Gaston 1999) and use the intrinsic physiological properties of species to predict their responses to climatic variables and how it may affect the geographic range of assemblages, species and populations (e.g., Addo-Bediako *et al.* 2000; Somero 2005; Compton *et al.* 2007; Calosi *et al.* 2010). Consequently, in this chapter, we evaluate what is known of how geographic range size and distributional limits can be shaped by physiological capacity, tolerance, and plasticity (see Box 17.1 for definition of terms). We also consider the importance of understanding the proximate mechanisms through which climatic variability interacts with the physiological function to generate macrophysiological patterns. We end the chapter with some thoughts on future directions that may lead to advances in our understanding of the physiological basis for geographic distributions.

LIMITS TO GEOGRAPHIC RANGES: EXAMPLES INVOLVING CAPACITIES AND TOLERANCES

One of the most commonly invoked explanations for a link between climatic variability and geographic ranges is Rapoport's rule (see Stevens 1989), which states that when the latitudinal extent of the geographic ranges of organisms is plotted against the latitude middle

Box 17.1.

Physiological tolerance: Organism exhibit different patterns in dealing with environmental variability. They may avoid them by behavioral traits or they may allow to be exposed to the environment and thus tolerate it. They are two groups of tolerators: (a) some may conform to changes by follows the external environmental conditions, and (b) some may regulate their internal state to maintain independence (within limits) of the environment. This may occur as a result of acclimatization/acclimation (see below) but also by selection of more tolerant or capable genotypes. For instance, animals vary in their tolerance of varying body temperatures and consequently in the temperatures that elicit acclimatory response

Physiological range: A measure of the overall spread of physiological variation in relation to environmental factors.

Physiological plasticity: is the capability of an organism to change its physiology in response to changes in the environmental variables; when these changes are reversible in time the concept of physiological flexibility is commonly used. Plasticity is a pervasive trait that can be extended equally to both physiological limits (tolerance) and performances (capacities)

Acclimation: is the physiological adjustment to cope with variable environmental conditions under laboratory conditions.

Acclimatization: is the process in which the physiology under field conditions is adjusted to cope with change in environmental variables, allowing to maintain performance and homeostasis.

point, a simple positive correlation is found. Stevens (1989)– based on previous work done by Janzen in 1967 (see below) – suggested that the variability among species' physiological tolerance level is the possible explanation for this pattern, what became later known as the climatic variability hypothesis (CVH). Specifically, this hypothesis posits that: (1) as the range of climatic fluctuation experienced by terrestrial animals increases with latitude (or altitude), individuals at higher latitudes (or altitudes) require a broader range of tolerances or greater acclimation abilities to persist at a site (see Box 17.1); (2) the broader range of tolerance or the greater acclimation abilities, in turn, allow species to occupy more habitats and then to become more widely distributed (Janzen 1967; Stevens 1989; Gaston 2003; Chown *et al.* 2004). In other words, for terrestrial animals, there is expected to exist a match between levels of climatic variability experienced (typically increasing with increased latitude or altitude) and the width of their thermal tolerance range and/or their acclimation abilities (Compton *et al.* 2007). Note that the assumption to link the CVH with the Rapoport's rule is that geographic range is maintained only by differences in physiological tolerances or acclimation abilities. However, in most cases it is difficult to determine whether current physiological tolerance ranges and/or acclimation ability are the byproduct of local adaptation after colonization to a new climate or the driving force allowing changes in species distribution (Gaston 2003).

Thermal environmental conditions hold considerable significance for most levels of biological hierarchies, and thus, thermal physiology may be a significant factor underpinning the ecological and evolutionary success (see Chapter 14, this volume). Not surprisingly, thermoregulatory capacities have often been invoked to explain the distribution of endothermic animals. For instance, Root (1988) suggests that the Northern boundary of the winter distribution of several passerines species is related with the annual minimum temperature isotherm, in such a way that the metabolic rate at this distribution boundary is close to 2.5 times the basal metabolic rate (BMR). More recently, Humphries *et al.* (2002) suggest that the distribution of the little brown bat in North America is constrained by thermal effects on hibernation energetics, and provide a mechanistic explanation of how energetics, climate, and distribution are related. In addition, Lovegrove (2000) showed that Nearctic and Palearctic mammals had higher BMR than, their Afrotropical, Australasian, Indomalayan, and Neotropical counterparts. This author explained these patterns with a model describing geographical variance in BMR in terms of the influence of climate variability along geographic ranges and zones (see also Speakman 2000; Lovegrove 2003; Naya *et al.* 2013a). The relationships between geographic distributions and physiological traits have also been complimented by comparative physiological studies of species from different environments. For example, Rezende *et al.* (2004) and Bozinovic and Rosenmann (1989) demonstrated a significant correlation between maximum cold-induced metabolic rate (MMR) of rodent species and the environmental temperature from which they came from. Also, Rodriguez-Serrano and Bozinovic (2009) analyzed the diversity of physiological responses in nonshivering thermogenesis (NST) among rodents from different biogeographic zones, and found a negative correlation between NST and ambient temperature, suggesting that selection may act to shape thermoregulatory performance (Figure 17.1). Collectively, these results put forward that temperature is an important selective pressure not only on basal metabolic rates in endotherms, but also on maximum thermoregulatory capabilities. In line with this, biogeographic analyses of passerine birds demonstrate that suboscines are dominant in South America, whereas oscines are dominant elsewhere. One hypothesis for oscine "superiority" is that oscines hold higher metabolic capacities than suboscines, and this favors oscines in most geographic areas (Swanson & Bozinovic 2011). Specifically, Swanson and Bozinovic (2011) tested the hypothesis by comparing MMR between oscines and suboscines using conventional and phylogenetically corrected approaches. They found that oscines have generally higher cold-induced maximum metabolism than suboscines, which suggests that metabolic capacity is one factor influencing the evolution of broad biogeographical patterns in passerines (Swanson & Bozinovic 2011). Lastly, it is important to note that since higher or lower values of metabolic rates among endothermic species are associated with acclimation to temperature and photoperiod, a component of phenotypic plasticity may partially account for all these

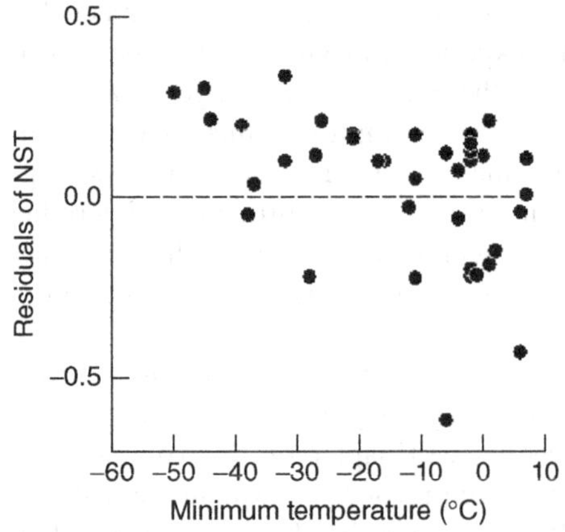

Figure 17.1. Correlation between mean minimum habitat temperature and the residuals of the relationships between body mass and nonshivering thermogenesis or NST (mass-independent NST) among rodents species. Statistical parameters are $r = -0.41$, $P = 0.009$. *Source*: Modified from Rodriguez-Serrano, E. & Bozinovic, F. (2009) Interplay between global patterns of environmental temperature and variation in non-shivering thermogenesis of rodents species across large spatial scales. *Global Change Biology*, **15**, 2116–22.

large-scale biogeographical patterns. However, a gross comparison of current data on intra and inter-specific variability on BMR indicates that the inter-specific component of variation is greater than the intra-specific component (Naya *et al.* 2013b). This suggests that all of the above mentioned patterns are likely due to differences among species in their physiological capacities more than in their physiological plasticity (see also below).

The macrophysiological patterns between climate, temperature, and the thermal physiology of ectotherms provides an interesting and complimentary contrast to those of endotherms. The classic metabolic cold adaptation hypothesis (Scholander *et al.* 1953) posits that at equivalent environmental temperature the metabolic rate of species from cold climates is greater than that of their warm climates relatives. The increase in metabolic rate is considered to be adaptive, that is, there is compensation for the short period of favorable environmental conditions for development, growth and reproduction (Clarke 1993; Addo-Bediako *et al.* 2002). The metabolic cold adaptation hypothesis has also been extended to explain metabolic differences across altitude, latitude and habitat conditions, since it is thought to be a general evolutionary physiological adaptation among ectotherms to compensate for low environmental temperature. Even though several studies support this hypothesis (Chown *et al.* 1997; Addo-Bediako *et al.* 2002), some authors have failed to find an increase in metabolic rate at lower environmental temperature in marine organisms (e.g., Clarke 1991). Nevertheless, it is important to point out that there is also a cost to the higher metabolic rate in terms of greater food requirements. In theory the evolution of higher metabolic rates should reflect both the costs and benefits, and may explain why there are exceptions to this hypothesis (see also Chapter 18, this volume).

Although factors such as acclimatization and acclimation (Box 17.1), activity level, sex, reproductive status, food intake and experimental technique employed may strongly influence measurements of metabolic rate, the general pattern of increasing metabolic rates with lower environmental temperature, seems to provide a physiological explanation in support of Rapoport's rule. That is, as the mean temperature experienced by species decreases with latitude, animals require a higher level of metabolic expenditure and hence a broader tolerance to survive and persist in the habitat, and as consequence species can become more widely distributed. One underlying assumption of this hypothesis is that metabolic plasticity lies within a limited range, which is characteristic of each species or population. Yet, the role of physiological plasticity in these kinds of studies has been largely ignored (see below).

LIMITS TO GEOGRAPHIC RANGES: EXAMPLES INVOLVING TOLERANCES AND PLASTICITY

As mentioned above, the first part of the CVH posits that as the range of climatic variability increases with latitude, individuals from populations inhabiting higher latitudes require a broader range of physiological tolerances and/or greater acclimation abilities (i.e., greater plasticity) to persist at that site. With regard to the range of physiological tolerance, studies of ectothermic animals do show a strong positive relationship between thermal tolerance range and geographic latitude for several taxa (e.g., genus *Drosophila* (Levins 1969; Kimura 1988; Hoffman & Watson 1993; Overgaard *et al.* 2011), diving beetles (Calosi *et al.* 2008, 2010), insects (Addo-Bediako *et al.* 2000; Deutsch *et al.* 2008), porcelain crabs (Stillman & Somero 2000), fish (Brett 1970), amphibians (Brattstrom 1968; Snyder & Weathers 1975), and reptiles (van Berkum 1988; Cruz *et al.* 2005; Deutsch *et al.* 2008)). By contrast, comparative studies of endotherms examining latitudinal variation in physiological ranges are still very scarce. Endothermic species can cope with lower temperature, if they have enough food and also the ability to convert this food into heat, which in turn is related with their digestive and metabolic attributes (Naya *et al.* 2012). Thus, macrophysiological patterns for endothermic animals could be more noticeable for a different set of traits than in ectothermic animals. In line with this, Naya *et al.* (2012) evaluated the correlation between metabolic scope (cold-induced MMR-BMR) in rodent species and different abiotic variables such as latitude, altitude, rainfall, and temperature. The main result of this work is that, as predicted from the CVH, absolute metabolic scope is positively correlated with geographic latitude (Naya *et al.* 2012).

The prediction that acclimation abilities should increase with latitude or climatic variability also has mixed support, however analyzing physiological flexibility for a population or species tends to be a more complex goal than analyzing their physiological tolerance ranges (see Box 17.1). Consequently, studies designed to evaluate latitudinal patterns in physiological flexibility are scarce in comparison to those analyzing physiological ranges of tolerance. In addition, the former kinds of studies usually have a reduced taxonomic (e.g., one genus or one family) and geographic (e.g., one continent) scope in relation to the latter ones. Probably and as a consequence, latitudinal patterns in physiological flexibility do not support any clear pattern of variation. That is, while some studies suggest a positive relationship between physiological flexibility in metabolic and thermal traits and geographic latitude (e.g., amphibians (Feder 1982), lizards (Tsuji 1988)) others do not (e.g., genus *Drosophila* (Mitchell *et al.* 2011; Overgaard *et al.* 2011); diving beetles (Calosi *et al.* 2010)). Naya *et al.* (2008) applied a meta-analytical approach to data of rodents' small intestine length flexibility distributed along the five continents, under the rationale that flexibility in digestive traits should be connected with metabolic flexibility, and then with thermal tolerance (see above). Naya *et al.* (2008) found a positive correlation between small intestine length flexibility and latitude, a result that comprises important evidence on the adaptive value of physiological flexibility in small mammals, which is also consistent with the CVH. Finally, latitudinal patterns in plasticity have been recently analyzed for three eco-physiological and three life-history traits in five populations of a plant species (*Taraxacum officinale*), which inhabit a latitudinal range from 0° to 53°. Again, a strong correlation between latitude and both plasticity for each specific trait and overall mean was observed (Molina-Montenegro & Naya 2012).

The second part of the CVH states that broader ranges of tolerance or greater acclimation abilities allow species to occupy more habitats and then to become more widely distributed. Studies on this topic are still very scarce, but some works suggest that tolerance ranges and physiological flexibility are positively related with species distributional range extension (e.g., genus *Drosophila* (Kellerman *et al.* 2009), diving beetles (Calosi *et al.* 2008), and amphibians (Brattstrom 1970)) or with the number of habits occupied by different species (Naya *et al.* 2008). For instance, recent works done by Calosi *et al.* (2008, 2010) found a positive relationship between diving beetles acclimatory ability to heat tolerance and their geographical range, thus showing that widespread species appear to possess a higher degree of plasticity. By contrast, Stillman and Somero (2000) and Stillman (2003) have shown that in porcelain crab species tolerance levels and their related plastic response are negatively related, thus indicating the existence of an physiological trade-off between these traits, both for tolerance to heat and cold. A potential explanation for these contrasting results could be due differences in the evolutionary trajectory of thermal tolerance in diving beetles and crabs (J.H. Stillman, personal communication, in Bozinovic *et al.* 2011). In any case, it is clear that much more work needs to be done before we understand the relationship between physiological range, physiological flexibility, and species distributional ranges.

PROXIMATE CAUSES BEHIND MACROPHYSIOLOGICAL PATTERNS

Interestingly, most of the studies analyzing latitudinal patterns in physiological ranges or physiological flexibility document a greater association with geographic latitude, rather than with climatic variables (Naya *et al.* 2012). There are three reasons commonly invoked to explain this result. First, latitude is probably a better predictor of long-term regimens of climatic variables than current climate values provided by weather stations (Speakman 2000; Rezende *et al.* 2004). Second, latitude is correlated with several other environmental factors that are often not assessed, such as climatic (e.g., wind speed), ecological (e.g., day length, environmental productivity), and historical factors (e.g., geographical boundaries) that could affect physiological traits (Rezende *et al.* 2004; Naya *et al.* 2008). Third, species included in physiological studies are not present only at the collection site(s), but occur over large areas. Thus, we should not expect adaptation to environmental conditions to be more fine-tuned than adaptation to the conditions taking place over these large areas. Consequently, given the smooth variation of climatic variables in the space, latitude may represent a weighted variable of climatic conditions acting over spatial scales that are much closer to the scale at which adaptation is expected to occur than the conditions at the collection sites (Naya *et al* 2012; Figure 17.2). These three, not mutually exclusive, arguments may be the reason why many times geographical latitude is a better predictor of physiological ranges and flexibility than local climatic conditions, despite the fact that climatic variables (and not latitude) should be the ultimate factors underlying physiological variation.

On the other hand, in some cases physiological patterns are more clearly related with mean values of climatic variables than with variability indexes (e.g., Molina-Montenegro & Naya 2012; Naya *et al.* 2012). This result appears to be contradictory with the CVH since temperature variability and not temperature mean should be the putative cause of physiological ranges and flexibility. We believe that these differences between theoretical predictions and observed data could be due to several possible explanations. First, the indices of climatic variability that are usually used may be too coarse to reflect the proper variability that organisms have to cope with. For instance, the fact that diurnal variation in temperature at high altitudes in the tropics is close to seasonal variation in the temperate north (Ghalambor *et al.* 2006) is probably not captured by the climatic variability indexes commonly used, such as annual temperature range or temperature seasonality. Second, behavior may buffer variation in environmental temperature, allowing animals to avoid exposure to extreme temperatures (Huey *et al.* 2003). For instance, if behavior isolates an organism from one extreme temperature (e.g., maximum temperatures) more than the other one (e.g., minimum temperatures), a better correlation with the variable from which the organism is less isolated than with the temperature range is expected (Figure 17.3). Interestingly, the hypothetical example depicted in Figure 17.3 also illustrates the fact that the final impact of environmental variation on organisms is usually mediated by their behavior, and hence, illustrates how different phenotypic traits are integrated (see below). In any case, current data indicate that it remains unclear to what extent climatic variability or other factors related with latitude, act on the physiological phenotype and account for any observed geographical gradient in physiological range flexibility (Maldonado *et al.* 2011; Naya *et al.* 2011).

In recent years, some interesting attempts have been made to explore the proximate causes behind geographical patterns in physiological flexibility. For example, Maldonado *et al.* (2011) analyzed the physiological flexibility within the CVH framework by using a within species comparative approach and tested if there was a positive relationship between digestive flexibility, latitude, climate and diet variability in populations of the rufous-collared sparrows (*Zonotrichia capensis*) inhabiting desert, Mediterranean, and

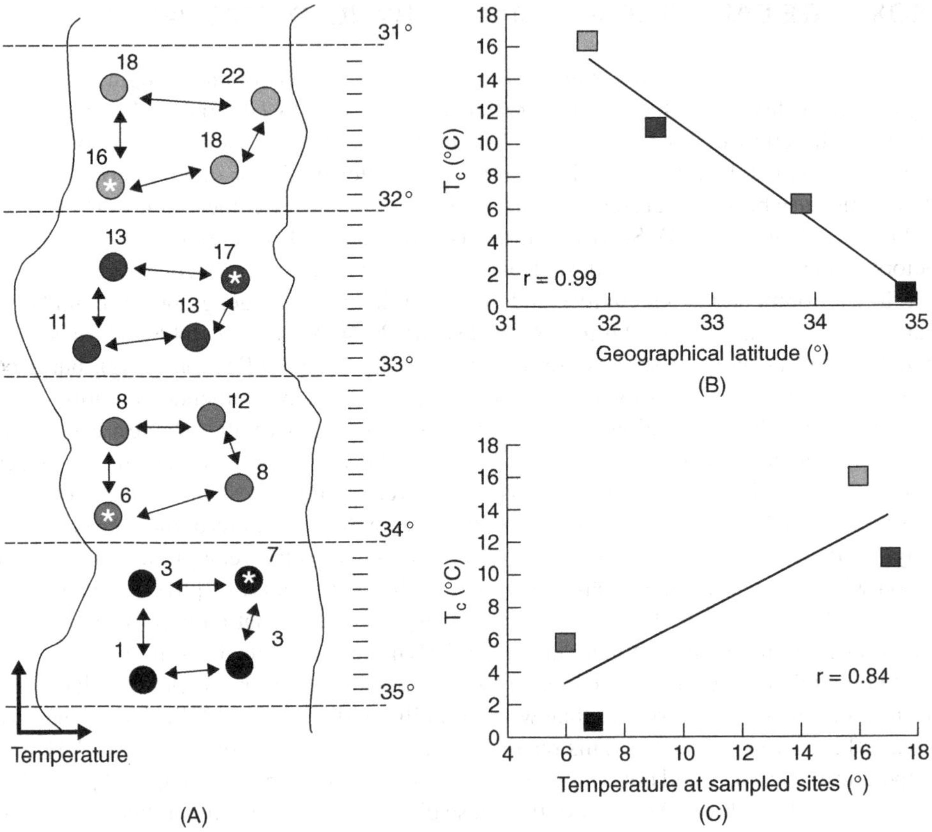

Figure 17.2. A hypothetical example showing the "spatial scale problem" (see text). Four different species (shown in different colors) each inhabit four different localities (circles) that are connected by migration (arrows); the population of each species that is sampled is marked with an asterisk, and the data above each locality represent a climatic variable, e.g., minimum temperature (A). If the physiological variable under study (e.g., minimum critical temperature, T_c) is fitted to the entire distributional range – for example, it is equal to the minimum temperature recorded over the entire range – it will be better correlated with latitude (B) than with minimum temperatures at the sampled localities (C). (See plate section for color version.)

cold-temperate habitats. In agreement with the CVH, they found a positive relationship between the magnitude of digestive flexibility and environmental variability but not latitude. The greatest digestive flexibility was observed in birds at middle latitudes, which experience the most environmental variability (Mediterranean climate), whereas individuals from the most stable climates (desert and cold-temperate) exhibited little or no digestive flexibility in response to experimental diets. Maldonado *et al.* (2012) also investigated the relationship between the magnitude of BMR flexibility in response to inter-annual rainfall variability in the same three populations. These authors addressed the question of whether birds from a desert environment are more flexible in BMR than those from nondesert habitats, and found a positive trend between BMR flexibility and rainfall variability. At the same time and following the same rationale, Naya *et al.* (2011) tested for physiological flexibility in digestive and metabolic traits for three populations

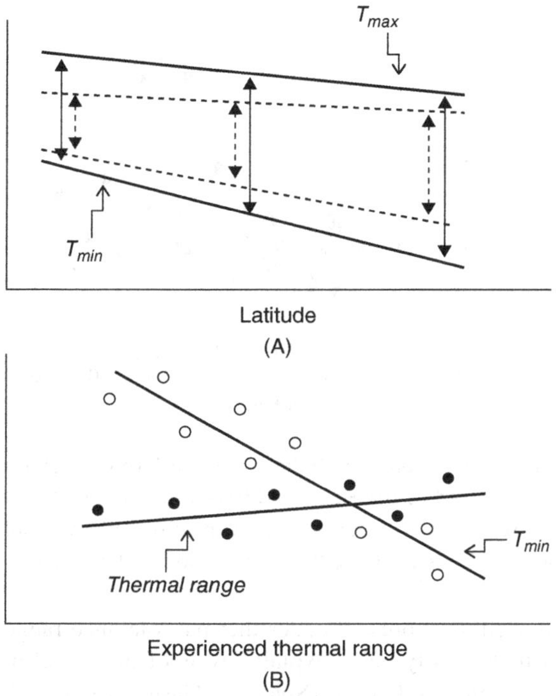

Figure 17.3. An hypothetical example showing the latitudinal change in minimum and maximum temperatures obtained from weather stations (continuous lines) and minimum and maximum temperatures experienced by an organism (dotted lines) (A). Given that, in this example, a behavioral mechanism allows the organism to isolate from maximum temperature more than from minimum temperature, the real thermal range experienced by the organism is better correlated with minimum temperature (open circles) than with the temperature range obtained from weather stations (closed circles) (B).

of the snail *C. aspersum* for which climatic variability (both in temperature and rainfall) and latitude were inversely correlated. As expected, they found that digestive flexibility was positively related with climatic variability and negatively correlated with latitude. An additional perspective on the relationship between latitude and climatic variability was taken by Naya and Bozinovic (2012) within aquatic systems. They rationed that in aquatic systems, physiological ranges should be related with species distributional range but not with latitude. This is because thermal variability in water is fairly constant along latitudinal gradients, and thus, the only way to increase thermal variability is through increasing distributional range. To test this idea with real data they evaluated metabolic scope (i.e., routine metabolic rate minus standard metabolic rate) in 38 fish species, finding that, as expected, metabolic scope was positively correlated with species' distributional range (and also marginally correlated with the thermal range within species' distributional area), but not with species distribution middle point (Figure 17.4). Taken together all these results strongly suggest that, as predicted by the CVH, climatic variability and not latitude is the ultimate factor affecting both physiological flexibility and physiological ranges.

To summarize, numerous organisms are able to adjust their physiological abilities following daily and seasonal fluctuations of environmental variables along geographic gradients. Acclimatization can thus allow individual organisms to exploit a broad thermal

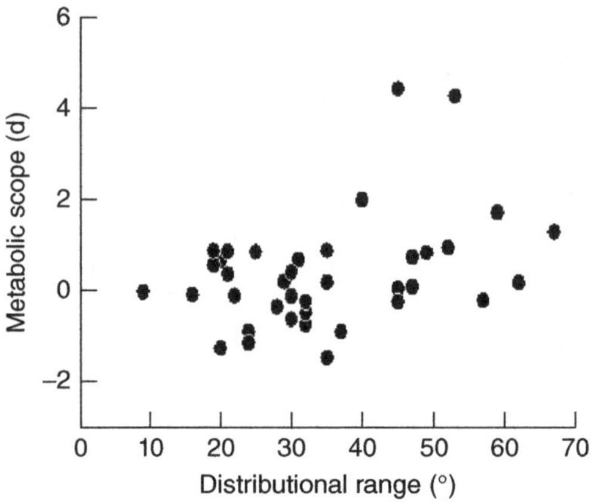

Figure 17.4. Correlation between metabolic scope (difference between routine and standard metabolic rates, expressed as the Hedge's difference) and distributional range (difference between northern and southern distribution limit, expressed in latitudinal degrees) among fish species. Statistical parameters are $r = 0.38$, $P < 0.001$. Modified from Naya and Bozinovic (2012).

capacity window, as well as express broader thermal tolerance ranges. This represents a form of plasticity, which clearly helps explain levels of intra- and inter-individual physiological variation over time and space (Spicer & Gaston 1999). Where present it could clearly modulate species and populations resilience to climatic variability (Stillman 2003). As suggested by Bozinovic *et al* (2011) plasticity is a pervasive phenomenon, which can be equally extended to physiological limits (tolerance), performances (capacities) and affecting or perhaps determining geographic ranges and its limits.

CONCLUSIONS AND FUTURE DIRECTIONS

In conclusion what seems to be clear is that we need to advance our current understanding of the relationship between species and populations physiological responses and how they are linked to their distribution. Ultimately, this will lead to a better understanding of the relationship among tolerance, capacity and plasticity, and guide the integration of these traits in the investigation of "what determines a species range size?" The dispute on how to integrate physiological traits, how to better parameterize each and every trait considered fundamental for the modeling of species range limits is a vigorous one (see Gaston *et al.* 2009; Bozinovic *et al.* 2011). This complex problem requires an understanding of the regional impact of climate, the physiological mechanisms (tolerance, capacity and plasticity) by which animals cope to climate during their life cycles, and the nature of the differential effects of climate on life history of species and populations along geographic gradients.

Perhaps more importantly is the fact that Earth's climate is changing rapidly. The data of the ICPP 2007-2008 signal an evident trend toward increasing global temperatures. The causes of these changes are complex and associated with a network of events in which human action appears to have a determinant role. Climate change reflects changes at the global, regional, and local levels. The observed responses to recent climate change are many and important but no doubts include shifts in geographic ranges (Parmesan 2006). Although these trends are expected to cause pervasive effects on the biota, the nature, span, and final consequences of climate change on biogeography must differ among taxa. In fact, the diversity in physiological traits and ecological associations are evident in the fauna and flora, even among individuals. For instance, the response of endotherms distributional range to climate change likely would be very different from that found in ectotherms, as it would be between mobile and sedentary species. Thus, understanding the nature of differential effects of climate change on animal and plant species geographic range is one of the many urgent interdisciplinary challenges faced by contemporary science (McNab 2012; Kirkpatrick & Barton 1997). Indeed, recently, Araújo *et al.* (2013) asked if species physiologically adapt to climate warming. These authors analyzed thermal tolerances of a large number of terrestrial ectotherms, endotherms, and plant species across the world. Araújo *et al.* (2013) observed that that tolerance to heat is largely conserved across lineages, but tolerance to cold varies between and within species. This pattern challenges the classic view that physiological tolerances of species change continuously across climatic gradients, and portrays an alternative view in which the thermal component of climatic niches would overlap across species more than expected. One consequence of conservatism of upper tolerances is that estimated niches for cold-adapted species will be likely to underestimate their upper thermal limits, so potentially inflating assessments of risk from climate change. On the contrary, species whose climatic preferences are close to their upper thermal limits will unlikely evolve physiological tolerances to increased heat, and could be predictably more affected by global warming.

To end this chapter, we wish to point out three very simple questions that, in our opinion, should be taken into account more seriously in future studies aimed to disentangle the complex relationship between integrative biology and species distribution. First, what is the mechanistic link between the focal trait and organism fitness? This question is critical in establishing a link between individual physiological responses and population viability. Second, to what degree is a trait under selection integrated with other organismal traits (e.g., behavior), in such a way that alternative solutions could emerge from this integration? This theme of understanding trait integration is shared throughout many chapters in this book (e.g., Chapters 1 and 19), and will require a more multivariate approach that simultaneously examines how multiple traits change in response to climatic variability. And finally, to what extent does the environmental variables being tested reflect the real environments in which organisms are living? This last point may seem obvious, but if we ultimately want to predict geographic range limits and responses to changes in climate, we will need much more environmental data collected at the scale organisms live. Our hope is that a focus on these questions will allow for progress to be made on the links between climate, physiology, and patterns of distribution.

ACKNOWLEDGMENTS

The editors of this book and Piero Calosi made relevant contributions and comments. Funded by FONDECYT 1130015 to FB and Programa Iberoamericano de Ciencia y Tecnología para el Desarrollo (CYTED 410RT0406) to FB and DEN. The authors thank Fondo Basal FONDECYT FB-002 for support.

REFERENCES

Addo-Bediako, A., Chown, S.L. and Gaston, K.J. (2000) Thermal tolerance, climatic variability and latitude. *Proceedings of the Royal Society of London B*, **267**, 739–45.

Addo-Bediako, A., Chown, S.L. and Gaston, K.J. (2002) Metabolic cold adaptation in insects: a large-scale perspective. *Functional Ecology*, **16**, 332–8.

Araújo, M.B., Ferri-Yañez, F., Bozinovic, F. *et al.* (2013) Heat freezes niche evolution. *Ecology Letters*, **16**, 1206–19.

Block, B.A. and Finnerty, J.R. (1994) Endothermy in fishes: A phylogenetic analysis of constraints, predispositions, and selection pressures. *Environmental Biology of Fish*, **40**, 283–302.

Bozinovic, F., Calosi, P. and Spicer, J.I. (2011) Physiological correlates of geographic range in animals. *Annual Review of Ecology, Evolution and Systematics*, **42**, 155–79.

Bozinovic, F. and Rosenmann, M. (1989) Maximum metabolic rate of rodents: physiological and ecological consequences on distributional limits. *Functional Ecology*, **3**, 173–81.

Brattstrom, B.H. (1968) Thermal acclimation in anuran amphibians as a function of latitude and altitude. *Comparative Biochemistry and Physiology*, **24**, 93–111.

Brattstrom, B.H. (1970) Thermal acclimation in Australian amphibians. *Comparative Biochemistry and Physiology*, **35**, 69–103.

Brett, J.R. (1970) Temperature. Animals. Fishes, in *Marine Ecology* (ed O. Kinne), John Wiley & Sons, Inc., New York, pp. 515–60.

Calosi, P., Bilton, D.T. and Spicer, J.I. (2008) Thermal tolerance, acclimatory capacity and vulnerability to global climate change. *Biology Letters*, **4**, 99–102.

Calosi, P., Bilton, D.T., Spicer, J.I. *et al.* (2010) What determines a species' geographical range? Thermal biology and latitudinal range size relationships in European diving beetles (Coleoptera: Dytiscidae). *Journal of Animal Ecology*, **79**, 194–204.

Chown, S.L., van der Merwe, M. and Smith, V.R. (1997) The influence of habitat and altitude on oxygen uptake in sub-Antarctic weevils. *Physiological Zoology*, **70**, 116–24.

Chown, S.L., Gaston, K.J. and Robinson, D. (2004) Macrophysiology: large-scale patterns in physiological traits and their ecological implications. *Functional Ecology*, **18**, 159–67.

Clarke, A. (1991) What is cold adaptation and how should we measure it? *American Zoologist*, **31**, 81–92.

Clarke, A. (1993) Seasonal acclimatization and latitudinal compensation in metabolism: do they exist? *Functional Ecology*, **7**, 139–49.

Compton, T.J., Rijkenberg, M.J.A., Drent, J.S. and Piersma, T. (2007) Thermal tolerance ranges and climate variability: a comparison between bivalves from differing climates. *Journal of Experimental Marine Biology and Ecology*, **352**, 200–11.

Cruz, F.B., Fitzgerald, L.A., Espinoza, R.E. and Schulte, J.A. II (2005) The importance of phylogenetic scale in tests of Bergmann's and Rapoport's rules: lessons from a clade of South American lizards. *Journal of Evolutionary Biology*, **18**, 1559–74.

Deutsch, C.A., Tewksbury, J.J., Huey, R.B. *et al.* (2008) Impacts of climate warming on terrestrial ectotherms across latitude. *Proceedings of the National Academy of Science USA*, **105**, 6668–72.

Feder, M.E. (1982) Environmental variability and thermal-acclimation of metabolism in tropical anurans. *Journal of Thermal Biology*, **7**, 23–8.

Futuyma, D.J. (1998) *Evolutionary Biology*, 3rd edn, Sinauer Associates, Inc, p. 763.

Gaston, K.J. (2003) *The structure and dynamics of geographic ranges*, Oxford University Press, Oxford, p. 276.

Gaston, K.J., Chown, S.L., Calosi, P. *et al.* (2009) Macrophysiology: a conceptual reunification. *American Naturalist*, **174**, 595–612.

Ghalambor, C.K., Huey, R.B., Martin, P.R. *et al.* (2006) Are mountain passes higher in the tropics? Janzen's hypothesis revisited. *Integrative and Comparative Biology*, **46**, 5–17.

Hoffman, A.A. and Watson, M. (1993) Geographical variation in the acclimation response of *Drosophila* to temperature extremes. *American Naturalist*, **142**, S93–S113.

Huey, R.B., Hertz, P.E. and Sinervo, B. (2003) Behavioral drive versus behavioral inertia in evolution: a null model approach. *American Naturalist*, **161**, 357–66.

Humphries, M.M., Thomas, D.W. and Speakman, J.R. (2002) Climate-mediated energetic constraints on the distribution of hibernating mammals. *Nature*, **418**, 313–16.

IPCC (2007) *Climate Change 2007: The Physical Science Basis. Summary for Policymakers.* http://www.ipcc.ch/SPM2feb07.pdf

Jacobsen, D. and Brodersen, K.P. (2008) Are altitudinal limits of equatorial stream insects reflected in their respiratory performance? *Freshwater Biology*, **53**, 2295–2308.

Janzen, D.H. (1967) Why mountain passes are higher in the tropics. *American Naturalist*, **101**, 233–49.

Kellerman, V., van Heerwaarden, B., Sgro, C.M. and Hoffmann, A.A. (2009) Fundamental evolutionary limits in ecological traits drive *Drosophila* species distributions. *Science*, **325**, 1244–6.

Kimura, M.T. (1988) Adaptations to temperate climates and evolution of overwintering strategies in the *Drosophila melanogaster* species group. *Evolution*, **42**, 1288–97.

Kirkpatrick, M. and Barton, N.H. (1997) Evolution of a species' range. *American Naturalist*, **46**, 954–71.

Levins, R. (1969) Thermal acclimation and heat resistance in *Drosophila* species. *American Naturalist*, **103**, 483–99.

Lovegrove, B.G. (2000) The zoogeography of mammalian basal metabolic rate. *American Naturalist*, **156**, 201–19.

Lovegrove, B.G. (2003) The influence of climate on the basal metabolic rate of small mammals: a slow-fast metabolic continuum. *Journal of Comparative Physiology*, **73B**, 87–112.

Maldonado, K., Bozinovic, F., Rojas, J.M. and Sabat, P. (2011) Within-species digestive tract flexibility in rufous-collared sparrow and the climatic variability hypothesis. *Physiological and Biochemical Zoology*, **84**, 377–84.

Maldonado, K., Bozinovic, F., Cavieres, G. *et al.* (2012) Phenotypic flexibility in basal metabolic rate is associated with rainfall variability among populations of rufous-collared sparrow. *Zoology*, **115**, 128–33.

McNab, B.K. (2002) *The Physiological Ecology of Vertebrates. A View from Energetics*, Comstock, Cornell.

McNab, B.K. (2012) *Extreme Measure: The Ecological Energetics of Birds and Mammals*, The University of Chicago Press, Chicago and London.

Mitchell, K.A., Sgro, C.M. and Hoffmann, A.A. (2011) Phenotypic plasticity in upper thermal limits is weakly related to *Drosophila* species distributions. *Functional Ecology*, **25**, 661–70.

Molina-Montenegro, M.A. and Naya, D.E. (2012) Latitudinal patterns in phenotypic plasticity and fitness-related traits: assessing the climatic variability hypothesis (CVH) with an invasive plant species. *PloS One*, **7**, e47620.

Naya, D.E. and Bozinovic, F. (2012) Metabolic scope of fish species increase with distributional range. *Evolutionary Ecology Research*, **14**, 769–77.

Naya, D.E., Bozinovic, F. and Karasov, W.H. (2008) Latitudinal trends in digestive flexibility: testing the climatic variability hypothesis with data on the intestinal length in rodents. *American Naturalist*, **172**, 122–34.

Naya, D.E., Catalan, T.P., Artacho, P. *et al.* (2011) Exploring the functional association between physiological flexibility, climatic variability and geographical latitude: lesson from land snails. *Evolutionary Ecology Research*, **13**, 647–59.

Naya, D.E., Spangenberg, L., Naya, H. and Bozinovic, F. (2012) Latitudinal patterns in rodent metabolic flexibility. *American Naturalist*, **179**, 172–9.

Naya, D.E., Spangenberg, L., Naya, H. and Bozinovic, F. (2013a) How does evolutionary variation in basal metabolic rates arise? A statistical assessment and a mechanistic model. *Evolution*, **67**, 1463–76.

Naya, D.E., Spangenberg, L., Naya, H., & Bozinovic, F. (2013b) Thermal conductance and basal metabolic rate are part of a coordinated system for heat transfer regulation. *Proceedings of the Royal Society of London B,* **280**, 20131629, 10.1098/rspb.2013.169.

Overgaard, J., Kristensen, T.N., Mitchell, K.A. and Hoffmann, A.A. (2011) Thermal tolerance in widespread and tropical *Drosophila* species: does phenotypic plasticity increase with latitude? *American Naturalist*, **178**, S80–S96.

Parmesan, C. (2006) Ecological and evolutionary responses to recent climate change. *Annual Review of Ecology Evolution and Systematic*, **37**, 637–69.

Pither, J. (2003) Climate tolerance and interspecific variation in geographic range size. *Proceedings of the Royal Society of London B*, **270**, 475–81.

Rezende, E.L., Bozinovic, F. and Garland, T. Jr. (2004) Climatic adaptation and the evolution of basal and maximum rates of metabolism in rodents. *Evolution*, **58**, 1361–74.

Rodriguez-Serrano, E. and Bozinovic, F. (2009) Interplay between global patterns of environmental temperature and variation in non-shivering thermogenesis of rodents species across large spatial scales. *Global Change Biology*, **15**, 2116–22.

Root, T. (1988) Energy constraints on avian distributions and abundances. *Ecology*, **69**, 330–9.

Scholander, P.F., Flagg, W., Walters, V. and Irving, L. (1953) Climatic adaptation in Arctic and tropical poikilotherms. *Physiological Zoology*, **26**, 67–92.

Snyder, G.K. and Weathers, W.W. (1975) Temperature adaptations in amphibians. *American Naturalist*, **109**, 93–101.

Somero, G. (2005) Linking biogeography to physiology: evolutionary and acclimatory adjustments of thermal limits. *Frontiers in Zoology*, **2**, 1–9.

Speakman, J.R. (2000) The cost of living: field metabolic rates of small mammals, in *Advances in Ecological Research* (eds A.H. Fisher and D.G. Raffaelli), Academic Press, San Diego, California, pp. 178–294.

Spicer, J.I. and Gaston, K.J. (1999) *Physiological Diversity and Its Ecological Implications*, Blackwell Scientific Publications, Oxford.

Stevens, G.C. (1989) The latitudinal gradient in geographical range: how so many species coexist in the tropics. *The American Naturalist*, **133**, 240–56.

Stillman, J.H. (2003) Causes and consequences of thermal tolerance limits in rocky intertidal porcelain crabs, genus *Petrolisthes. Integrative and Comparative Biology*, **42**, 790–6.

Stillman, J.H. and Somero, G.N. (2000) A comparative analysis of the upper thermal tolerance limits of Eastern Pacific porcelain crabs, genus *Petrolisthes*: Influences of latitude, vertical zonation, acclimation, and phylogeny. *Physiological and Biochemical Zoology*, **73**, 200–8.

Swanson, D.L. and Bozinovic, F. (2011) Metabolic capacity and the evolution of biogeographic patterns in oscine and suboscine passerine birds. *Physiological and Biochemical Zoology*, **84**, 185–94.

Tsuji, J.S. (1988) Thermal-acclimation of metabolism in sceloporus lizards from different latitudes. *Physiological Zoology*, **61**, 241–53.

van Berkum, F.H. (1988) Latitudinal patterns of the thermal sensitivity of sprint speed in lizards. *American Naturalist*, **132**, 327–43.

18

TRADE-OFFS AND BIOLOGICAL DIVERSITY: INTEGRATIVE ANSWERS TO ECOLOGICAL QUESTIONS

Paul R. Martin

Department of Biology, Queen's University, Kingston, ON, Canada

INTRODUCTION

What determines the abundance and distribution of species? This question is paramount to ecology because it encompasses the interactions of individuals, populations, and species with each other, and with their environments – interactions that ultimately create the patterns of biological diversity that we see in nature. Ecological approaches and frameworks have successfully addressed this question across diverse species and contexts, and yet the broader rules that underlie these patterns across environments and taxonomic groups remain elusive (Lawton 1999; Simberloff 2004). In this chapter, I will argue that our difficulties in finding broad answers to this question are, in part, because the answers are not strictly ecological, but broadly biological. Understanding patterns of diversity requires the integration of diverse traits, processes, and selective pressures that span fields, from physiology to biophysics, genetics to biogeography. Importantly, the answers to why species vary in their abundance and distribution benefit from an evolutionary framework that tests ideas from the perspective of the fitness costs and benefits to individuals, and their consequences for populations and species.

An integrative approach within an evolutionary framework helps us to understand species distributions and abundances because of its role in elucidating fundamental trade-offs – adaptations that confer a fitness advantage for performing one function, and simultaneously confer a fitness disadvantage for performing another function (Bohannan *et al.* 2002; Figure 18.1). These trade-offs can be understood from the level of genetics (e.g., antagonistic pleiotropy, linkage; Falconer & McKay 1996) to physiology and

Integrative Organismal Biology, First Edition.
Edited by Lynn B. Martin, Cameron K. Ghalambor, and H. Arthur Woods.

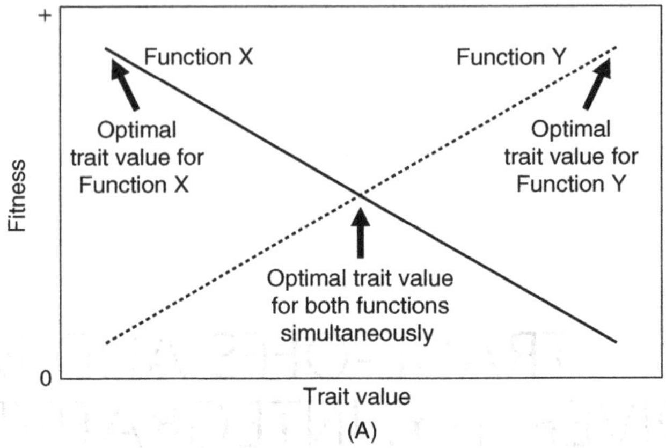

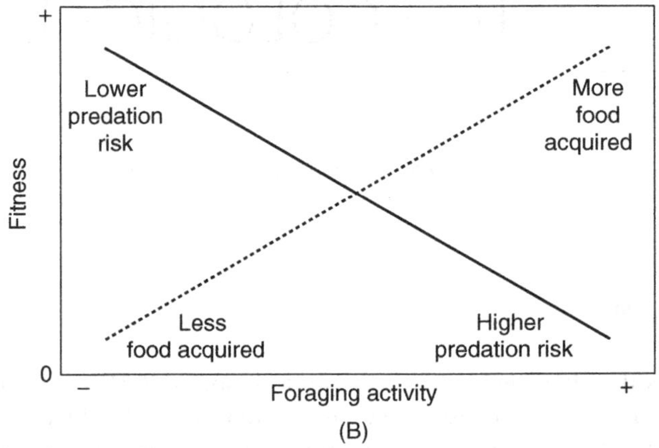

Figure 18.1. (A) A trade-off, where a trait that confers a fitness advantage for performing one function (X), simultaneously confers a fitness disadvantage for performing another function (Y). In the case of allocative trade-offs, the focal trait controls the allocation of limited resources between two (or more) other phenotypic traits. If the selection pressures that create the trade-off vary in space or time, different species may coexist with adaptations that optimize trait values for functions X or Y. For example (B), increased foraging activity increases the rate of food acquisition, but comes at a cost of increased predation. Variation in the risk of predation or availability of resources can lead to coexistence of species whose foraging activity is adapted to either reduce predation risk at the expense of food acquisition, or increase food acquisition at the expense of susceptibility to predation.

development (e.g., trade-offs in allocation of energy, resources; Sibly & Calow 1986) and developed phenotypes (Agrawal *et al.* 2010), and have fundamentally shaped the way biologists think about the evolution of life histories (Stearns 1992), physiologies (Sibly & Calow 1986), behaviors (Krebs & Davies 1997), and adaptation in general (Kawecki & Ebert 2004). Trade-offs are central to the study of diversity because without them, we should expect one species to dominate in all environments (Dobzhansky 1964; Tilman 1982). By forcing the choice between evolutionary paths that cannot be simultaneously taken, trade-offs allow different species to partition energy and other resources, and thus coexist, creating (e.g., Duffy & Forde 2009) and maintaining (e.g., Kneitel & Chase 2004) diversity in nature (Figure 18.1).

The framework of trade-offs for understanding species distributions, interactions, and diversity has a long and productive history in both plant and animal ecology (e.g., Grime 1977; Tilman 1982; Chapin 1980; Holt 1995; McPeek 1996; Chesson 2000). This framework, however, differs from historically common and pervasive approaches to addressing ecological questions that highlight, among other things, the importance of different ecological processes such as competition, predation or parasitism, realized versus fundamental niches, biotic or abiotic factors, and top-down versus bottom-up constraints. By focusing on trade-offs, we no longer emphasize the relative importance of these ecological processes or how species are different, but instead ask why the traits of one species cannot do it all, or in other words, why a trait that is adaptive in one context cannot also excel in another context simultaneously (Box 18.1). By focusing on fundamental trade-offs confronting organisms, we are also forced to understand the mechanisms creating the trade-offs (e.g, Chapin *et al.* 1993; Angert *et al.* 2009), thus promoting the integration of diverse fields often lost in our tendency to specialize in science.

Box 18.1. Trade-offs and the Niche

Niches define different ecological strategies of organisms, and thus have direct relevance to the framework of trade-offs. For example, a trade-off in competitive ability versus ability to detoxify heavy metals may result in two different ecological strategies or niches for plants – one adapted to heavy metal soils, and another competitively dominant on soils lacking heavy metals (Macnair *et al.* 2000). A descriptive characterization of the niche, however, tells only part of the story, and misses the most important part: distinct niches occur because of underlying trade-offs confronting organisms, and these trade-offs can be understood from the level of genetics, development, and physiology through to the consequences for species interactions and distributions. Niches also create a narrow view of interactions among species and patterns of coexistence. For example, a recent study suggests that within-species trade-offs involving sexual selection can lead to the coexistence of species without ecological (niche) differentiation (M'Gonigle *et al.* 2012). In this case, the key to coexistence is a trade-off between the costs of searching for suitable mates (that increase with distance) and the costs of competition (that decrease with distance) within a heterogeneous environment. Such patterns of coexistence without ecological divergence challenge a central ecological premise that species can only coexist if they have different ecological niches.

Niches have an inherent appeal because they describe what we can all see – species with different ecological strategies living together, or apart. Understanding these patterns, however, will require more than a study of ecological strategies. It will require an understanding of why one species cannot do it all, and thus the trade-offs that force distinct ecological strategies and create and maintain biodiversity in nature.

Below, I provide select examples to illustrate the importance of key trade-offs and the integration of diverse fields that promote a more mechanistic understanding of the factors underlying the distributions of species and interactions between them. I focus on ecologically similar, closely related populations or species that share most traits in common through recent shared ancestry. Similar trade-offs extend to distantly related species, albeit in more complicated ways as species accumulate divergent traits over evolutionary time. While I focus on trade-offs among populations or species, trade-offs acting within populations can have important consequences for species interactions, distributions, and patterns of coexistence (e.g., Tessier *et al.* 2000; Agrawal 2001; M'Gonigle *et al.* 2012; see also Clark 2010), and thus understanding within-species trade-offs and variation is also important for our understanding of patterns of diversity (Box 18.2). I begin with case studies and finish with a discussion of why an integrative approach that focuses on trade-offs will advance our understanding of community ecology and biodiversity.

Box 18.2. Trade-offs, Phenotypic Plasticity, and Specialization among Individuals

The trade-offs that confront populations and species can act similarly on individuals within populations, with consequences for individual phenotypes. For example, phenotypic plasticity allows individual organisms to match their phenotypes to environmental conditions, such that one individual can settle a trade-off in different ways that corresponds to environmental variation (e.g., McCollum & Van Buskirk 1996). This environmental variation may be current, or predicted to occur in the future (e.g., predicted adaptive responses; Gluckman *et al.* 2005). Phenotypic plasticity frequently mirrors heritable variation among populations and species (e.g., Ruell *et al.* 2013), and can have similar consequences for the abundances and distributions of species because it can determine the relative fitness of individuals in different environments. For example, if individuals in a population cannot adjust their phenotypes to match environmental challenges, then these individuals may be more likely to experience lower fitness, reduced abundance, and a higher risk of population extinction. While phenotypic plasticity can clearly benefit individuals, evidence suggests that plasticity itself may come at a cost (DeWitt *et al.* 1998), presenting its own trade-off that varies with environmental conditions (e.g., Relyea 2002).

While plasticity can allow an individual to settle trade-offs in different ways, variation among individuals within a population can allow different individuals to settle trade-offs in different ways (i.e., individual specialization), with similarly important consequences for community ecology and diversity (Bolnick *et al.* 2003, 2011). Sex-specific and age-related variation in ecological and other traits are common; however, variation among individuals of the same sex and age class is also widespread within populations (Bolnick *et al.* 2003). While gene flow may restrict the evolution of distinct traits within a population, competitive interactions among conspecific individuals are usually more intense (Clark 2010), and may cause the divergence of traits within a population (specialization), even in the face of gene flow. Altogether, trade-offs confront organisms within populations, and commonly favor phenotypic plasticity or individual specialization that parallels divergence among populations and species. These adaptive responses within populations may play an important and underappreciated role in structuring diversity (Agrawal 2001; Bolnick *et al.* 2003; 2011; see also Clark 2010).

Resource Availability and Environmental Gradients: The Importance of Body Size

All organisms must obtain resources for growth, survival and reproduction, and these resources vary in their availability. Traits that maximize fitness when resources are abundant are often different from those that maximize fitness when resources are scarce (e.g., Chapin 1980), and these traits frequently involve variation in body size. Large body size provides advantages in competitive interactions among diverse groups, including both animals and plants, allowing larger organisms to preferentially acquire limiting resources when interacting with smaller organisms (e.g., Morse 1974; Grime 1977; Schoener 1983; Keddy 2001). Small-bodied organisms, however, require fewer resources for development, growth, survival, and reproduction, providing advantages when resources are too scarce to sustain large body sizes (Peters 1983; Schmidt-Nielsen 1984; Bonner 2006). These different patterns of resource requirement and use create an important body size trade-off that varies with resource abundance.

Body size trade-offs extend to many other selective pressures beyond resource availability (Peters 1983; Schmidt-Nielsen 1984; Bonner 2006). These selective pressures can be traced back to fundamental biophysical constraints on body size that ultimately help to shape most aspects of an organism's biology (Peters 1983; Schmidt-Nielsen 1984; Bonner 2006). These diverse body size trade-offs create variation in the selective advantage of different body sizes along equally diverse gradients of climatic, chemical, physical, and biotic variation. The important and diverse consequences of body size make size one of the most influential traits for determining the distributions, interactions, and patterns of coexistence of organisms (West *et al.* 1997; Bonner 2006).

Body Size Trade-offs and the Segregation of Rodents along a Resource Gradient Desert gerbil species (*Gerbillus*) use similar resources but occupy distinct microhabitats and forage at different times of the night in the Western Negev Desert, Israel. Food abundance (seeds) differs among microhabitats and declines over the course of the night as rodents consume seeds made available by strong afternoon winds (Ziv *et al.* 1993). Field experiments reveal a trade-off between two species involving body size. The larger *G. pyramidum* excludes the smaller *G. allenbyi* from preferred seed-rich habitat (semistabilized sand dunes), particularly during seed-rich foraging times (early night). *G. allenbyi*, however, performs well in seed-poor conditions (stabilized sand dunes, foraging later at night) where it typically occurs, because it can harvest seeds profitably at low resource densities. The larger *pyramidum* is a less efficient forager at low seed densities, and prefers high resource habitat and foraging times (Ziv *et al.* 1993). Field and laboratory experiments suggest a trade-off involving size, with large size conferring an advantage to *pyramidum* in aggressive interactions, and small size allowing *allenbyi* to survive and reproduce in low food environments because of reduced energetic demands and thermoregulatory costs (Kotler & Brown 1990; Ziv *et al.* 1993). The two species also differ in other aspects of foraging efficiency (Kotler & Brown 1990), but whether these traits are functionally linked to body size is not clear.

Similar dominance hierarchies among closely related species are common in animals, with larger species typically dominant in aggressive contests for resources (e.g., Morse 1974; Donadio & Buskirk 2006). While gerbils reach adult size relatively early in development, many other organisms (e.g., fish) grow over extended periods through their lifetimes. These changes in size with development can create dynamic interactions and trade-offs that shift over time, complicating interactions among species and between organisms and their environments, and playing an important role in structuring communities (e.g., Werner & Hall 1988).

Ecological Defense: Guarding against Natural Enemies

Organisms must guard against being eaten and infected, and most have evolved adaptations for reducing these risks. Adaptations that deter natural enemies are diverse (Edmunds 1974; Coustau *et al.* 2000; Berenbaum 2001) and often costly, creating a trade-off between adaptations that reduce the likelihood of being consumed versus allocation to growth and reproduction (e.g., Strauss *et al.* 2002; Knowles *et al.* 2009; van der Most *et al.* 2011). Trade-offs also occur when organisms encounter multiple threats, such as risk from different predators or threats from predation and disease. For example, behavioral adaptations to reduce predation risk in damselflies (*Enallagma* spp.) vary depending on the predator, and present a trade-off between reducing predation by fish (e.g., remaining still) versus reducing predation by dragonflies (e.g., swimming away) (McPeek 1990). The importance of natural enemies in nature creates widespread trade-offs confronting hosts and prey, with diverse interactions and investments in adaptations that vary with risk (Raffel *et al.* 2008). These trade-offs can allow species with different strategies to coexist in diverse ways (Raffel *et al.* 2008), increasing diversity.

Trade-offs between Viral Defense and Competitive Ability in Bacteria

Viruses are abundant predators of bacteria in nature (Suttle 2005). Studies of the interactions between *E. coli* and T-type bacteriophage virus have identified *E. coli* that are resistant to viral predation, but their resistance comes at a cost to their growth and competitive ability (Lenski 1988a; Bohannan *et al.* 2002; Figure 18.2). The trade-off between competitive ability and resistance to viral predation typically involves changes to the receptor molecules on the bacteria's cell surface to which the virus initially binds. The changes to the receptor molecules simultaneously compromise aspects of metabolism, such as nutrient uptake, thereby reducing the competitive ability of virus-resistant bacteria (Figure 18.2; Lenski 1988a; Bohannan & Lenski 2000b; Bohannan *et al.* 2002). The magnitude of the trade-off between competitive ability and virus resistance varies (Bohannan & Lenski 2000b), and declines as modifier alleles evolve to mitigate, but not eliminate, the costs of resistance (Lenski 1988b). The cost of viral resistance, and thus the size of the trade-off, also varies with environmental factors such as resource abundance and temperature (Bohannan & Lenski 2000a; Quance & Travisano 2009; Gómez & Buckling 2011), which would alter the distributions of virus-resistant and virus-susceptible bacteria by influencing when they can coexist along environmental gradients.

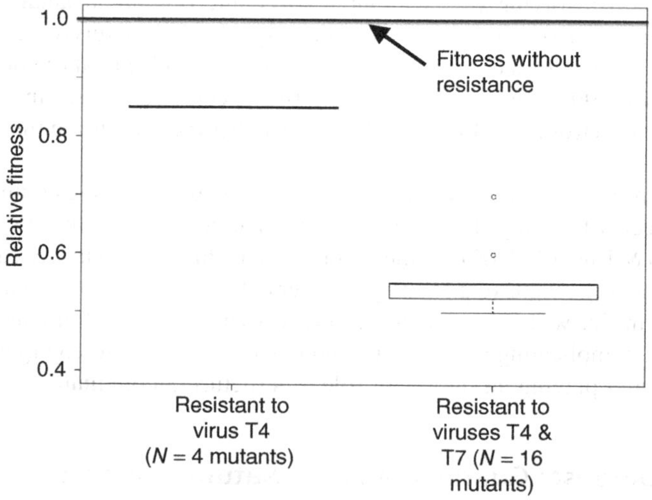

Figure 18.2. The relative fitness of 20 independent *Escherichia coli* mutants that were resistant to virus T4 (but not to virus T7) or to both viruses T4 and T7 (data from Lenski 1988a). The fitness of virus-sensitive *E. coli* is equal to 1. Mutations that altered receptors on *E. coli*'s cell surface prevented viral infection, but compromised metabolism, growth, and competitive ability, illustrating an evolutionary trade-off. This trade-off allows both virus-resistant and virus-susceptible strains of *E. coli* to coexist when viruses are present (Lenski 1988a).

While microbial interactions in the lab provide insight into evolutionary trade-offs, do they inform patterns in nature? Studies of both ocean cyanobacteria (Avrani *et al.* 2011, 2012) and soil bacteria (Gómez & Buckling 2011) and their viruses provide evidence that the evolutionary trade-offs demonstrated in the lab also play an important role in determining the distributions and coexistence of microbial diversity in nature. In nature, however, the costs of resistance are likely to be higher with resource limitation, and extend beyond reduced competitive abilities to trade-offs involving different viruses. In addition, the spatial and temporal complexities of selection, not evident in simplified laboratory settings, may play pivotal roles in maintaining diversity in nature (Avrani *et al.* 2011, 2012; Gómez & Buckling 2011).

Abiotic Challenges: The Costs of Adaptation

Abiotic conditions across most of the Earth's surface significantly challenge the majority of organisms (Russell 1990). Yet some species persist, and even thrive, in the face of these challenges as a result of specific adaptations. These adaptations often come at a competitive cost because allocative or functional trade-offs are required to address the environmental challenge (e.g., Coustau *et al.* 2000). Examples of adaptations to abiotic challenges include ion excretion organs in aquatic organisms that inhabit extremely hyperosmotic environments (e.g., Shimizu & Bradley 1994), physiological adaptations providing resistance to herbicides, pesticides, and antibiotics (e.g., Vila-Aiub *et al.* 2009), and well-developed thermal insulation for homeothermic animals confronting low temperatures (e.g., Scholander 1955).

Trade-offs during Plant Adaptation to Serpentine Soils Serpentine soils represent a challenge to plant growth (Figure 18.3), particularly due to low Ca:Mg ratios (Kruckeberg 2002; Brady *et al.* 2005). Absorption of Ca typically leads to greater absorption of Mg when Ca:Mg ratios are low (Madhok & Walker 1969), with elevated levels of Mg being detrimental to photosynthesis (Marschner 2001). Ca is a vital component of cell walls, and is thus essential for plant growth (Marschner 2001), leaving plants with little option but to maintain adequate Ca uptake while minimizing the problem of Mg. These and other challenges of serpentine soils have led to low productivity, high endemism, and distinct vegetation types in serpentine plant communities (Figure 18.3; Brady *et al.* 2005; Harrison & Rajakaruna 2011).

Despite the challenges of serpentine soils, plants have repeatedly and independently adapted to their hostile conditions (Harrison & Rajakaruna 2011). Several adaptations address the challenge of low Ca:Mg ratios by discriminating Ca and Mg ions during absorption, translocating Mg and other toxins to minimize impacts, and sequestrating or chelating Mg and other toxins within the plant (Rajakaruna *et al.* 2003; Brady *et al.* 2005). These adaptations to serpentine soils, however, come at a competitive cost. Many serpentine-adapted populations and species are limited to serpentine soils because they are unable to compete with plant species in nonserpentine environments (Kruckeberg 1954; Moore & Elmendorf 2011) and show reduced growth rates relative to nonserpentine plants, even in the absence of competitors (Brady *et al.* 2005; Wright & Stanton 2011). The reduced competitive ability of serpentine-adapted plants suggests that adaptations to serpentine soils compromise their ability to recolonize and persist in their ancestral habitats (Brady *et al.* 2005). This trade-off allows different plant species to persist on different soil types, increasing local and regional diversity.

Figure 18.3. Plant communities on nonserpentine (background) and serpentine (foreground) soils, Snell Valley, Napa County, California. Many different species of plants have adapted to live on serpentine soils, however, adaptations frequently compromise the ability of serpentine plants to compete on nonserpentine soils, illustrating a trade-off. This trade-off allows different species of plants to coexist by using different types of soil. Photo by John McKay. (See plate section for color version.)

EVOLUTIONARY TRADE-OFFS AND THE DISTRIBUTIONS OF SPECIES

These select studies illustrate integrative approaches to the study of trade-offs and their consequences for diversity. While our understanding of each trade-off is incomplete, the studies collectively illustrate the strength of integrative approaches for the study of evolutionary trade-offs of broad importance – trade-offs that should impact the ecology and evolution of most organisms in nature. The groupings are somewhat artificial (i.e., examples will often fit into multiple categories), but the challenges confronting organisms are real. Still other trade-offs are widespread in nature, such as trade-offs in life history strategies (Stearns 1992), ecological breadth (Futuyma & Moreno 1988), competitive versus colonization abilities (e.g., Violle *et al.* 2010), digestive physiologies (Afik & Karasov 1995), and physiological plasticity (DeWitt *et al.* 1998; Chapter 17, this volume), with important consequences for the distributions of species (Chapter 17, this volume). Furthermore, some evidence suggests that a diversity of trade-offs, rather than a few, may be the key to explaining the coexistence of rich biodiversity (Clark *et al.* 2007). We may expect such diversity of trade-offs if we consider trade-offs in the context of coevolution among species (cf. Thompson 2005).

Trade-offs allow species to coexist because they prevent any one species from monopolizing energy and other resources, particularly when selective pressures vary over space and time. For example, trade-offs between competitive dominance and herbivory defense may promote the coexistence of diverse plant species on two distinct soil types in Amazonian forest (Fine *et al.* 2004). Without the trade-off (herbivory removed), species adapted to clay soils prevailed across both types of soil (Fine *et al.* 2004). Similarly, trade-offs between minimizing energetic costs of foraging versus minimizing predation allow rodent species to coexist in different microhabitats within the same community (Kotler 1984). Overall, variation in selective pressures, coupled with the inability of species to adapt to all selective pressures simultaneously (trade-offs), create varying opportunities for diverse species. Understanding the mechanistic basis of these trade-offs and their consequences for individual fitness, population dynamics, species interactions, and coexistence will help us to understand how diversity is structured and why it varies (e.g., Bohannan *et al.* 2002; Kneitel & Chase 2004; Angert *et al.* 2009).

The same factors that determine the distributions of species also influence their abundances within their distributions. For example, the size of the trade-off between competitive ability and viral resistance influences the abundance of virus-resistant and virus-susceptible bacteria and their predatory virus. The costs of resistance, and thus the size of the trade-off, vary with both temperature and resources, creating variation in the abundance of coexisting bacteria and their viruses (Bohannan & Lenski 2000a; Quance & Travisano 2009; Gómez & Buckling 2011). Similarly, coexisting species of plants in the Sonoran Desert exhibit temporal variation in their abundance from year to year. This variation is best explained by a fundamental trade-off between the ability to photosynthesize and grow rapidly in wetter years versus the ability to withstand the low resource abundance in drier years. While these species coexist together, their demographic responses, and thus abundances, vary with environmental variation from year to year (Angert *et al.* 2009).

ARE TRADE-OFFS UNIVERSAL?

Trade-offs confront organisms everywhere we look, and yet when we look for specific or expected trade-offs, we don't always find them (e.g., Strauss *et al.* 2002). Conflicting evidence for trade-offs may be a function of the difficulties of measuring weak selection, the masking of trade-offs by other traits or variation (genetic, environmental), or delayed fitness benefits of different allocation strategies (Van Buskirk 2000; Agrawal *et al.* 2010). Despite these difficulties, simple experiments, such as reciprocal transplants, have been effective at identifying phenotypic traits involved in trade-offs and their consequences for abundances and distributions. Furthermore, the expanding fields of genomics and proteomics will provide new opportunities to understand the mechanistic basis of trade-offs, from genes to phenotypes.

Beyond the mechanics of identifying trade-offs in nature, we should also expect the occurrence of specific trade-offs to vary. Natural selection should strongly favor the evolution of traits that mitigate the costs of trade-offs, especially in the case of functional trade-offs (e.g., Lenski 1988b; Thaler *et al.* 2012). For example, the evolution of weapons (e.g., talons) in smaller species can offset the advantage larger species have in aggressive contests (Martin & Ghalambor, unpublished data). Allocative trade-offs are less likely to be directly mitigated by other traits because they are dependent upon resource abundance (Lenski 1988b). Allocative trade-offs, however, become more prominent when resources are limiting (e.g., Gómez & Buckling 2011), either in the environment, or because of the poor condition of the individual. In these latter cases, we may find condition-dependent trade-offs that can lead to distinct phenotypes within species (e.g., Emlen *et al.* 2012).

Evolutionary forces, such as gene flow, can constrain divergent evolution in response to trade-offs, potentially explaining why some species respond differently when confronted with similar challenges. For example, trade-offs commonly favor distinct phenotypes, but the evolution of distinct phenotypes is impeded by gene flow. Reproductive isolation creates the evolutionary independence that can free species to settle trade-offs in different ways (Coyne & Orr 2004). This evolutionary independence is a key reason why species formation is so important for adaptation and biodiversity (Coyne & Orr 2004).

In contrast, evolutionary forces, such as character displacement, can promote divergent evolution in response to trade-offs. The presence of other species can act to shift patterns of evolution from cases where one species optimizes its phenotype in response to a trade-off, to cases where two species optimize their phenotypes to settle a trade-off in different ways (Schluter 2000; Pfennig & Pfennig 2012; Figure 18.1). An illustrative example comes from comparisons of the many *Geospiza* finches found on several of the Galápagos Islands to the one species found on Cocos Island. Similar selective pressures and trade-offs may confront finches on Cocos Island, but without geographic isolation, the evolution of reproductive isolation, or character displacement, the one species has not diverged into distinct, sympatric ecological forms (Grant & Grant 2008; Losos & Ricklefs 2009).

Overall, we should expect variation in the occurrence and importance of specific trade-offs, and expect different trade-offs to determine the distributions and abundances of species under different circumstances (McPeek 1996). Understanding when specific trade-offs occur, including their prevalence at different scales of space and time (Kneitel & Chase 2004), would provide insight into how trade-offs are maintained or lost over time. Some trade-offs may prove to be fundamental across species and difficult to overcome (Tilman 2011), while others may prove fleeting.

TRADE-OFFS, INTEGRATIVE BIOLOGY, AND MOVING FORWARD

How is a trade-off framework different from other approaches to studying community ecology, given that both approaches focus on the same processes (competition, predation, parasitism, climate) (see also Box 18.1)? To illustrate, I use an example from my own work on wood warblers (Parulidae; Martin & Martin 2001a,b; Figure 18.4), the same family that was the focus of MacArthur's classic study of competition (MacArthur 1958).

Orange-crowned (*Oreothlypis celata*) and Virginia's (*O. virginiae*) warblers are closely related species that segregate their distributions along a temperature and moisture gradient, repeated across multiple spatial scales, but coexist in some areas of sympatry (Figure 18.4). *O. virginiae* occupies hotter, drier regions of North America, and, within regions of sympatry with *celata*, occupies lower elevations that are also hotter and drier. At even finer spatial scales, the two species completely overlap their breeding territories in a zone of geographic and elevational sympatry in central Arizona, USA, where *virginiae* typically nest in hotter, drier nest sites under oak (*Quercus*) or locust (*Robinia*), while *celata* nest in moister regions under maple (*Acer*). The larger *celata* is aggressive to the smaller *virginiae*, but does not fully exclude *virginiae* from its territories. Reciprocal removal experiments revealed high fitness costs of co-occurrence for both species caused by density-dependent nest predation.

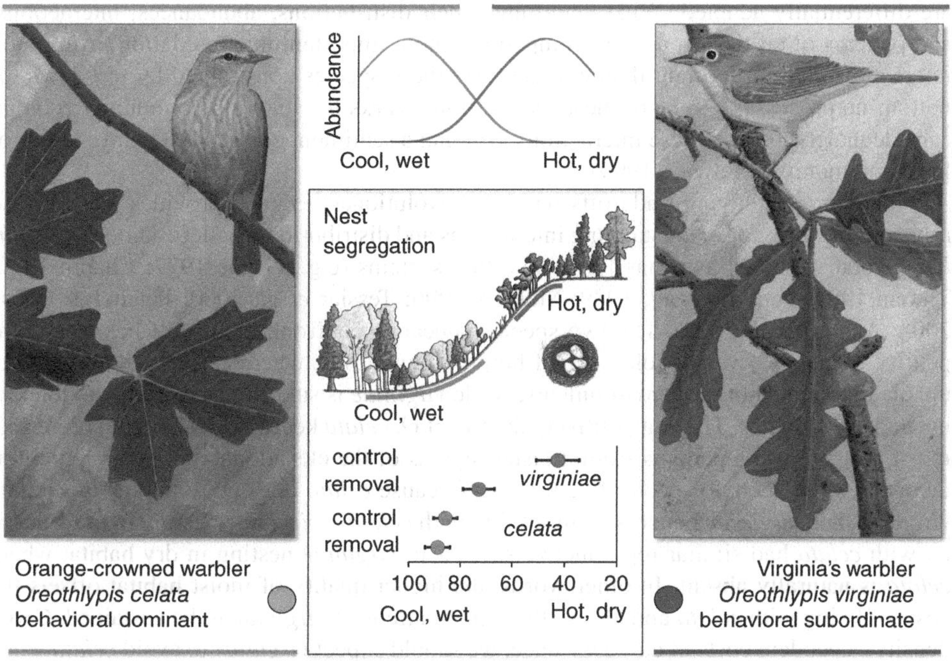

Figure 18.4. Segregation of Orange-crowned (*Oreothlypis celata*, blue) and Virginia's (*O. virginiae*, red) warblers (Parulidae) along gradients of temperature and moisture during the breeding season. These two species appear to face a trade-off, where the larger *celata* is dominant in aggressive interactions and excludes the smaller *virginiae* from preferred nest sites in cool, moist regions of their breeding territories. The smaller *virginiae*, however, can tolerate hot, dry nesting conditions that *celata* cannot tolerate, leading to spatial partitioning of nest sites where their breeding territories overlap. Experimental removal of the opposite species resulted in *virginiae* shifting their nest sites into cool, wet regions of the gradient in the absence of *celata*, but no shift in *celata* nest sites in the absence of *virginiae* (x-axis is the mean percentage of maple stems near the nest +/– SE). The temperature and moisture gradients that characterize overlapping territories also characterize patterns of elevational and geographic segregation, and temporal variation in abundance, suggesting that these trade-offs may also influence their distributions on broader spatial and temporal scales (Martin 1998, 2001; Martin & Martin 2001a,b). Similar trade-offs between competitive ability and environmental tolerance face barnacles (Connell 1961) and many other closely-related species, from chipmunks to salamanders, cattails to crayfish (Martin & Martin 2001a). (See plate section for color version.)

When *celata* was experimentally removed, *virginiae* shifted their nest sites to sites indistinguishable from *celata*; however, *celata* did not shift their nest sites when *virginiae* had been removed, and appear to be unable to nest successfully in hot, dry conditions. In addition, *O. virginiae* increased their feeding rates at nests with the removal of *celata*, which could reflect competition for food, or shifted time budgets that allow more time for foraging when *virginiae* nest in optimal sites (Martin 1998, 2001; Martin & Martin 2001a,b). The take home message from this study is this: interactions between these species are significant and complicated, involving asymmetric competition for nest sites, high costs of predation, and segregation along a climatic gradient of temperature and moisture to which the species are differentially adapted. What determines their distributions, abundances, interactions, and patterns of co-occurrence? Is competition more important than predation? Are biotic interactions more important than climate? Are these species constrained by top-down, or bottom-up processes? None of these classic frameworks for studying community ecology can adequately explain these interactions, echoing a sentiment common to many studies of local communities (Lawton 1999).

In contrast, studying trade-offs within an evolutionary context provides a useful and robust framework for understanding interactions and distributional patterns among *Oreothlypis* wood warblers, as it has for many other systems (e.g., Grime 1977; Chapin 1980; Tilman 1982; Petraitis *et al.* 1989; McPeek 1996; Tessier *et al.* 2000; Kneitel & Chase 2004; Angert *et al.* 2009). The two species appear to confront a trade-off between competitive dominance versus tolerance of hot, dry conditions, where *celata* is dominant but unable to tolerate hot, dry environments, while *virginiae* is subordinate but able to tolerate hot and dry climates. The competitively dominant *O. celata* keeps *virginiae* from preferred, moist nest sites, and perhaps from moister regions of the elevational and geographic gradients as well. The two species live together because *celata* does not completely exclude *virginiae*, and *virginiae* benefits from settling with *celata* – *virginiae* nesting in moist habitat with *celata* had similar reproductive success to *virginiae* nesting in dry habitat where *celata* is naturally absent. In other words, the higher quality of moist habitat offsets the costs of living with *celata* and allows these populations of *virginiae* to be sustained. Given density-dependent costs of co-occurrence, we should expect *virginiae* to avoid *celata* when the densities of *celata* reach a threshold (Martin 1998, 2001; Martin & Martin 2001a,b; Clark & Martin 2007; Figure 18.4). Questions remain, such as what adaptations allow *virginiae*, but not *celata*, to use hot and dry environments, and what are the ecological and fitness consequences of aggression by *celata*?

Trade-offs represent evolutionary challenges that play an integral role in ecology. Reviewing examples of widespread trade-offs reveals repeated patterns that suggest fundamental constraints. Among closely-related species, we repeatedly see a trade-off between competitive ability and ability to tolerate diverse environmental challenges, including low resources, high risk of predation, herbivory, parasitism, or pathogens, or hostile conditions such as poisonous soils or extreme climate (Connell 1961; Colwell & Fuentes 1975; Chapin 1980; Wisheu 1998; Keddy 2001; Martin & Martin 2001a; McGill *et al.* 2006). This broad trade-off makes sense from an evolutionary perspective – selective pressures that present significant challenges to a group of organisms require adaptations that should compromise organismal performance and competitive abilities in the context of environments that lack these challenges (e.g., Coustau *et al.* 2000). Importantly, competition does not structure these relationships. Without a trade-off involving diverse selective pressures, these species would be unlikely to coexist (e.g., Lenski & Levin 1985).

We should expect other rules to structure biodiversity, and many have been identified. For example, the extreme variation in body size, from bacteria to large trees and mammals, reveals fundamental scaling laws (West *et al.* 1997) and reflects distinct evolutionary strategies in response to trade-offs in organismal complexity, rates of evolution mediated by generation time, and inertial versus viscous forces (Peters 1983; Bonner 2006). Trade-offs are evident between endothermic and ectothermic strategies – in endotherms, maintaining high body temperatures allows consistently high metabolic rates and performance independent of the thermal environment, but comes at a high energetic cost of maintaining body temperatures (Schmidt-Nielsen 1997). Such a trade-off may explain the success of bird pollinators and bird-pollinated plants on high neotropical mountains where rain and low temperatures constrain bee pollination activity (Cruden 1972).

The evolution of specialized biotic interactions and coevolution could create an abundance of trade-offs, where adaptations that are specific and dependent on another species come at a cost to their function in the absence of that species. For example, adaptations in *Prochlorococcus* cyanobacteria to one set of viruses come at a cost of enhanced susceptibility to a different set of viruses (Avrani *et al.* 2011; 2012). Variation in the occurrence of virus types can lead to the coexistence of bacteria with different patterns of resistance (Avrani *et al.* 2011; 2012). This evolution of traits in response to new and diverse biotic trade-offs could promote speciation and the coexistence of even more species, particularly in species-rich environments (Schemske 2009).

Other trade-offs may also play important roles in the formation of new species. For example, Dobzhansky-Muller incompatabilities – where mutations that function well in the genetic background of their own population, function poorly in the genetic background of another population – can be viewed as a simple trade-off, but one potentially playing an important role in the formation of species by creating costs to hybridizing among populations (Coyne & Orr 2004). Overall, these few examples of fundamental trade-offs represent generalities that can partly explain distributions and diversity – the "rules" for which community ecologists search (Lawton 1999; Simberloff 2004).

In sum, trade-offs provide a productive framework for understanding the distributions and abundance of species in nature, integrating and uniting diverse fields, levels of study, and perspectives in the search to understand broad ecological patterns. This integrative approach yields richer stories, and elucidates mechanisms that help us to identify causal factors shared across diverse taxa and environments. These integrative stories will link together to provide a more complete understanding of how and why organisms live where they do, and ultimately, the factors that structure biological diversity.

ACKNOWLEDGMENTS

Thanks to Rose Barcarse, Fran Bonier, Chloe Boynton, Alice Domalik, Sarah Mangin, Mark McPeek, Bill Nelson, Chris Ng, Vanya Rohwer, the editors, and an anonymous reviewer for constructive and insightful feedback on this work. Thanks to John McKay for his photo of plants.

REFERENCES

Afik, D. and Karasov, W.H. (1995) The trade-offs between digestion rate and efficiency in warblers and their ecological implications. *Ecology*, **76**, 2247–57.

Agrawal, A.A. (2001) Phenotypic plasticity in the interactions and evolution of species. *Science*, **294**, 321–6.

Agrawal, A.A., Conner, J.K. and Rasmann, S. (2010) Tradeoffs and negative correlations in evolutionary ecology, in *Evolution since Darwin. The First 150 Years* (eds M.A. Bell, D.J. Futuyma, W.F. Eanes and J.S. Levinton), Sinauer Associates, Sunderland, pp. 243–68.

Angert, A.L., Huxman, T.E., Chesson, P. and Venable, D.L. (2009) Functional tradeoffs determine species coexistence via the storage effect. *Proceedings of the National Academy of Sciences USA*, **106**, 11641–5.

Avrani, S., Wurtzel, O., Sharon, I. *et al.* (2011) Genomic island variability facilitates *Prochlorococcus* – virus coexistence. *Nature*, **474**, 604–8.

Avrani, S., Schwartz, D.A. and Lindell, D. (2012) Virus-host swinging party in the oceans. Incorporating biological complexity into paradigms of antagonistic coexistence. *Mobile Genetic Elements*, **2**, 88–95.

Berenbaum, M.R. (2001) Plant – herbivore interactions, in *Evolutionary Ecology: Concepts and Case Studies* (eds C.W. Fox, D. Roff and D.J. Fairbairn), Oxford University Press, Oxford, pp. 303–14.

Bohannan, B.J.M., Kerr, B., Jessup, C.M. *et al.* (2002) Trade-offs and coexistence in microbial microcosms. *Antonie van Leeuwenhoek*, **81**, 107–15.

Bohannan, B.J.M. and Lenski, R.E. (2000a) Linking genetic change to community evolution: insights from studies of bacteria and bacteriophage. *Ecology Letters*, **3**, 362–77.

Bohannan, B.J.M. and Lenski, R.E. (2000b) The relative importance of competition and predation varies with productivity in a model community. *American Naturalist*, **156**, 329–40.

Bolnick, D.I., Amarasekare, P., Araújo, M.S. *et al.* (2011) Why intraspecific trait variation matters in community ecology. *Trends in Ecology and Evolution*, **26**, 183–92.

Bolnick, D.I., Svanbäck, R., Fordyce, J.A. *et al.* (2003) The ecology of individuals: incidence and implications of individual specialization. *American Naturalist*, **161**, 1–28.

Bonner, J.T. (2006) *Why Size Matters*, Princeton University Press, Princeton.

Brady, K.U., Kruckeberg, A.R. and Bradshaw, H.D. (2005) Evolutionary ecology of plant adaptation to serpentine soils. *Annual Review of Ecology, Evolution and Systematics*, **36**, 243–66.

Chapin, F.S. III (1980) The mineral nutrition of wild plants. *Annual Review of Ecology and Systematics*, **11**, 233–60.

Chapin, F.S. III, Autumn, K. and Pugnaire, F. (1993) Evolution of suites of traits in response to environmental stress. *American Naturalist*, **142**, S78–S92.

Chesson, P. (2000) Mechanisms of maintenance of species diversity. *Annual Review of Ecology and Systematics*, **31**, 343–66.

Clark, J.S. (2010) Individuals and the variation needed for high species diversity in forest trees. *Science*, **327**, 1129–1132.

Clark, J.S., Dietze, M., Chakraborty, S. *et al.* (2007) Resolving the biodiversity paradox. *Ecology Letters*, **10**, 647–59.

Clark, M.E. and Martin, T.E. (2007) Modeling tradeoffs in avian life history traits and consequences for population growth. *Ecological Modelling*, **209**, 11–120.

Colwell, R.K. and Fuentes, E.R. (1975) Experimental studies of the niche. *Annual Review of Ecology and Systematics*, **6**, 281–310.

Connell, J.H. (1961) The influence of interspecific competition and other factors on the distribution of the barnacle *Chthamalus stellatus*. *Ecology*, **42**, 710–23.

Coustau, C., Chevillon, C. and ffrench-Constant, R. (2000) Resistance to xenobiotics and parasites: can we count the cost? *Trends in Ecology and Evolution*, **15**, 378–83.

Coyne, J.A. and Orr, H.A. (2004) *Speciation*, Sinauer Associates, Sunderland.

Cruden, R.W. (1972) Pollination in high elevation ecosystems: relative effectiveness of birds and bees. *Science*, **176**, 1439–40.

DeWitt, T.J., Sih, A. and Wilson, D.S. (1998) Costs and limits of phenotypic plasticity. *Trends in Ecology and Evolution*, **13**, 77–81.

Dobzhansky, T. (1964) Biology, molecular and organismic. *American Zoologist*, **4**, 443–52.

Donadio, E. and Buskirk, S.W. (2006) Diet, morphology, and interspecific killing in Carnivora. *American Naturalist*, **167**, 524–36.

Duffy, M.A. and Forde, S.E. (2009) Ecological feedbacks and the evolution of resistance. *Journal of Animal Ecology*, **78**, 1106–12.

Edmunds, M. (1974) *Defence in Animals: A Survey of Anti-predator Defences*, Longman Inc., New York.

Emlen, D.J., Warren, I.A., Johns, A. *et al.* (2012) A mechanism of extreme growth and reliable signaling in sexually selected ornaments and weapons. *Science*, **337**, 860–4.

Falconer, D.S. and Mackay, T.F.C. (1996) *Quantitative Genetics, 4th Edition* edn, Pearson Education, Harlow, UK.

Fine, P.V.A., Mesones, I. and Coley, P.D. (2004) Specialization by trees in Amazonian forests. *Science*, **305**, 663–5.

Futuyma, D.J. and Moreno, G. (1988) The evolution of ecological specialization. *Annual Review of Ecology and Systematics*, **19**, 207–33.

Gluckman, P.D., Hanson, M.A. and Spencer, H.G. (2005) Predictive adaptive responses and human evolution. *Trends in Ecology and Evolution*, **20**, 527–33.

Gómez, P. and Buckling, A. (2011) Bacteria-phage antagonistic coevolution in soil. *Science*, **332**, 106–9.

Grant, P.R. and Grant, B.R. (2008) *How and Why Species Multiply: The Radiation of Darwin's Finches*, Princeton University Press, Princeton.

Grime, J.P. (1977) Evidence for the existence of three primary strategies in plants and its relevance to ecological and evolutionary theory. *American Naturalist*, **111**, 1169–94.

Harrison, S. and Rajakaruna, N. (eds) (2011) *Serpentine: The Evolution and Ecology of a Model System*, University of California Press, Berkeley.

Holt, R.D. (1995) Linking species and ecosystems: where's Darwin?, in *Linking Species & Ecosystems* (eds C.G. Jones and J.H. Lawton), Chapman & Hall, London, pp. 273–9.

Kawecki, T.J. and Ebert, D. (2004) Conceptual issues in local adaptation. *Ecology Letters*, **7**, 1225–41.

Keddy, P.A. (2001) *Competition, 2nd Edition* edn, Kluwer Academic Publishers, Dordrecht.

Kneitel, J.M. and Chase, J.M. (2004) Trade-offs in community ecology: linking spatial scales and species coexistence. *Ecology Letters*, **7**, 69–80.

Knowles, S.C.L., Nakagawa, S. and Sheldon, B.C. (2009) Elevated reproductive effort increases blood parasitaemia and decreases immune function in birds: a meta-regression approach. *Functional Ecology*, **23**, 405–15.

Kotler, B.P. (1984) Risk of predation and the structure of desert rodent communities. *Ecology*, **65**, 689–701.

Kotler, B.P. and Brown, J.S. (1990) Rates of seed harvest by two species of gerbilline rodents. *Journal of Mammalogy*, **71**, 591–6.

Krebs, J.R. and Davies, N.B. (1997) *Behavioral Ecology: An Evolutionary Approach, 4th Edition* edn, Blackwell Science, Oxford.

Kruckeberg, A.R. (1954) Plant species in relation to serpentine soil. *Ecology*, **35**, 267–74.

Kruckeberg, A.R. (2002) *Geology and Plant Life: The Effects of Landforms and Rock Types on Plants*, University of Washington Press, Seattle.

Lawton, J.H. (1999) Are there general laws in ecology? *Oikos*, **84**, 177–92.

Lenski, R.E. (1988a) Experimental studies of pleiotropy and epistasis in *Escherichia coli*. I. Variation in competitive fitness among mutants resistant to virus T4. *Evolution*, **42**, 425–32.

Lenski, R.E. (1988b) Experimental studies of pleiotropy and epistasis in *Escherichia coli*. II. Compensation for maladaptive effects associated with resistance to virus T4. *Evolution*, **42**, 433–40.

Lenski, R.E. and Levin, B.R. (1985) Constraints on the coevolution of bacteria and virulent phage: a model, some experiments, and predictions for natural communities. *American Naturalist*, **125**, 585–602.

Losos, J.B. and Ricklefs, R.E. (2009) Adaptation and diversification on islands. *Nature*, **457**, 830–6.

MacArthur, R.H. (1958) Population ecology of some warblers in northeastern coniferous forests. *Ecology*, **39**, 599–619.

Macnair, M.R., Tilstone, G.H. and Smith, S.E. (2000) The genetics of metal tolerance and accumulation in higher plants, in *Phytoremediation of Contaminated Soil and Water* (eds N. Terry and G. Bañuelos), CRC Press, Boca Raton, pp. 235–50.

Madhok, O.P. and Walker, R.B. (1969) Magnesium nutrition of two species of sunflower. *Plant Physiology*, **44**, 1016–22.

Marschner, H. (2001) *Mineral Nutrition of Plants, 2nd Edition* edn, Academic Press, San Diego.

Martin, P.R. and Martin, T.E. (2001a) Ecological and fitness consequences of species coexistence: a removal experiment with wood warblers. *Ecology*, **82**, 189–206.

Martin, P.R. and Martin, T.E. (2001b) Behavioral interactions between coexisting species: Song playback experiments with wood warblers. *Ecology*, **82**, 207–18.

Martin, T.E. (1998) Are microhabitat preferences of coexisting species under selection and adaptive? *Ecology*, **79**, 656–70.

Martin, T.E. (2001) Abiotic vs. biotic influences on habitat selection of coexisting species: climate change impacts? *Ecology*, **82**, 175–88.

McCollum, S.A. and Van Buskirk, J. (1996) Costs and benefits of a predator-induced polyphenism in the Gray Treefrog *Hyla chrysoscelis*. *Evolution*, **50**, 583–93.

M'Gonigle, L.K., Mazzucco, R., Otto, S.P. and Dieckmann, U. (2012) Sexual selection enables long-term coexistence despite ecological equivalence. *Nature*, **484**, 506–9.

McGill, B.J., Enquist, B.J., Weiher, E. and Westoby, M. (2006) Rebuilding community ecology from functional traits. *Trends in Ecology and Evolution*, **21**, 178–85.

McPeek, M.A. (1990) Behavioral differences between *Enallagma* species (Odonata) influencing differential vulnerability to predators. *Ecology*, **71**, 1714–26.

McPeek, M.A. (1996) Trade-offs, food web structure, and the coexistence of habitat specialists and generalists. *American Naturalist*, **148**, S124–S138.

Moore, K.A. and Elmendorf, S.C. (2011) Plant competition and facilitation in systems with strong environmental gradients, in *Serpentine: The Evolution and Ecology of a Model System* (eds S. Harrison and N. Rajakaruna), University of California Press, Berkeley, pp. 223–36.

Morse, D.H. (1974) Niche breadth as a function of social dominance. *American Naturalist*, **108**, 818–30.

Peters, R.H. (1983) *The Ecological Implications of Body Size*, Cambridge University Press, Cambridge.

Petraitis, P.S., Latham, R.E., & Niesenbaum, R.A. (1989) The maintenance of species diversity by disturbance. *Quarterly Review of Biology*, 393–418.

Pfennig, D.W. and Pfennig, K.S. (2012) *Evolution's Wedge: Competition and the Origins of Diversity*, University of California Press, Berkeley.

Quance, M.A. and Travisano, M. (2009) Effects of temperature on the fitness cost of resistance to bacteriophage T4 in *Escherichia coli*. *Evolution*, **63**, 1406–16.

Raffel, T.R., Martin, L.B. and Rohr, J.R. (2008) Parasites as predators: unifying natural enemy ecology. *Trends in Ecology and Evolution*, **23**, 610–18.

Rajakaruna, N., Siddiqi, M.Y., Whitton, J. *et al.* (2003) Differential responses to Na^+/K^+ and Ca^{2+}/Mg^{2+} in two edaphic races of the *Lasthenia californía* (Asteraceae) complex: a case for parallel evolution on physiological traits. *New Phytologist*, **157**, 93–103.

Relyea, R.A. (2002) Costs of phenotypic plasticity. *American Naturalist*, **159**, 272–82.

Ruell, E.W., Handelsman, C.A., Hawkins, C.L. *et al.* (2013) Fear, food and sexual ornamentation: plasticity of colour development in Trinidadian guppies. *Proceedings of the Royal Society B*, **280**, 2012–19.

Russell, N.J. (1990) Cold adaptation of microorganisms. *Philosophical Transactions of the Royal Society of London B*, **326**, 595–611.

Schemske, D.W. (2009) Biotic interactions and speciation in the tropics, in *Speciation and Patterns of Diversity* (eds R.K. Butlin, J.R. Bridle and D. Schluter), Cambridge University Press, Cambridge, pp. 219–39.

Schluter, D. (2000) *The Ecology of Adaptive Radiation*, Oxford University Press, Oxford.

Schmidt-Nielsen, K. (1984) *Scaling: Why Is Animal Size So Important?*, Cambridge University Press, Cambridge.

Schmidt-Nielsen, K. (1997) *Animal Physiology: Adaptation and Environment*, Cambridge University Press, Cambridge.

Schoener, T.W. (1983) Field experiments on interspecific competition. *American Naturalist*, **122**, 240–85.

Scholander, P.F. (1955) Evolution of climate adaptation in homeotherms. *Evolution*, **9**, 15–26.

Shimizu, M. and Bradley, T.J. (1994) Concentrated excretion of sulfate by the anterior colon of the brine fly, *Ephydra hians*. *Physiological Zoology*, **67**, 54–67.

Sibly, R.M. and Calow, P. (1986) *Physiological Ecology of Animals: An Evolutionary Approach*, Blackwell Scientific, Oxford.

Simberloff, D. (2004) Community ecology: is it time to move on? *American Naturalist*, **163**, 787–99.

Stearns, S.C. (1992) *The Evolution of Life Histories*, Oxford University Press, Oxford.

Strauss, S.Y., Rudgers, J.A., Lau, J.A. and Irwin, R.E. (2002) Direct and ecological costs of resistance to herbivory. *Trends in Ecology and Evolution*, **17**, 278–85.

Suttle, C.A. (2005) Viruses in the sea. *Nature*, **437**, 356–61.

Tessier, A.J., Leibold, M.A. and Tsao, J. (2000) A fundamental trade-off in resource exploitation by *Daphnia* and consequences to plankton communities. *Ecology*, **81**, 826–41.

Thaler, J.S., McArt, S.H. and Kaplan, I. (2012) Compensatory mechanisms for ameliorating the fundamental trade-off between predator avoidance and foraging. *Proceedings of the National Academy of Sciences USA*, **109**, 12075–80.

Thompson, J.N. (2005) *The Geographic Mosaic of Coevolution*, University of Chicago Press, Chicago.

Tilman, D. (1982) *Resource Competition and Community Structure*, Princeton University Press, Princeton.

Tilman, D. (2011) Diversification, biotic interchange, and the universal trade-off hypothesis. *American Naturalist*, **178**, 355–71.

Van Buskirk, J. (2000) The costs of an inducible defense in anuran larvae. *Ecology*, **81**, 2813–21.

van der Most, P.J., de Jong, B., Parmentier, H.K. and Verhulst, S. (2011) Trade-off between growth and immune function: a meta-analysis of selection experiments. *Functional Ecology*, **25**, 74–80.

Vila-Aiub, M.M., Neve, P. and Powles, S.B. (2009) Fitness costs associated with evolved herbicide resistance alleles in plants. *New Phytologist*, **184**, 751–67.

Violle, C., Pu, Z. and Jiang, L. (2010) Experimental demonstration of the importance of competition under disturbance. *Proceedings of the National Academy of Sciences USA*, **107**, 12925–9.

Werner, E.E. and Hall, D.J. (1988) Ontogenetic niche shifts in bluegill: the foraging rate – predation risk trade-off. *Ecology*, **69**, 1352–66.

West, G.B., Brown, J.H. and Enquist, B.J. (1997) A general model for the origin of allometric scaling laws in biology. *Science*, **276**, 122–6.

Wisheu, I.C. (1998) How organisms partition habitats: different types of community organization can produce identical patterns. *Oikos*, **83**, 246–58.

Wright, J.W. and Stanton, M.L. (2011) Local adaptation in heterogeneous landscapes, in *Serpentine: The Evolution and Ecology of a Model System* (eds S. Harrison and N. Rajakaruna), University of California Press, Berkeley, pp. 155–79.

Ziv, Y., Abramsky, Z., Kotler, B.P. and Subach, A. (1993) Interference competition and temporal and habitat partitioning in two gerbil species. *Oikos*, **66**, 237–46.

19

CONCLUSIONS: THE CENTRAL ROLE OF THE ORGANISM IN BIOLOGY

H. Arthur Woods[1], Lynn B. Martin[2], and Cameron K. Ghalambor[3]

[1]*Division of Biological Sciences, University of Montana, Missoula, MT, USA*
[2]*Department of Integrative Biology, University of South Florida, Tampa, FL, USA*
[3]*Department of Biology and Graduate Degree Program in Ecology, Colorado State University, Fort Collins, CO, USA*

> "The task of science is to make use of the world's redundancy to describe that world simply."
>
> – *(H.A. Simon, 1962)*

> "A new model of scientific investigation to understand complex systems would require shifting the perspective from the parts to the whole and back again."
>
> – *(F. Mazzocchi, 2008)*

INTRODUCTION

In thinking about the chapters in this book, and grappling with their synergisms and contradictions, we have identified an emerging set of themes. These themes appear, to us, to be on the verge of coalescing into a novel view of what organisms are: how they function, how they interact with their environments, and how they evolve. Here we attempt to give this view a push – first, by briefly outlining the historical shoulders on which our book stands; second, by reexamining several ideas from the Introduction so that we can more explicitly define the key characteristics of organisms; and, third, by showing how these ideas illuminate a number of interesting, modern problems in biology.

Integrative Organismal Biology, First Edition.
Edited by Lynn B. Martin, Cameron K. Ghalambor, and H. Arthur Woods.

In 1987, George Bartholomew contributed a chapter to *New Directions in Ecological Physiology* (Feder *et al.* 1987), writing, "It is the intact and functioning organism on which natural selection operates. Organisms are therefore the central element of concern to the biologist who aspires to a broad and integrated understanding of biology." The evolutionary emphasis in the first sentence fits most naturally into the overall themes of *New Directions*, which advocated a particular, evolutionary point of view – that the future of the field lay in integrating physiological ecology more fully with evolutionary ideas, and in focusing explicitly on individual variation and the phylogenetic contexts in which lineages evolve. By contrast, Bartholomew's emphasis on organismal biology (which characterizes his writing generally; for a summary, see Huey & Bennett 2008) played a less significant a role in *New Directions*. Indeed, in the past few decades, organisms themselves *have not* been the central element of concern. They have largely been contexts for other, more suborganismal pursuits or the units in higher-level ecological and evolutionary investigations.

It is time for the study of organisms to reassume its central role, and we think that this will require an approach to organismal biology that is integrative in new ways. The phrase "integrative biology" has been used so frequently over the past few decades that it has become unremarkable. Nevertheless, the idea remains fuzzy (Wake 2003), which reflects that there have been ongoing, even tectonic, shifts in how biologists think about integration. Twenty years ago, being integrative meant drawing on multiple conceptual fields, using diverse technical approaches, or spanning levels of organization in the course of a single project. We fully support this kind of integration. Nevertheless, a different kind of integration has emerged in the past decade – one that focuses more on the actual biological integration occurring within organisms, rather than on the conceptual integration done by biologists.

Below we identify four core themes that characterize integrative organismal biology, with emphasis on this second sense of the word integrative. Many more themes could be added, but our goal is to pose a simple set that is necessary and also sufficient for capturing a significant swath of interesting problems identified by various chapters in this book. Such a minimal approach risks too much abstraction. We believe, however, that a productive organismal biology will emerge only from conceptual breadth and integration, which requires a certain distance from the grit of individual studies and organisms.

1. *Organisms are collections of tradeoffs*. Tradeoffs permeate biology (Williams 1966; Noordwijk & de Jong 1986; Stearns 1989; Zera & Harshman 2001; Ricklefs & Wikelski 2002; Kneitel & Chase 2004) (chapters 5, 8, 10, 11, 12, 13, 14, 17, 18, this volume). Phenotypes are strongly shaped by linkage and pleiotropy, which constrain evolutionary trajectories by coupling changes in one trait to changes in others. Constraints also arise from physiological tradeoffs, which occur because physiological systems draw on common pools of energy and materials, and because evolved regulatory networks share common elements of sensing and control (Martin *et al.* 2011; Cohen *et al.* 2012).
2. *Organisms are complex*. Even the smallest metazoans contain fantastically complex systems. Work in many fields now suggests that biological complexity has significant structure (Kauffman 1993; Goldenfeld & Kadanoff 1999; Alm & Arkin 2003; Mazzocchi 2008). Moreover, the structure of complexity may be the key to understanding the position of the organism in the ill-defined relationships among genes, environments, and the evolutionary process (Chapters 2, 3, 6, 7, & 9, this volume). At organismal and even physiological levels, general descriptions of complexity's structure and consequences are not yet mature.

3. *Organisms are modular but integrated.* Networks within bodies are organized into modules, within which there are intensely interacting parts but among which the connections are weaker (Wagner & Altenberg 1996; Ayroles *et al.* 2009; Martin *et al.* 2011; Cohen *et al.* 2012). From physiological and organismal points of view, fitness depends on coordinating the actions of modules, which in turn requires sensing and signaling connections among them, and central mechanisms to coordinate them (Chapters 4, 8, 9, 10, 15, & 16, this volume). Networks within bodies, however, are also integrated wholes – because the modules are coordinated, because information must pass effectively among parts (Woods & Wilson 2013), and because modules draw on so many common resources (Mykles *et al.* 2010).
4. *Organisms are plastic.* Most physiological traits change over multiple time scales, from seconds to lifetimes. Indeed, plasticity is now the expectation rather than the exception, and even traits that are canalized often stem from plasticity in other traits (Chapter 8, this volume). As a concept, plasticity is stimulating – and central to biology – because it depends on so many integrated factors. It occurs in response to environments filtered through complex, interconnected sets of sensory and physiological systems; it emerges from coordinated, level-spanning changes in different modules, all subject to the constraints and tradeoffs that characterize integrated systems; and it is the organismal phenomenon that connects physiological, behavioral, and morphological phenotypes most strongly to key problems in evolution and ecology (Pigliucci 2001; Agrawal 2001; West-Eberhard 2003; Miner *et al.* 2005; Nussey, Wilson & Brommer 2007). Plasticity thus provides a natural conceptual bridge between organisms and the higher-level ecological and evolutionary processes that they are shaping (and that are shaping them; Chapters 1–5, 7–9, 12, 14, & 16, this volume).

Similar themes have been recognized and discussed already in the developmental and evolutionary literature (Gottlieb 2001; Pigliucci & Muller 2010). The novelty of this volume stems from (1) focusing on the integrated systems that underlie flexible changes in physiology, performance, and behavior across a variety of time scales, and (2) viewing the individual parts as a conceptual whole. Such an approach should generate fresh approaches to long-standing problems in biomedicine (Chapters 7, 8, 9, 11, 12, & 16, this volume) and other disciplines (Chapters 2, 4, 5, 14, 17, & 18, this volume), and it should provide new ways to define the roles of individual organisms in ecology and evolution. The four sections below outline broad examples of this approach. The first two are about basic research questions; the third and fourth focus on the implications of the core themes for how we, as organismal biologists, think about biology and formulate projects.

HOW DO COMPLEX TRAITS INTERACT WITH COMPLEX ENVIRONMENTS?

In biology, a common, implicit view is that organisms are collections of traits all working more-or-less autonomously – and, likewise, that environments are collections of factors each varying in its own way. Such a view underlies the typical experiments and observations made in, e.g., physiological ecology and evolutionary physiology, which focus on one or two traits that interact in some way with one or two environmental variables. Clearly, however, organism-environment interactions are significantly more complicated than this. First, single traits do not stand alone. Each is linked to multiple other traits by shared evolutionary interests and history, resources and information, underlying genetic components, and systems for coordination and control (core themes 1–3 above). Second, organisms do not interact with environmental traits one at a time; they interact with multivariate sets of environmental factors that covary over multiple scales. At first glance, such complexity-meets-complexity seems impenetrable; not only are organisms complex, so are the environments in which they live!

How should organismal biologists approach such complexity? One path forward is emerging from molecular biology, and specifically from recent progress in high-throughput approaches to assessing genomic and transcriptomic states of organisms. In a study of complex traits in *Drosophila*, Ayroles *et al.* (Ayroles *et al.* 2009) write, "Determining the genetic architecture of complex traits is challenging because phenotypic variation arises from interactions between multiple, environmentally sensitive alleles." To analyze their particular genotype-phenotype map in *Drosophila*, Ayroles *et al.* propose a "systems genetics" approach (see also Nadeau & Dudley 2011), which they illustrate by evaluating the transcriptomic states of thousands of genes across a set of phenotypes. Their results suggest that connections among genes are highly modular and that variation in several hundred candidate genes is most closely related to variation in phenotypes.

For integrative organismal biologists, such an approach could be expanded in two productive ways. First, we'd like to see stronger connections between the transcriptome and *physiological* measures of multiple phenotypes. The connections should go both ways: we need better understanding of how complex physiological traits emerge from the collective action of many genes *and* how physiological states feedback onto the transcriptome. Such assessments could be used to identify key patterns of physiological covariation and constraint. Second, we need realistic characterizations of the actual, complex patterns of temporal and spatial variation in environments. Here too there are many pre-existing methods developed by ecologists (Lepš & Šmilauer 2003). Like the systems genetics approach, these ecological methods are aimed at delivering statistical descriptions of variation of multiple factors simultaneously. The outstanding hard problem is to bridge the gap – between the complexity of phenotypes and the complexity of the environments with which they interact. Integrative organismal biology is uniquely positioned to take on this problem.

WHERE DOES ORGANISMAL DIVERSITY COME FROM AND HOW DOES IT EVOLVE?

A long-standing problem in biology has been to explain the origins of traits, and many of the chapters in this volume examine this problem (Chapters 4, 6, & 9). Very generally, this problem includes: (1) explaining how single cells, containing single genomes, differentiate into the many cells types contained in (most) multicellular organisms; (2) understanding the roles that developmental noise plays in establishing the diversity of phenotypes shown by cells and tissues making up physiological systems (Woods 2014); (3) explaining how organismal phenotypes change as they interact with their environments (Pigliucci 2001); and (4) explaining the evolutionary origins of complex novelty (Muller & Wagner 1991; Shubin *et al.* 2009). So far, progress on these questions has come primarily from molecular genetic analyses. For example, we know now that many apparently unrelated morphological structures share "deep homologies" – their development is supported by homologous genetic circuits coopted in different contexts for novel purposes (Shubin *et al.* 2009). Our organismal perspective suggests that higher, physiological levels also play important roles in the origins of diversity and novelty: the actions of lower-level genetic circuitry all are filtered through the complex set of organismal characteristics above. That is, the core organismal characteristics – tradeoffs, complexity, integration, modularity, and plasticity – probably play roles that are just as important, if not more so, than the actions of the genes operating at lower levels of organization; physiological traits filter and integrate the actions of genes before those actions are manifest as organismal phenotypes visible to their environments and to selection.

A related problem is to predict how existing complex traits will evolve. In general, trajectories of complex traits under selection are difficult to predict because of the multitude of physical, physiological, and genetic linkages underlying them. These connections can be formalized within a quantitative genetic framework as variance-covariance matrices (G matrices) (Arnold *et al.* 2008) or, for traits that naturally occur as curves (e.g., performance curves), as function-valued traits (Kingsolver *et al.* 2001; Chapter 3, this volume). Physiological, morphological, and behavioral traits also exhibit patterns of correlations that arise from having to perform particular functions, and the matrix describing these relationships may in some cases be more important than the underlying genetic architecture in determining the responses to selection (Ghalambor *et al.* 2003; Walker 2007). In all these cases, when traits are correlated, selection on one trait generates changes in others and this has consequences for organismal performance.

HOW USEFUL IS REDUCTIONISM?

Reductionism has been a key driver of progress in biology during the 20th century (Mazzocchi 2008; Noble 2012). Nevertheless, we now recognize many problems that resist its power. These include the mechanistic steps leading from genes to phenotypes; the origins and maintenance of plasticity; redundancy in many biological systems; and the origins of variation and evolutionary trajectories of complex phenotypes. Responding to the rise of molecular biology and the pervasiveness of genetic determinism – an idea that stems directly from reductionist approaches to genetics – Strohman wrote more than 15 years ago (1997), "In promising to penetrate and reveal the secrets of life, it has extended to itself a level of complexity where, as a paradigm, it has little power and must eventually fail." The more general statement of the problem is that reductionism's weaknesses reflect the key themes in organismal biology – tradeoffs, complexity, modularity, integration, and plasticity – that we lay out above. In a real sense, reductionist biology is to integrative biology as Newtonian physics is to relativity and quantum mechanics (Strohman 1997).

Such a view has deep implications for future progress in industries based on biology, including medicine. Many practical uses of biology involve moving genes between organisms, or changing levels of expression, in attempts to engineer phenotypes. Such an approach is reductionist in that it ignores the complex, integrated systems in which individual genes operate (Nicholson *et al.* 2004). In medicine, much of the effort of modern pharmaceutics goes to identifying *single* compounds that can affect disease phenotypes. That the industry continues to find new drugs is proof that reductionism works. Nevertheless, the rate of new drug discovery has slowed markedly, perhaps because the single-gene, single-compound approach largely ignores the integrated complexity of physiological systems (Nicholson *et al.* 2004) and the power of the environment to induce changes that are often larger than those stemming from variation in genes (Horrobin 2003). This ignorance is, in part, what leads to the prevalence of unintended side effects from modern drugs. If medicine could really deal explicitly with this problem, perhaps we'd make more rapid progress toward subtler manipulations of whole physiological systems, which would simultaneously give more power over target phenotypes and fewer side effects.

WHAT ARE THE CONSEQUENCES OF EMERGENCE FOR ORGANISMAL BIOLOGY?

In organismal biology, reductionism is also partly foiled by emergence. Emergent properties describe phenomena that are not readily decomposable into, or even understandable from, descriptions of their component parts (de Haan 2006; Bizzarri *et al.* 2013). As a physical example, consider the life-giving properties of liquid water – high heat capacity, high boiling point for its molecular size, surface tension, etc. All of these properties emerge from the fact that individual water molecules are electric dipoles that interact promiscuously, via hydrogen bonds, with other water molecules. In retrospect, it is possible to see how water's properties emerge from the bulk interactions of millions of tiny dipoles, but those properties would be difficult to predict from information about hydrogen and oxygen alone.

Many phenotypes can be understood in the same way. A trait at some level of organization emerges from complex processes at other levels of organization: action potentials emerge from interactions among voltages, membrane properties, and ion channels; body temperature from interactions among pathways of heat gain and heat loss, and the transport of heat within the body; size and shape of the cranium from interactions between growing bony plates, feedback loops in gene networks, and environmental conditions such as nutrition and social context. It can be useful to divide emergent properties (Noble 2012) into those that are "weakly emergent," which refer to phenomena at a particular level that can, with some effort, be explained by analysis at the level below; and those that are "strongly emergent," which refer to phenomena that are practically impossible to predict. Weakly emergent traits include such traits as skin thickness and eye color (e.g., they emerge in a relatively straightforward way from number of skin cells and identity of eye pigments). Strongly emergent traits include many developmental, physiological, and performance phenotypes, as well as the difficult-to-define states of health, stress, and disease (Chapters 9, 11, 12, & 14, this volume).

Emergence also has practical consequences for how we do biology. Biologists frequently invoke the metaphor of "levels of biological organization" because it provides a shorthand for associating phenomena with particular fields. This metaphor has been distorted in various ways, such that most biologists view causation as something rising up from below, and that is ultimately traceable to genes (Strohman 1997; Noble 2012; Bizzarri *et al.* 2013), or even to physics and chemistry (Schrödinger 1944). The idea of emergence sharpens the utility of level-thinking by providing a real justification for biological hierarchies; each level can be defined as the emergent set of traits arising from interactions at the next lower level. In this view, organisms are nested sets of Russian dolls (and in which the smaller dolls occur in increasingly large numbers). The lowest biological level, the smallest dolls, contains the physical and chemical traits of atoms and molecules; the highest biological level is "the organism" (including superorganisms) – the set of morphological, behavioral, and performance traits that interact directly with the environment (Chapters 2, 6, 7, 8, 11, 14, & 15, this volume).

Practically, emergence *reduces* the total scope of particular projects. It does so by relieving the researcher from having to be so integrative that he or she feels obliged to understand and examine the entire set of nested phenomena related to some problem. Instead, it may be enough to focus on a few levels right around the one of interest. In this vein, Bartholomew wrote (Bartholomew 1966), "This is the idea that there are a number of levels of biological integration and that each level offers problems and insights that are unique to it; further, that each level finds its explanations of mechanism in the levels below, and its significances in the levels above it." In practice, emergence *hides* the other higher and lower levels from making strong immediate contributions to the one of interest. In Goldenfeld and Kadanoff's (1999) words, "Use the right level of description to catch the phenomena of interest. Don't model bulldozers with quarks." Such a view provides a counterpoint to the prevailing, and often uncritical, view that the way forward in biology is always to maximize the vertical integration of projects.

CONCLUSIONS TO THE CONCLUSIONS

Advocating a renewed focus on individual organisms brings with it a certain historical irony. Most working biologists were inspired early in their lives by organisms (e.g., Wilson 2006) rather than by nutrients cycling through ecosystems, or by the genetic architectures underlying evolutionary trajectories, or by how genes map onto phenotypes. Growing up scientifically has come to mean expanding one's horizons beyond organisms, to higher or lower levels of organization. Indeed, the trajectory of science over the past two hundred years has followed a similar path. Whereas taxonomy and physiology, two overtly organismal sciences, were ascendant in the 19th and early 20th centuries, they have been eclipsed ever since by work at higher and lower levels of organization. What we advocate here may thus seem retrograde: a return to the organism. But the return we envision is equipped with a fresh set of perspectives focused on systems, connections, and complexity. These perspectives lead toward thinking of organisms in much richer terms, and as key integrators across biological disciplines. Indeed, this sentiment is not even new, as the writings of Bartholomew over the past 50 years attest. Still this view has had little effect on the trajectory of biology outside restricted enclaves working on explicitly organismal problems. It is time for this to change. The integrative biology we envision reasserts the primacy of the organism while also serving as a conceptual hub around which the biological sciences are organized.

ACKNOWLEDGMENTS

Thanks to Mihaela Pavličev for comments on this chapter.

REFERENCES

Agrawal, A.A. (2001) Phenotypic plasticity in the interactions and evolution of species. *Science*, **294**, 321–6.

Alm, E. and Arkin, A. (2003) Biological networks. *Current Opinion in Structural Biology*, **13**, 193–202.

Arnold, S.J., Bürger, R., Hohenlohe, P.A. *et al.* (2008) Understanding the evolution and stability of the G-matrix. *Evolution*, **62**, 2451–61.

Ayroles, J.F., Carbone, M.A., Stone, E.A. *et al.* (2009) Systems genetics of complex traits in Drosophila melanogaster. *Nature Genetics*, **41**, 299–307.

Bartholomew, G.A. (1966) Interaction of physiology and behavior under natural conditions, in *The Galapagos* (ed R.I. Bowman), University of California Press, Berkeley, pp. 39–45.

Bizzarri, M., Palombo, A. and Cucina, A. (2013) Theoretical aspects of systems biology. *Progress in Biophysics and Molecular Biology*, **112**, 33–43.

Cohen, A.A., Martin, L.B., Wingfield, J.C. *et al.* (2012) Physiological regulatory networks: ecological roles and evolutionary constraints. *Trends in Ecology & Evolution*, **27**, 428–35.

Feder, M.E., Bennett, A.F., Burggren, W.W. and Huey, R.B. (1987) in *New Directions in Ecological Physiology* (eds M.E. Feder, A.F. Bennett, W.W. Burggren and R.B. Huey), Cambridge University Press, Cambridge.

Ghalambor, C.K., Walker, J.A. and Reznick, D.N. (2003) Multi-trait selection, adaptation, and constraints on the evolution of burst swimming performance. *Integrative and Comparative Biology*, **43**, 431–8.

Goldenfeld, N. and Kadanoff, L.P. (1999) Simple lessons from complexity. *Science*, **284**, 87–9.

Gottlieb, G. (2001) *Individual Development & Evolution*, Psychology Press.

Horrobin, D.F. (2003) Modern biomedical research: an internally self-consistent universe with little contact with medical reality? *Nature Reviews. Drug Discovery*, **2**, 151–4.

Huey, R.B. and Bennett, A.F. (2008) Bart's familiar quotations: the enduring biological wisdom of George A. Bartholomew. *Physiological and Biochemical Zoology*, **81**, 519–25.

Kauffman, S.A. (1993) *The Origins of Order: Self-Organization and Selection in Evolution*, Oxford University Press, New York.

Kingsolver, J.G., Gomulkiewicz, R. and Carter, P.A. (2001) Variation, selection and evolution of function-valued traits. *Genetica*, **112–13**, 87–104.

Kneitel, J.M. and Chase, J.M. (2004) Trade-offs in community ecology: linking spatial scales and species coexistence. *Ecology Letters*, **7**, 69–80.

Lepš, J. and Šmilauer, P. (2003) *Multivariate Analysis of Ecological Data Using CANOCO*, Cambridge University Press.

Martin, L.B., Liebl, A.L., Trotter, J.H. *et al.* (2011) Integrator networks: illuminating the black box linking genotype and phenotype. *Integrative and Comparative Biology*, **51**, 514–27.

Mazzocchi, F. (2008) Exceeding the limits of reductionism and determinism using complexity theory. *EMBO Reports*, **9**, 10–14.

Miner, B.G., Sultan, S.E., Morgan, S.G. *et al.* (2005) Ecological consequences of phenotypic plasticity. *Trends in Ecology & Evolution*, **20**, 685–92.

Muller, G. and Wagner, G. (1991) Novelty in evolution: restructuring the concept. *Annual Review of Ecology and Systematics*, **22**, 229–56.

Mykles, D.L., Ghalambor, C.K., Stillman, J.H. and Tomanek, L. (2010) Grand challenges in comparative physiology: integration across disciplines and across levels of biological organization. *Integrative and Comparative Biology*, **50**, 6–16.

Nadeau, J.H. and Dudley, A.M. (2011) Systems genetics. *Science*, **331**, 1015–16.

Nicholson, J.K., Holmes, E., Lindon, J.C. and Wilson, I.D. (2004) The challenges of modeling mammalian biocomplexity. *Nature Biotechnology*, **22**, 1268–74.

Noble, D. (2012) A theory of biological relativity: no privileged level of causation. *Interface focus*, **2**, 55–64.

Noordwijk, A.V. and Jong, G.D. (1986) Acquisition and allocation of resources: their influence on variation in life history tactics. *The American Naturalist*, **128**, 137–42.

Nussey, D.H., Wilson, A.J. and Brommer, J.E. (2007) The evolutionary ecology of individual phenotypic plasticity in wild populations. *Journal of Evolutionary Biology*, **20**, 831–44.

Pigliucci, M. (2001) *Phenotypic Plasticity: Beyond Nature and Nurture*, Johns Hopkins University Press.

Pigliucci, M. and Muller, G. (eds) (2010) *Evolution – the Extended Synthesis*, MIT Press.

Ricklefs, R.E. and Wikelski, M. (2002) The physiology/life-history nexus. *Trends in Ecology & Evolution*, **17**, 462–8.

Schrödinger, E. (1944) *What Is Life?: With Mind and Matter and Autobiographical Sketches*, Cambridge University Press.

Shubin, N., Tabin, C. and Carroll, S. (2009) Deep homology and the origins of evolutionary novelty. *Nature*, **457**, 818–23.

Simon, H. (1962) The architecture of complexity. *Proceedings of the American Philosophical Society*, **106**, 467–82.

Stearns, S.C. (1989) Trade-offs in life-history evolution. *Functional Ecology*, **3**, 44–50.

Strohman, R. (1997) The coming Kuhnian revolution in biology. *Nature Biotechnology*, **15**, 194–200.

Wagner, G.P. and Altenberg, L. (1996) Complex adaptations and the evolution of evolvability. *Evolution*, **50**, 967–76.

Wake, M. (2003) What is "Integrative Biology"? *Integrative and Comparative Biology*, **241**, 239–41.

Walker, J.A. (2007) A general model of functional constraints on phenotypic evolution. *The American Naturalist*, **170**, 681–9.

West-Eberhard, M. (2003) *Developmental Plasticity and Evolution*, Oxford University Press, Oxford.

Williams, G. (1966) Natural selection, the costs of reproduction, and a refinement of Lack's principle. *The American Naturalist*, **100**, 687–90.

Wilson, E.O. (2006) *Naturalist*, Island Press.

Woods, H.A. (2014) Mosaic physiology from developmental noise: within-organism physiological diversity as an alternative to phenotypic plasticity and phenotypic flexibility. *The Journal of Experimental Biology*, **217**, 35–45.

Woods, H.A. and Wilson, J.K. (2013) An information hypothesis for the evolution of homeostasis. *Trends in Ecology & Evolution*, **28**, 283–9.

Zera, A. and Harshman, L. (2001) The physiology of life history trade-offs in animals. *Annual Review of Ecology and Systematics*, **32**, 95–126.

INDEX

Note: Page numbers in *italics* indicate figures; tables are denoted with '*t*'.

Integrative Organismal Biology, First Edition.
Edited by Lynn B. Martin, Cameron K. Ghalambor, and H. Arthur Woods.

Printed in the USA/Agawam, MA
September 17, 2026

910661.219

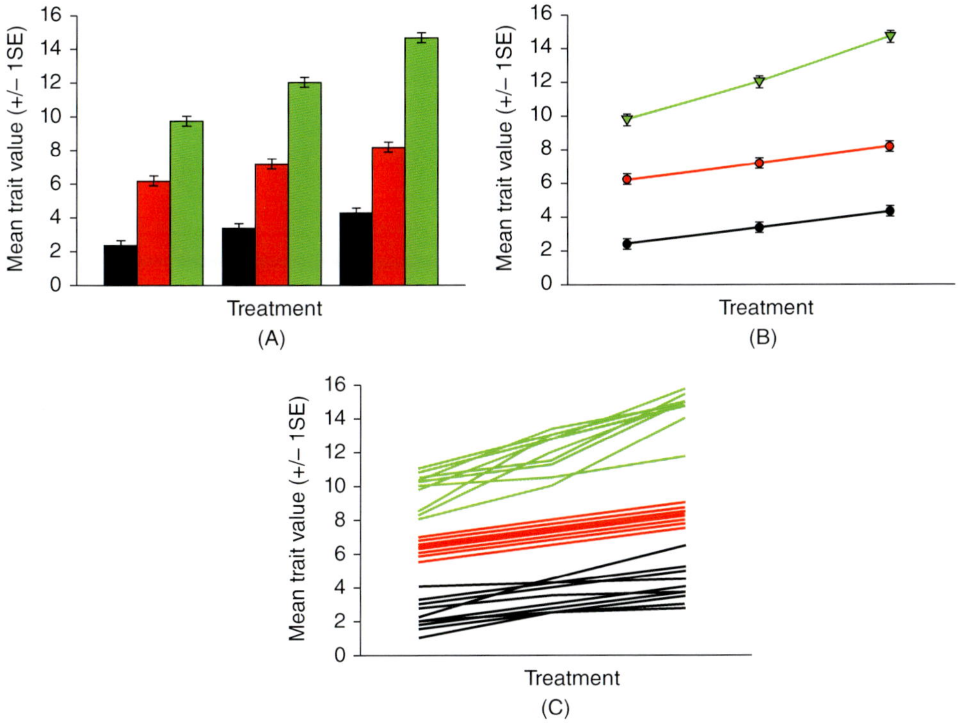

Figure 1.1. Three ways of representing the same data. Shown are three populations under three different environmental treatments. (A) A traditional bar graph showing the mean and variance for each population and each treatment. (B) The mean reaction norm for each population. (C) The individual reaction norms within each population.

Integrative Organismal Biology, First Edition.
Edited by Lynn B. Martin, Cameron K. Ghalambor, and H. Arthur Woods.

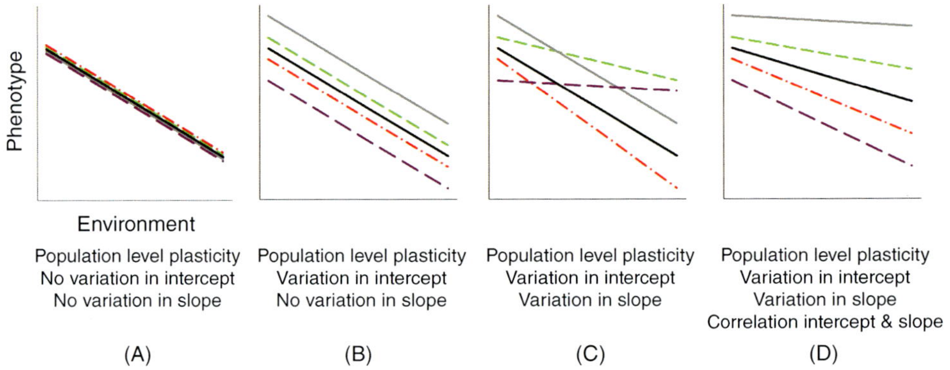

Figure 2.1. Some potential patterns of individual variation in a simple linear reaction norm, illustrating situations in which (A) neither elevations nor slopes vary among individuals, (B) elevations but not slopes vary, (C) both elevations and slopes vary, and (D) elevations and slopes vary and are positively correlated with one another.

Figure 5.1. Classic examples of polyphenism. (A) Horn polyphenism in *Onthophagus* beetles (image reproduced from Valena & Moczek 2012); (B) seasonal polyphenism in *Bicyclus* butterflies (image copyright H. Piel and Antónia Monteiro); (C) phase polyphenism in *Schistocerca* locusts (image copyright Tom Fayle); (D) wing polyphenism in pea aphids; (E) feeding polyphenism in *Spea* tadpoles (image copyright David Pfennig).

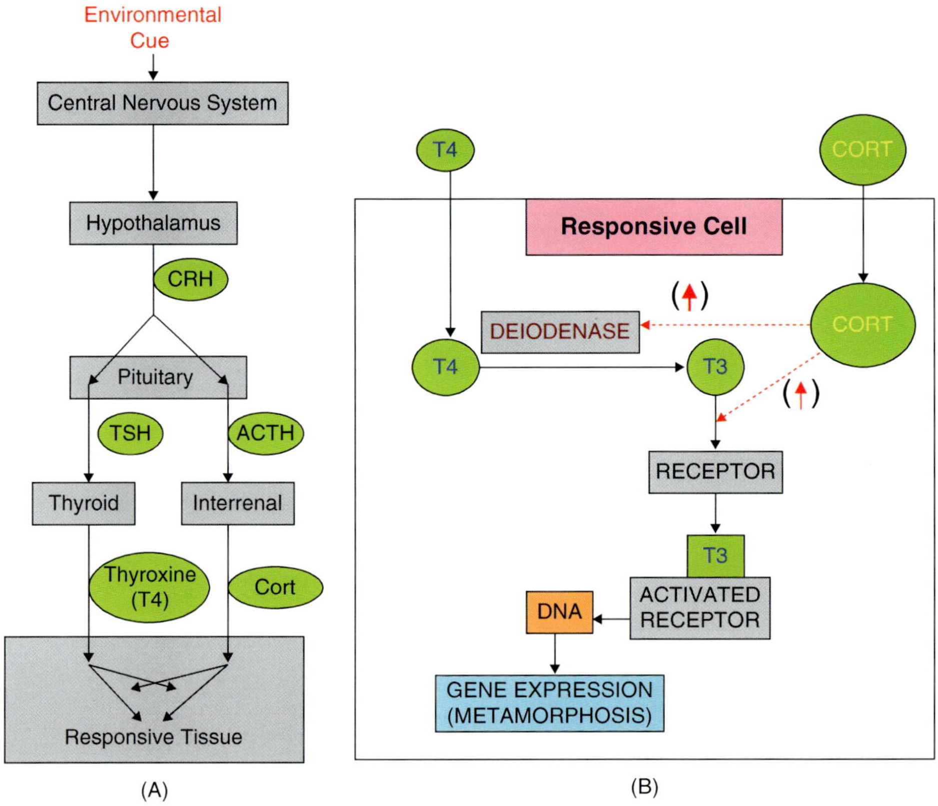

Figure 5.2. Panel (A) Simplified diagram of the endocrine regulation of the timing of metamorphosis in anurans, illustrating various aspects of endocrine regulation. CRH = corticotropin releasing hormone, TSH = thyroid-simulating hormone, ACTH = adrenocorticotropic hormone, CORT = corticosterone. As described in the text, an environmental signal is transduced through the brain into an internal endocrine signal that induces metamorphosis. Endocrine control is hierarchical with multiple control steps. Common regulators (e.g., CRH) coordinate the release of multiple hormones which interact in multiple ways to cause a tissue response (see Panel B). Panel B) Hormonal interactions in a peripheral tissue undergoing metamorphosis. T4 = thyroxine, T3 = triiodothyronine. CORT potentiates the effect of thyroxine in responsive cells by increasing the activity of the enzyme (deiodenase) that converts T4 (the prohormone, thyroxine) to the more active T3 (hormone, triidothyronine), and by upregulating the T3 nuclear receptor. There are other interactions between these hormones in peripheral tissues that are not shown.

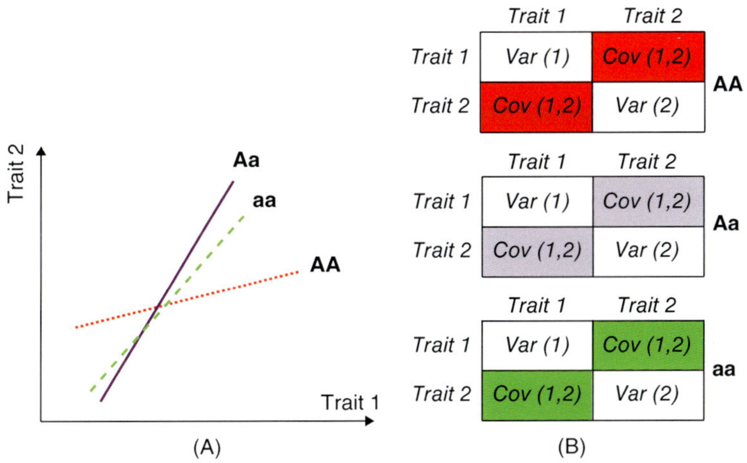

Figure 6.2. A scheme of relationship QTL. (A) the regression slope of the values of one trait on the values of a second trait differs among the genotypes at this locus (here genotypes AA, Aa and aa). (B) the phenotypic covariance also differs between the subpopulations sharing a genotype at this locus.

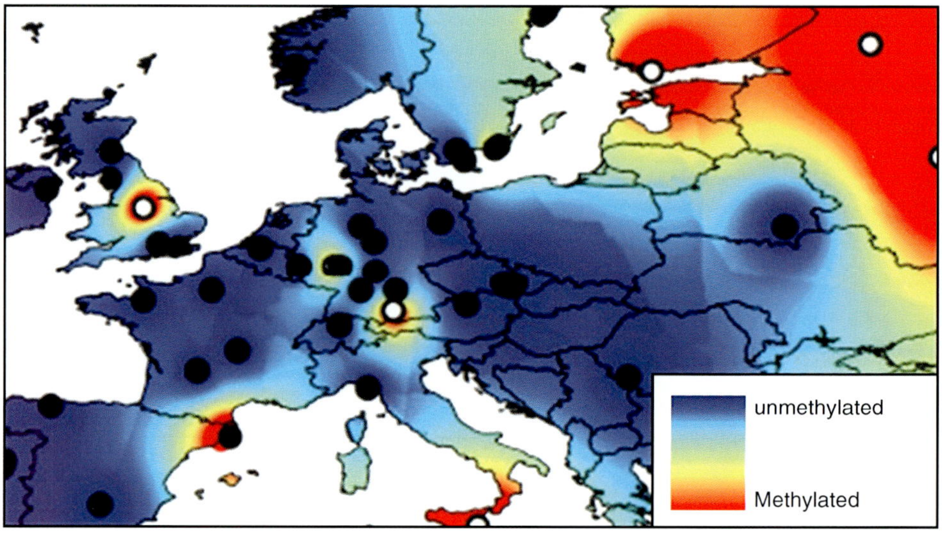

Figure 7.1. The chromosomal fragment around the gene *MEE57* in *Arabidopsis thaliana* has two different, presumably stable epi-alleles (methylated versus unmethylated) that segregate in the wild. The spatially interpolated frequencies of the two contrasting epi-alleles across the landscape are represented here, based on the locations of collection of the 76 accessions in our study and their epi-genotypes at the *MEE57* fragment. Circles shaded black represent accessions with the methylated epi-allele, and circles shaded white represent accessions with the unmethylated epi-allele. Black areas indicate a high frequency of the methylated epi-allele, and white areas indicate a high frequency of the unmethylated epi-allele. Many points are obscured by other points.

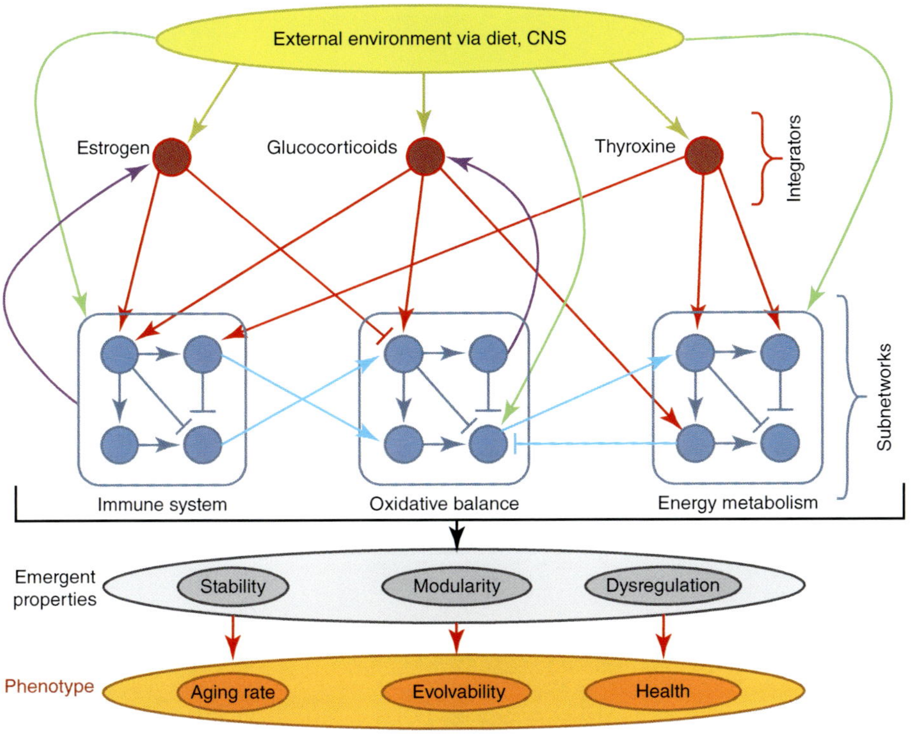

Figure 9.1. A simplified schematic of a physiological regulatory network (PRN). Green arrows depict direct effects of the environment on subnetworks, such as content of antioxidants in the diet. Light-blue arrows indicate direct interactions among subnetworks via party hubs (see Figure 9.2), such as immune regulation by dietary antioxidants. Red arrows indicate hierarchical control within PRNs via key nodes (i.e., integrators; see Figure 9.3). Yellow arrows indicate environmental regulation of integrators, usually via the central nervous system (CNS). Purple arrows indicate feedback effects of subnetworks on integrators, such as antioxidant effects on glucocorticoids. Emergent, system-level properties of the PRN (grey oval) manifest at different levels, depending on PRN state and structure (see text). Likewise, phenotypic traits (e.g. health and evolvability, respectively; orange oval) arise via individual (state) or species-level (structure) PRN configurations. Connectivity is the number of edges among nodes (total arrows), modularity is the number of possible light-blue arrows relative to all other possible arrows, and hierarchy is the distribution of arrows (edges) among nodes. As edge distributions in bionetworks tend to follow a power law and as integrators appear common in metazoans (see Martin *et al*. 2011), we expect a branching hierarchy in PRNs. *Source*: Adapted from Alan A. Cohen, Lynn B. Martin, John C. Wingfield, Scott R. McWilliams, Jennifer A. Dunne (2012) Physiological regulatory networks: ecological roles and evolutionary constraints. *Trends in Ecology & Evolution*, **28**(8), 428–35. Copyright © 2012, Elsevier.

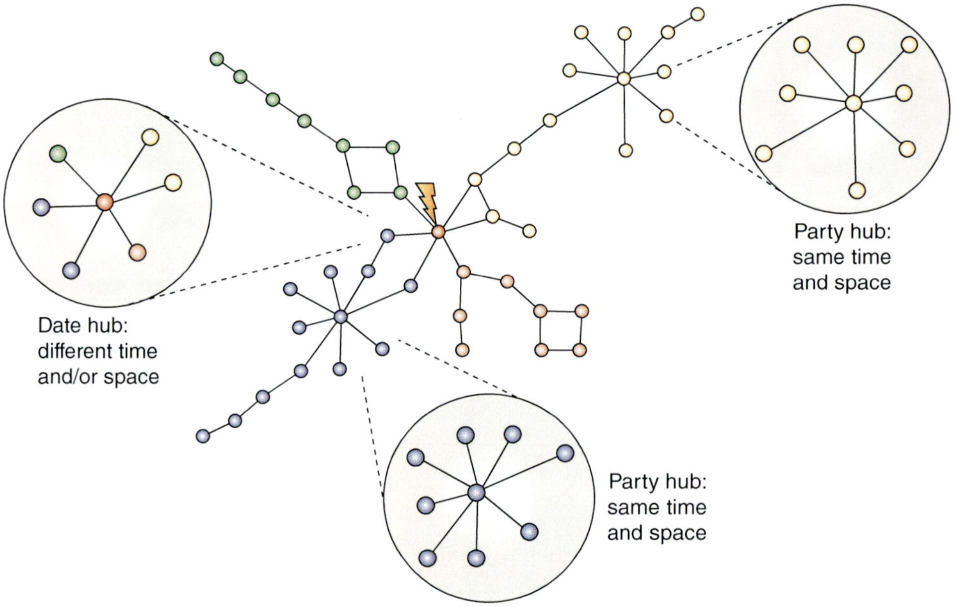

Figure 9.2. A bionetwork with two types of highly connected nodes. Network nodes are colored circles; edges are lines linking nodes. 'Party' hubs (i.e., central blue or yellow nodes in zoomed figure elements) interact simultaneously with many partners whereas 'date' hubs (i.e., red node in figure center; equivalent to integrators in Figure 9.3) interact with many other nodes at many times or locations. Colors depict subnetwork membership. Lightning bolt cartoon depicts particular environmental stimulus inducing this particular network state. *Source*: Adapted from Wagner *et al*. 2007; reprinted with permission.

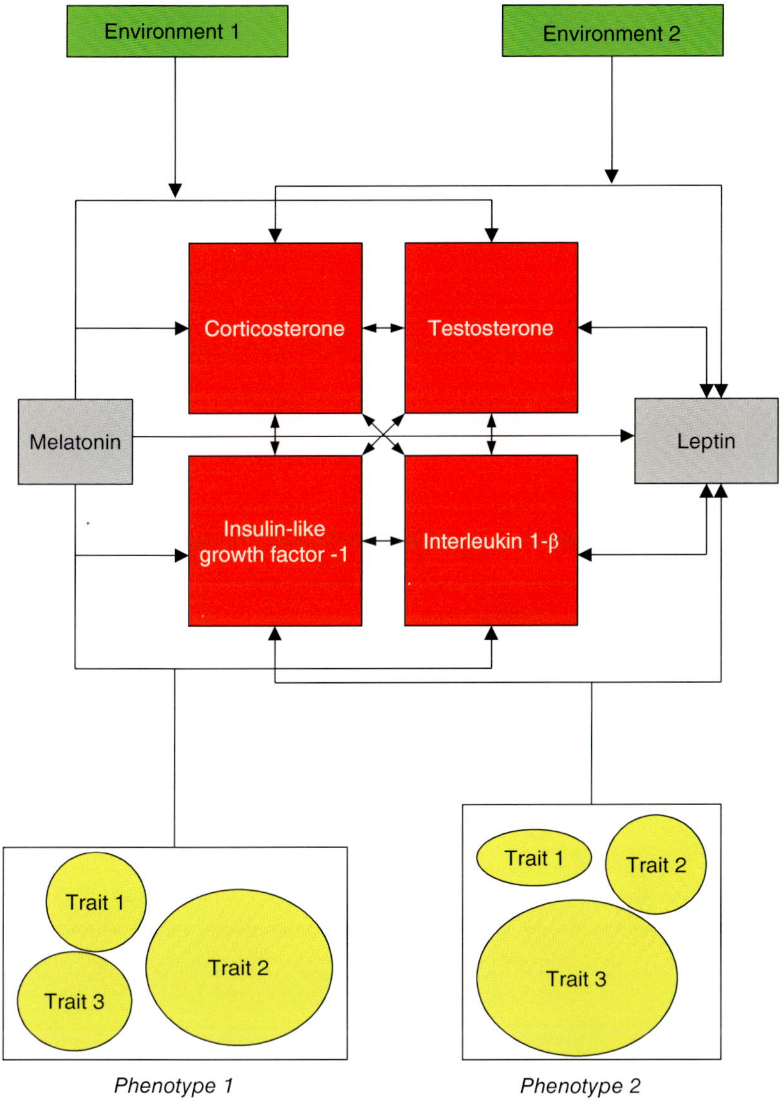

Figure 9.3. Integrators encode present and past environment to produce individual phenotypic variation from one genotype. Red boxes represent integrators (i.e., nodes acting as PRN date hubs) that regulate performance such as coping with stressors (corticosterone), reproduction (testosterone), growth (IGF-1), and interactions with parasites (IL-1β). Arrows between nodes depict edges of network, the configuration of which determines the specific and coordinated phenotypic response. Two integrators are set apart from the main network as their influences on PRNs may be unique. Melatonin impacts, but is not impacted by, other integrators; leptin, insulin and functionally similar molecules may have more labile connection to other date hubs, thus acting as mediators of resource-based trade-offs. The green boxes depict two different environments (e.g., conspecific density, latitude), and hence determine the specific phenotypic outcome (boxes with yellow circles) of the same genotype. Only a few integrators are depicted here, for clarity.

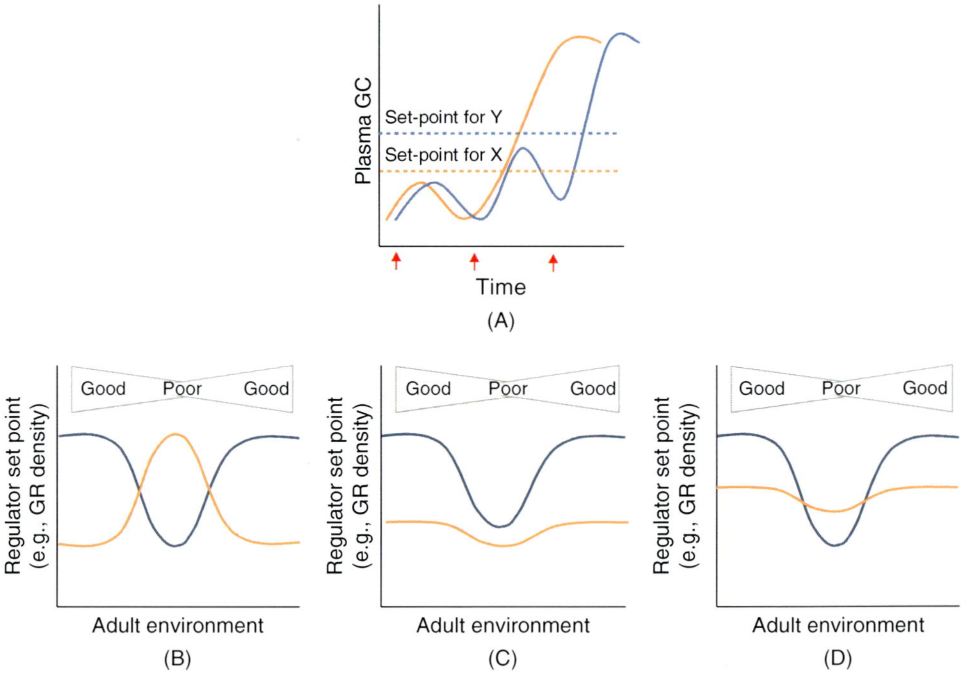

Figure 12.4. Schematic of how parental effort in Figure 12.2 can affect set points of regulators and plasma GC of offspring. To simplify, two offspring phenotypes are represented: low quality offspring phenotype X due to poor developmental environment (orange lines) and high quality offspring phenotype Y due to good developmental environment (blue lines). These phenotypes differ in a set point where if plasma GC surpasses in response to any stimulus (red arrow, e.g., low energy storage), neither behavioral and physiological modification nor negative feedback can restore reactive homeostasis (A). Two lines are off-set to increase visibility. In reality, the threshold between reactive homeostasis and homeostatic overload (in reactive scope model), or between physiological state B & C (in allostasis model) is likely to change with developmental environment, age, and interaction between developmental and adult environment. Figures B through D depict a change in regulator set point across developmental environment according to the environmental matching model (B), silver spoon model (C), and the combination of the two models (D). Gray box represents adult environments.

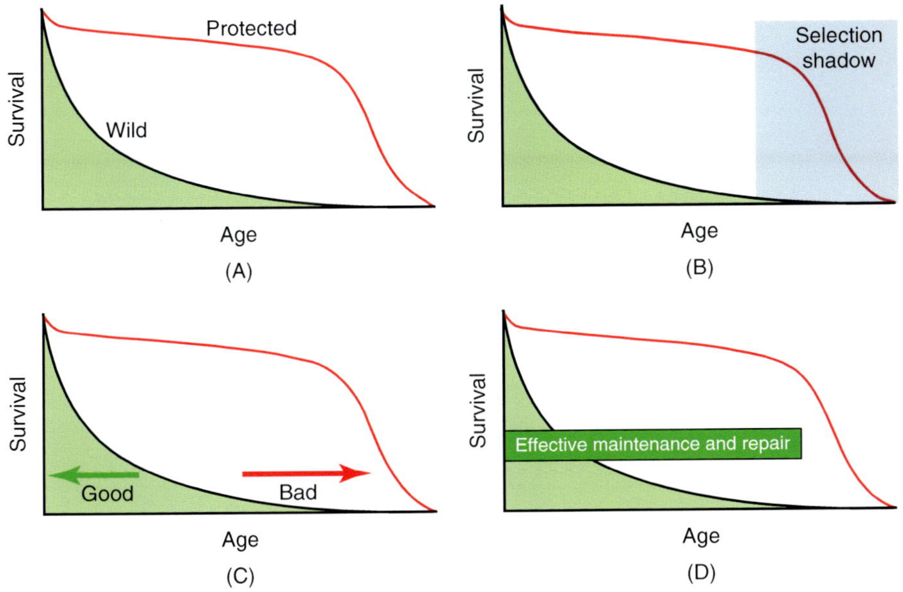

Figure 16.1. Survival curves illustrating the evolutionary theories of aging. *Source*: Thomas B.L. Kirkwood, Steven N. Austad (2000) Why do we age? *Nature*, **408**(6809). Reprinted by permission from Macmillan Publishers Ltd: *Nature*. Copyright 2000.

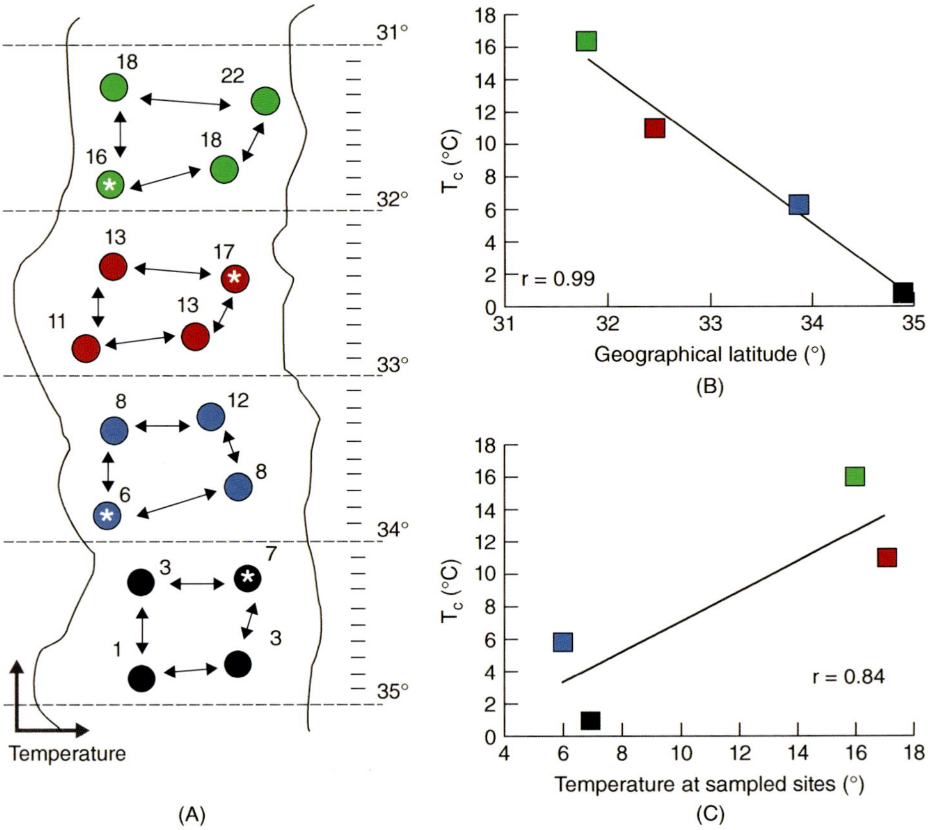

Figure 17.2. A hypothetical example showing the "spatial scale problem" (see text). Four different species (shown in different colors) each inhabit four different localities (circles) that are connected by migration (arrows); the population of each species that is sampled is marked with an asterisk, and the data above each locality represent a climatic variable, e.g., minimum temperature (A). If the physiological variable under study (e.g., minimum critical temperature, T_c) is fitted to the entire distributional range – for example, it is equal to the minimum temperature recorded over the entire range – it will be better correlated with latitude (B) than with minimum temperatures at the sampled localities (C).

Figure 18.3. Plant communities on nonserpentine (background) and serpentine (foreground) soils, Snell Valley, Napa County, California. Many different species of plants have adapted to live on serpentine soils, however, adaptations frequently compromise the ability of serpentine plants to compete on nonserpentine soils, illustrating a trade-off. This trade-off allows different species of plants to coexist by using different types of soil. Photo by John McKay.

Figure 18.4. Segregation of Orange-crowned (*Oreothlypis celata*, blue) and Virginia's (*O. virginiae*, red) warblers (Parulidae) along gradients of temperature and moisture during the breeding season. These two species appear to face a trade-off, where the larger *celata* is dominant in aggressive interactions and excludes the smaller *virginiae* from preferred nest sites in cool, moist regions of their breeding territories. The smaller *virginiae*, however, can tolerate hot, dry nesting conditions that *celata* cannot tolerate, leading to spatial partitioning of nest sites where their breeding territories overlap. Experimental removal of the opposite species resulted in *virginiae* shifting their nest sites into cool, wet regions of the gradient in the absence of *celata*, but no shift in *celata* nest sites in the absence of *virginiae* (x-axis is the mean percentage of maple stems near the nest +/– SE). The temperature and moisture gradients that characterize overlapping territories also characterize patterns of elevational and geographic segregation, and temporal variation in abundance, suggesting that these trade-offs may also influence their distributions on broader spatial and temporal scales (Martin 1998, 2001; Martin & Martin 2001a,b). Similar trade-offs between competitive ability and environmental tolerance face barnacles (Connell 1961) and many other closely-related species, from chipmunks to salamanders, cattails to crayfish (Martin & Martin 2001a).